Advances in Intelligent Systems and Computing

171

Editor-in-Chief

Prof. Janusz Kacprzyk
Systems Research Institute
Polish Academy of Sciences
ul. Newelska 6
01-447 Warsaw
Poland
E-mail: kacprzyk@ibspan.waw.pl

For further volumes:
http://www.springer.com/series/11156

Jorge Casillas, Francisco J. Martínez-López,
and Juan M. Corchado (Eds.)

Management Intelligent Systems

First International Symposium

 Springer

Editors
Dr. Jorge Casillas
Dept. Computer Science & A.I.
University of Granada
Granada
Spain

Prof. Dr. Juan M. Corchado
Depto. Informática y Automática
Universidad Salamanca
Salamanca
Spain

Dr. Francisco J. Martínez-López
Dept. Management
University of Granada
Granada
Spain

ISSN 2194-5357
ISBN 978-3-642-30863-5
DOI 10.1007/978-3-642-30864-2
Springer Heidelberg New York Dordrecht London

e-ISSN 2194-5365
e-ISBN 978-3-642-30864-2

Library of Congress Control Number: 2012939518

Printed on acid-free paper

Springer is part of Springer Science+Business Media (www.springer.com)

Preface

This symposium is believed to be the first international forum to present and discuss original, rigorous and significant contributions on Artificial Intelligence-based (AI) solutions—with a strong, practical logic and, preferably, with empirical applications—developed to aid the management of organizations in multiple areas, activities, processes and problem-solving; what we call Management Intelligent Systems (MiS).

Basically, the AI core focuses on the development of valuable, automated solutions (mainly by means of intelligent systems) to problems that would require the human application of intelligence. In an organizational context, there are problems that require human judgement and analysis to assess and solve these problems with a guarantee of success. These decisional situations frequently relate to strategic issues in organizations in general, and in firms in particular, where problems are far from being well-structured. Developing and applying *ad hoc* intelligent systems, due to their particular strengths in processing data and providing valuable information either with a data-driven or, especially, with a knowledge-driven approach, might be of interest to managers in their decision-making.

In essence, AI offers real opportunities for advancing the analytical methods and systems used by organizations to aid their internal and external managerial processes and decision-making. Indeed, well-conceived and designed intelligent systems are expected to outperform operational research- or statistical-based supporting tools in complex, qualitative and/or difficult-to-program managerial problems and decisional scenarios. However, these opportunities still need to be fully realized by researchers and practitioners. Therefore, more interdisciplinary and applied contributions are necessary for this research stream to really take off.

Each paper submitted to IS-MiS 2012 went through a stringent peer review process by members of the Program Committee comprising 51 internationally renowned researchers (including a dozen Editors-in-Chief of prestigious research journals on management, business and intelligent systems) from 16 countries.

From the 46 submissions received, a total of 31 papers have been accepted, and they address diverse Management and Business areas of application such as decision support, segmentation of markets, CRM, product design, service personalization,

organizational design, e-commerce, credit scoring, workplace integration, innovation management, business database analysis, workflow management, and location
of stores, among others. A wide variety of AI techniques have been applied to
these areas such as multi-objective optimization and evolutionary algorithms, classification algorithms, ant algorithms, fuzzy rule-based systems, intelligent agents,
Web mining, neural networks, Bayesian models, data warehousing, and Rough sets,
among others.

Furthermore, the symposium featured two distinguished keynote speakers.
Dr. Jay Liebowitz (University of Maryland University College), Editor-in-Chief of
Expert Systems With Applications, presented the talk "Intelligent Systems in Man-
agement: Challenges and Future Directions," while Dr. Vladimir Zwass (Fairleigh
Dickinson University), Editor-in-Chief of the *Journal of Management Information
Systems*, talked about "Information Systems, Electronic Commerce, and the Bene-
fits of Intelligence."

We believe our aim to promote, stimulate and publish high-quality contributions
on applied-intelligent systems to support management in tackling all kinds of issues
faced by organizations has been realized with this first symposium. However, this
is only the starting point for future events which we hope will help to establish this
promising interdisciplinary research field.

We wish to acknowledge the support of the sponsors IEEE Systems, Man and
Cybernetics Society (Spain Section Chapter) and the IEEE Spain Section. We would
also like to thank all the contributing authors, keynote speakers, members of the Program Committee and the rest of the Organizing Committee for their highly valuable
work in enabling the success of this first edition of IS-M*i*S. Thanks for your generous contribution–IS-M*i*S 2012 would not have been possible without you all.

Jorge Casillas
Francisco J. Martínez-López
Juan M. Corchado

Organization

General Chair

Jorge Casillas · · · · · · · · · · · University of Granada, Spain

Program Chair

Francisco J. Martínez-López · · · · University of Granada, Spain
Open University of Catalonia, Spain

Organizing Committee

Juan M. Corchado · · · · · · · · · · University of Salamanca, Spain
Fernando de la Prieta · · · · · · · · University of Salamanca, Spain
Antonio J. Sánchez · · · · · · · · · University of Salamanca, Spain
Sara Rodríguez · · · · · · · · · · · University of Salamanca, Spain

Program Committee

Andrea Ahlemeyer-Stubbe · · · · · antz21, Germany
Daniel Arias · · · · · · · · · · · · University of Granada, Spain
Barry J. Babin · · · · · · · · · · · Louisiana Tech University, USA
P.V. (Sundar) Balakrishnan · · · · University of Washington, USA
Malcolm J. Beynon · · · · · · · · · Cardiff University, UK
Min-Yuang Cheng · · · · · · · · · · National Taiwan University of Science and
Technology, Taiwan
Bernard De Baets · · · · · · · · · · University of Ghent, Belgium
Bob Galliers · · · · · · · · · · · · Bentley University, USA
Juan Carlos Gázquez-Abad · · · · · University of Almería, Spain
Dawn Iacobucci · · · · · · · · · · · Vanderbilt University, USA
Frank Klawonn · · · · · · · · · · · Ostfalia University of Applied Sciences,
Germany
Kemal Kiliç · · · · · · · · · · · · · Sabanci University, Turkey

Contents

Production and Operations Management

E-Business and E-Commerce

Software Applications and Prototypes

Marketing and Consumer Behavior

Risk Assessment and Management

Various Applications

Innovation in Management
and Organizational Design

A Strategic Perspective on Management Intelligent Systems

Zhaohao Sun and Sally Firmin

School of Science, Information Technology and Engineering
University Of Ballarat
P.O. Box 663, Ballarat, Vic 3353
{z.sun,s.firmin}@ballarat.edu.au

Abstract. Management intelligent systems (MIS) is a new paradigm integrating management with intelligent systems. What are the core components of MIS? What are intelligent systems for management? How can management integrate with intelligent systems? All these questions remain open in an MIS context. This article addresses these issues by examining MIS from a strategic perspective. More specifically, this article first examines management in information systems, management for intelligent systems and intelligent systems for management. Then this article provides a strategic model for MIS encompassing core components of MIS, through integrating main management functions with intelligent systems taking into account decision making of managers in organizations. The approach proposed in this article will facilitate the research and development of MIS, management and intelligent systems and information systems.

Keywords: Management intelligent systems (MIS), information systems, artificial intelligence (AI), decision making, intelligent systems.

1   Introduction

Management intelligent systems (MIS) is a new paradigm that integrates management and intelligent systems, and is different from management information systems, which has become a discipline and an academic area. More specifically, MIS is a new research area that focuses on applying artificial intelligence (AI) in general, and intelligent systems in particular to the management of organizations in any of their multiple areas, activities, processes and decisional problems [1]. This new paradigm has not drawn much attention either in management or in intelligent systems. It is unclear what the core components of MIS are. How to integrate intelligent systems with management is still a controversial problem. Many textbooks in information systems cover management activities based on information technology and information systems [4][12], and a variety of articles have contributed a detailed account of management activity of managers in organizations based on intelligent systems,

J. Casillas et al. (Eds.): Management Intelligent Systems, AISC 171, pp. 3–14.
springerlink.com   © Springer-Verlag Berlin Heidelberg 2012

for example, intelligent supply chain management [14] and intelligent customer relationship management [15]. There is little focus on main management functions and decision making of managers in organizations based on intelligent systems [2]. How to fill this gap is a complicated issue. This consideration leads to the following issues for developing MIS: How are intelligent systems managed? What are intelligent systems for management? How can management and intelligent systems be integrated in a unified way? This article will address these issues by examining the MIS from a strategic perspective and provide a strategic model for MIS through integrating the main management functions with intelligent systems taking into account decision making of managers in organizations. The organization of this article is as follows: Section 2 provides some background for our research. Section 3 proposes a multiperspective on MIS, which is the basis for this article. Section 4 looks at management in information systems. Section 5 and 6 examines management of intelligent systems and intelligent systems for management respectively. Section 7 discusses intelligent systems for management decision making, and Section 8 proposes a strategic model for MIS and the core components of MIS. The final section ends this article with some concluding remarks and future work. The approach proposed in this article will facilitate the research and development of MIS, management, intelligent systems and information systems.

2   Background

This section will provide a background on management, information systems, intelligent systems and briefly incorporates them in management intelligent systems (MIS).

Management is what managers do [5] (p.12). More specifically, management is the process of manager's coordinating and overseeing the work activities of others so that their activities are completed. The main management functions or activities of a manager consist of planning, organizing, leading and controlling. Every function a manager undertakes involves decision making, negotiation, resource allocation and disturbance handling [5] (p.16). There are three levels of management: operational management, tactical management, and strategic management which correspond to activities of operational managers, middle managers and top managers of organizations respectively [5] (pp.14-19).

Information systems as a field of academic study or discipline encompasses the concepts, principles, methodologies and processes for two broad areas of activity within organizations: 1) acquisition, deployment, management, and strategy for information technology resources and services; and 2) packaged system acquisition or system development, operation, and evolution of infrastructure and systems for use in organizational processes [6].

Information systems enhance management processes at all levels – operational, tactical, and strategic management. Information systems are vital to problem identification, analysis and decision making of managers [6]. Planning, organizing, leading and controlling have improved remarkably as a result of the development of information systems and information technology in the past half a century. Decision support systems (DSS) and intelligent systems have also aided managers in making decisions for business activities [3].

It should be noted that information systems as a discipline has many alternative names such as management information systems, computer information systems, business information systems, information science, and business informatics, to name a few. This reflects the historical development of the field and understanding of similar programs with different emphases since its inception in the 1960s [6]. For example, universities in Australia prefer to use business information systems or information systems rather than management information systems.

Intelligent systems (IS) as an applied field of artificial intelligence (AI) encompasses the principles, methodologies, techniques and processes of applying AI to real world problem solving. Currently, IS is a discipline that studies intelligent behaviors and their implementations as well as their impacts on human society [7]. An intelligent system is a system that can imitate, and/or automate intelligent behaviors of human beings [7]. Expert systems and knowledge-based systems are examples of intelligent systems. Intelligent techniques consist of case-based reasoning (CBR), genetic algorithms, neural networks, fuzzy logic, intelligent agents, data mining, and knowledge management, to name a few [4] (p. 371-380). These intelligent systems and techniques have also aided managers by capturing knowledge, discovering knowledge, and generating solutions to problems encountered in planning, organizing, leading, controlling, and corresponding decision making. These intelligent systems and techniques have also enhanced information systems by providing organizations with intelligent techniques to automate business activities of organizations, and improve human-machine activities. The interrelationship among management, intelligent systems and information systems can be illustrated by the following Fig. 1.

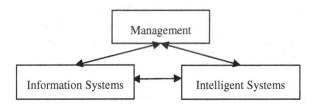

Fig. 1 The interrelationship among management, information systems and intelligent systems

3   Management Intelligent Systems: A Multiperspective

There are at least three different perspectives on management intelligent systems (MIS): A managerial perspective, a perspective from intelligent systems or AI, and a perspective from information systems.

From a managerial perspective, management is universal, because management is needed in all types and sizes of organizations, at all organizational levels, and in all organizational work areas, regardless of the country in which the organization is located [5] (p. 28). Furthermore, management is ubiquitous, because management is required on any occasion that human groups or intelligent agents are involved. A minimal human group may consist of two people, while a big human group may consist of all the people in a country like China. Management is also needed for any system whether small or large. Therefore, management for intelligent systems is necessary for managers, and an important component for MIS. However, what is management for intelligent systems? This remains open.

From an intelligent systems perspective, AI can be applied to improve everything existing in the world. This also implies that AI can be applied to management in order to improve effectiveness and efficiency of managers in organizations. Therefore, intelligent systems for management are an important field for AI or intelligent systems, just as information technology for management is an important topic for information technology (IT) and management [3]. Based on this perspective, intelligent systems for management is an important component of MIS. However, what are intelligent systems for management? Or what are the elements of intelligent systems for management? This also remains open.

From the perspective of information systems, MIS is a new paradigm that integrates management with intelligent systems. Management information systems have become an important discipline and research field over the past few decades [4]. As a related, evolved discipline, management intelligent systems will develop in coming decades. Therefore, MIS should be drawn more attentions from academic communities and industrial societies. Of course, we must note the viewpoints of some scholars that MIS is less necessary sometimes, because we have already had management information systems although management intelligent systems and management information systems have their own different emphases. The question is: What are their own different emphases? This also remains open.

Based on the above discussion, we find that MIS has social requirements from different academic communities and should be investigated in depth. In the following sections, we will address the above-mentioned issues and propose a strategic model for MIS.

4   Management in Information Systems

Management has played a significant role in information systems and information technology (IT) as a disciple. For example, management information systems has been developed as a discipline in USA, and offered as a course over the past decades in many universities throughout the world including Australia [4]. Information technology for management as a course has been offered by many universities including the University of Ballarat in Australia [3]. Management has also drawn lasting attention in e-commerce and e-business. For example, e-commerce for management as a course has been offered at many universities [2]. All these attempt to apply information systems, information technology and e-commerce to assist managers of organizations to solve their problems encountered in their business activities, processes and decision making. For example, sofware or information systems for customer relationships management (CRM) [4][11], business process management, human resource management, procurement management, supply chain management [4] and marketing analytics are all available on the market for acquistion from managers or organizations [2][3][12]. Major CRM application software vendors include Oracle's Siebel and SAP [4]. At the same time, the managers of organizations have also challenged information systems, information technology and e-commerce to provide new principles, methods and technologies to solve new problems that they encounter in the business environment. Therefore, management and information systems have had a relationship with each other in the past few decades, and facilitated continuing development of each over the past few decades. This relationship is depicted in Fig. 2.

Fig. 2 Management and Information Systems

Management of information systems and information security is also an important task for organizations, because both have become an inevitable part of many organizations in general and global organizations in particular [4][12]. Without information systems such as enterprise systems and ERP, large organizations such as airlines and banks could not be running their systems in an effective and efficient manner.

From a managerial perspective, management of information systems includes evaluation of information systems, investments to information systems, and location of the information systems functions including outsourcing. More generally, the following related to information systems in organizations must be managed: hardware platforms, network architecture, legacy information systems, operations management, database administration, user support and training, shared services, information systems staffing, services related to cloud computing for the organization, to name a few [12].

5 Management of Intelligent Systems

This section will examine management of intelligent systems. In other words, what is important for intelligent systems from a managerial perspective? From a managerial perspective, intelligent systems and information systems share commonality. The management of information systems discussed in the previous section is still valid for the management of intelligent systems. Furthermore, there are many methods, principles and techniques of management that are useful for managing and developing intelligent systems, for example, data management, information management, knowledge management, case management and experience management. These have been playing a significant role in intelligent systems, because data, information, knowledge, experience and cases are the foundation for intelligent systems and their applications in the real world [4][9], as shown in Fig. 3. In what follows, we only focus on case management and experience management, because data management [4], information management [4], knowledge management (KM) [11] are well-known in either information systems [2] or intelligent systems or business intelligence communities [3]. We do not go into the interrelationships among data, information, knowledge, experience and case, owing to space limitation. For detail see other references such as [9] and [11].

Case management (CM) is a part of case computing, which is an extended form of case based reasoning [9]. Case computing mainly consists of CBR, case management, case engineering and case systems. CM applies techniques of data base management and knowledge management to case bases, which are part of a case-based system. Case management usually consists of case repartition, case retrieval, case reuse, case retention, case revision, which are called R^5 for case management [8].

Experience management (EM) is a discipline that focuses on experience processing and corresponding management which is in each of following process stages: Discover, capture, gain and collect experience, model, store, evaluate, adapt, reuse, maintain experience, and transform experience into knowledge [10]. Furthermore, experience can be considered as a special knowledge, and then EM is a special kind of KM that is restricted to the management of experience. Therefore, methodologies, techniques and tools for KM can be directly reused for EM.

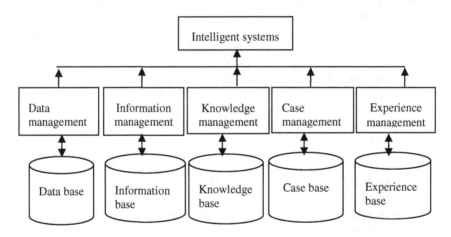

Fig. 3 A model of management for intelligent systems

6   Intelligent Systems for Management

This section will examine intelligent systems for management and the components of intelligent systems for management.

As mentioned earlier, intelligent systems have been an important topic for information systems in general and business management in particular. Business intelligence can be considered as a field of applying intelligent systems to business. For example, a review of three well-known textbooks illustrates common themes: The first explores business information systems [12] and the second, management information systems [4], and the third, information technology for management [3]. All of these three texts involve intelligent systems (including intelligent agents) and decision support systems although they have different classifications for intelligent systems. More specifically, Bocij et al, [12] introduce expert systems, business intelligence, data mining, AI, neural networks and knowledge management (including competitive intelligence as an example) as a part of decision support systems in their text [12] (pp. 260-262). Laudons [4] briefly introduces intelligent systems for decision support in their text. Their introduction covers expert systems, CBR, fuzzy logic systems, neural networks, genetic algorithms and intelligent agents [4] (pp. 380-387). Turban and Volonino [3] introduce intelligent systems as a part of business intelligence and decision support systems (DSS). None of these texts have considered the applications of intelligent systems in business or management at a detailed level. Therefore, intelligent systems for management still require investigation.

As we mentioned in Section 2, the main management functions or activities of a manager consist of planning, organizing, leading and controlling. Therefore, intelligent systems should be applied to each of the main management functions in order to aid managers to realize their organization objectives, that is, it is important for MIS to look at intelligent systems for planning, intelligent systems for organizing, intelligent systems for leading, and intelligent systems for controlling. These have not been explored in academic communities in terms of books and journal articles, to our knowledge. In what follows, we will look at each of these in some detail.

6.1 Intelligent Systems for Planning

Planning involves defining the organizations' goals, establishing an overall strategy for achieving those goals, and developing a comprehensive set of plans to integrate and coordinate organizational work [5] (p. 294). In the process of planning, managers define goals, establish strategies for achieving these goals, and develop plans to integrate and coordinate activities [5] (p. 14). To this end, the managers should define the nature and purpose of planning, classify the type of goals organizations use, and describe related types of plans organizations use, describe project management and discuss issues in planning. A comprehensive set of plans is the outcome of the planning process.

Intelligent systems for planning aim to imitate and automate some or all planning behaviors of managers of organizations such as supply chain planning [4] (p. 303). More specifically, they should imitate and automate definition of goals, establishment of strategies for achieving these goals, and development of plans to integrate and coordinate activities. To this end, data management, information management and knowledge management should be the basis for any intelligent systems for planning. Knowledge based systems, expert systems and intelligent agents have been developed to aid the process of planning. Case based systems as intelligent systems can also facilitate the intelligent planning, because "similar goals have similar strategies" for achieving these goals [9].

6.2 Intelligent Systems for Organizing

Organizing means arranging and structuring work to accomplish the organizations' plans and goals [5] (p.14). When managers organize, they determine what tasks need to get done, who is to do them, and how the tasks are to be decomposed and grouped, who reports to whom and at what level decisions to be made. They also allocate and deploy organizational resources during the organizing process [5] (p.368).

Intelligent systems for organizing aim to imitate and automate all or some organizing behaviors of managers of organizations. More specifically, they should imitate and automate decomposition of tasks, grouping of persons who complete the decomposed task, and allocation and deployment of organizational resources. Case based systems as intelligent systems can also facilitate intelligent organizing, because "similar decompositions of tasks have similar grouping strategies and similar allocation and deployment of organizational resources" [9]. ERP has been proved to be useful for automating allocation and deployment of organizational resources [3].

6.3 Intelligent Systems for Leading

Leading is to oversee and coordinate people to work so that organizational goals can be pursued and accomplished [5] (p. 467). When managers are leading, they motivate their subordinates, helping to resolve work group conflicts, influence individuals or work teams, select appropriate communication channels, or deal with individual or group behavior issues [5] (p. 14). Leading people involve understanding their attitudes, behaviors, personalities and motivations as an individual, or a group, or a community [5] (p. 469).

Intelligent systems for leading aim to imitate and automate all or some leading behaviors of managers of organizations. More specifically, they should imitate and automate how to motivate subordinates, help to resolve work group conflicts, influence individuals or work teams, select appropriate communication channels or deal with individual or group behavior issues, and understanding attitudes, behaviors, personalities and motivations of the individuals and teams. Intelligent systems for leading a team have been drawn some attention by the research group of the University of Technology Sydney (UTS) (http://www.uts.edu.au). Group DSS and executive DSS as intelligent systems have aided managers for leading to some extent [3]. Intelligent agents and multiagent systems also provide a better understanding of collaboration, coordination and cooperation of people within a team [9]. However, understanding attitudes, behaviors, personalities and motivations of individuals and teams is still a big challenge for research and development of intelligent systems.

6.4   Intelligent Systems for Controlling

Controlling involves monitoring, comparing and correcting work performance. When managers are in the process of controlling, they must monitor and evaluate the activities to make sure they are being done as planned and correct any significant deviations [5] (p. 645). Therefore, a control process consists of measuring actual performance, comparing actual performance against standards and taking managerial action, taking into account the goals and objectives of the organization.

Intelligent systems for controlling aim to imitate and automate all or some controlling behaviors of managers of organizations. More specifically, they should imitate and automate monitoring and evaluation of activities, measurement of actual performance, comparison of actual performance against standards and recommendations of managerial decisions. Digital surveillance and CCTV (closed circuit TV) camera, DSS, and intelligent agents have been used to monitor and evaluate activities and recommendations of managerial decisions. However, intelligent agents for controlling to replace the controlling behaviors of managers are still a big challenge for organizations and AI.

Even though we have briefly looked at intelligent systems for planning, organizing, leading and controlling, there is still a number of works to do to provide effective intelligent systems for main management functions in the real world. Furthermore, there are still no serious attempts towards unifying them into a comprehensive intelligent system to automate planning, organizing, leading and controlling at an organizational level although ERP software have realized the automation of business processes at enterprise levels to some extent [3]. Any attempt in this direction is significant for development of MIS.

7   Intelligent Systems for Management Decision Making

Management decision making (MDM) involves all the decision making of managers when they complete their management functions in an optimal way to achieve their business objectives. Much of a manager's work involves making decisions on planning, organizing, leading and controlling [12] (p.26). In other words, each management function of planning, organizing, leading and controlling involves management decision making as shown in Fig. 4. In this section we will look at a number of intelligent techniques and systems for supporting decision making of managers in organizations to some extent.

Similar to what we discussed in the previous section, we can also look at intelligent systems for planning decision making, intelligent systems for organizing decision making, intelligent systems for leading decision making and intelligent systems for controlling decision making.

Decision support systems (DSS) have been developed since the 1970s to combine models and data to aid decision makers to make decision with some semi-structured and unstructured problems. AI has been applied to DSS to provide automated decision support for solving repetitive managerial problems [3]. Therefore, intelligent systems for management decision making (MDM) can be replaced by DSS for planning, DSS for organizing, DSS for leading and organizing, DSS for controlling, that is, we have

Intelligent systems for MDM = DSS for planning + DSS for organizing
                    + DSS for leading and organizing + DSS for controlling

Therefore, it is fundamentally significant for MIS to develop DSS for planning, DSS for organizing, DSS for leading and organizing and DSS for controlling separately and then integrate them into a comprehensive intelligent DSS to aid managers of organizations to make decisions on planning, organizing, leading and controlling. We do not discuss this topic any more owing to the space limitation.

8 A Strategic Model for Management Intelligent Systems

This section provides a strategic model for management intelligent systems (MIS) through integrating main management functions with intelligent systems taking into account management decision making of managers in organizations. This model is shown in Fig. 4. This section also proposes the core components of MIS based on this model.

The four main management functions are located at middle level of the model. These are also what a manager of an organization does in order to realize organizational goals and improve performance. Therefore, they are core to MIS.

Every management function requires managers to make decisions. Decision making is at the top level which is involved in every management function, and then DSS for planning, DSS for organizing, DSS for leading and DSS for controlling are necessary for managers of organizations to make decisions on planning, organizing, leading and controlling. Therefore, as applications of intelligent systems, DSS for planning, DSS for organizing, DSS for leading and DSS for controlling are also core to MIS.

Any organization or business activity requires some, and to varying degree, planning, organizing, leading and controlling, if not all, as shown in Fig. 4. Therefore, MIS should seek intelligent systems and techniques to enhance either planning or organizing or leading or controlling, which lead to intelligent systems for planning, intelligent systems for organizing, intelligent systems for leading, and intelligent systems for controlling, all of them are as intelligent systems at bottom level. They support intelligent management in terms of intelligent planning, intelligent organizing, intelligent leading and intelligent controlling for managers of organizations. Thus, all of these four intelligent systems are also the core to MIS.

Therefore, from a strategic perspective, the foundation or core components of MIS forming a new paradigm or discipline should be

1. Planning, intelligent systems for planning and DSS for planning.
2. Organizing, intelligent systems for organizing and DSS for organizing.
3. Leading, intelligent systems for leading and DSS for leading.
4. Controlling, intelligent systems for controlling and DSS for controlling.

It should be noted that competitive analysis, current and potential markets' analyses, e-business, human resources, logistics and supply chain management, production and operations, sales management and strategic marketing and marketing-mix, to name a few, are interesting areas from a viewpoint of management or business, however, they are not the core of MIS, and can only be considered as the applications of management or business. For example, e-business is already a matured application field of business, and supply chain management can be considered as an application area of management and business or commerce [2][11]. Similarly, ad hoc machine learning approaches, artificial neural networks, association rules, bio-inspired optimization/learning algorithms, clustering, fuzzy logic and fuzzy systems, genetic algorithms and evolutionary computation, probabilistic graphical models, probabilistic logic, support vector machines, web mining belongs to intelligent systems or artificial intelligence [13]. They are not the core of MIS either, although each of them can be used to develop intelligent systems for planning, intelligent systems for organizing, intelligent systems for leading, and intelligent systems for controlling of MIS or DSS for planning, DSS for organizing, DSS for leading and DSS.

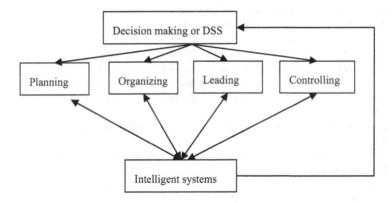

Fig. 4 A strategic model for Management Intelligent Systems

9   Conclusion

This article examined management intelligent systems (MIS) from a strategic perspective. More specifically, this article first examined management in information systems, management for intelligent systems and intelligent systems for management. Then this article provided a strategic model for MIS through integrating main management functions with intelligent systems taking into account decision making of managers in organizations and argued what are the core components of MIS. The approach will facilitate the research and development of MIS, management and intelligent systems and information systems.

In the future work we will elaborate intelligent systems for each of management functions based on more literature review; investigate intelligent techniques for management decision making and develop a prototype of an intelligent system for planning.

Business systems can be considered as a combination of business activities and business operations. Intelligent management systems in operations have been studied for over a decade. In the future work, we will integrate intelligent management systems with MIS.

References

[1] CFP, ISMiS,
    http://ismis.usal.es/sites/default/files/
    ismis-extended.pdf
[2] Chaffey, D.: E-Business and E-Commerce Management: Strategy, Implementation and Practice. Prentice Hall, England (2009)
[3] Turban, E., Volonino, L.: Information Technology for Management: Improving Performance in the Digital Economy, 7th edn. John Wiley & Sons, Hoboken (2010)
[4] Laudon, K., Laudon, J.: Essentials of Management Information Systems. Prentice Hall, Boston (2011)
[5] Robbins, S., et al.: Management, vol. 6. Pearson, Frenchs Forest (2012)
[6] ACM. IS 2010 Curriculum Guidelines (2010),
    http://www.acm.org/education/curricula/
    IS%202010%20ACM%20final.pdf (cited: February 16, 2012)
[7] Sun, Z., Zhang, P., Dong, D.: Customer Decision Making in Web Services. In: Kajan, E., Dorloff, F.-D., Bedini, I. (eds.) Handbook of Research on E-Business Standards and Protocls: Documents, Data, and Advanced Web Technologies. IGI, Hershey (2012)
[8] Finnie, G., Sun, Z.: R5 model for case-based reasoning. Knowl.-Based Syst. 16(1), 59–65 (2003)
[9] Finnie, G., Sun, Z.: Intelligent techniques in E-Commerce: A Case Based Reasoning Perspective, Berlin, Heidelberg (2004)
[10] Sun, Z., Finnie, G.R.: Experience Management in Knowledge Management. In: KES, Melbourne, vol. 1, pp. 979–986 (2005)
[11] Chaffey, D., White, G.: Business Information Management, 2nd edn. Prentice Hall, England (2011)
[12] Bocij, P., Greasley, A., Hickie, S.: Business Information Systems: Technology, Development & Management. Prentice Hall, England (2008)
[13] Russell, S., Norvig, P.: Artificial Intelligence: A Modern Approach, 3rd edn. Prentice-Hall, Upper Saddle River (2010)
[14] Khan, M.Z., et al.: Intelligent Supply Chain Management. Journal of Software Engineering and Applications 3(4), 404–408 (2010)
[15] Baxter, N., Collings, D., Adjali, I.: Agent-Based Modelling — Intelligent Customer Relationship Management. BT Technology Journal 21(2), 126–132 (2003)

Visualization of Agents and Their Interaction within Dynamic Environments

Elena García[1], Virginia Gallego[1], Sara Rodríguez[1], Carolina Zato[1], and Javier Bajo[2]

[1] Computer and Automation Department, University of Salamanca, Salamanca, Spain
{elegar,sandalia,srg,carol_zato}@usal.es
[2] Computer Department, Pontifical University of Salamanca, Salamanca, Spain
jbajope@upsa.es

Abstract. Many new technical systems are distributed systems that involve complex interaction between humans and machines, which notably reduces their usability. The properties of Agent Based Simulation make it especially suitable for simulating this kind of system. However, it is necessary to define new middleware solutions that allow the connection of simulation and visualization software. This paper describes the results achieved from a multiagent-based middleware for the behavior simulation and visualization of agents. The middleware modules presented in this study allow a complete integration of technologies for the development of Multiagent Systems and Agent Based Simulation, the construction of virtual organizations of agents, and the connection to external modules that represent the entities of the agents.

Keywords: Multi-agent systems, Simulation, Visualization, Virtual Organizations.

1 Introduction

Agents and Multiagent Systems (MAS) are adequate for developing applications in dynamic, flexible environments. Autonomy, learning and reasoning are especially important aspects for an agent. The development of open MAS, and of Virtual Organizations (VO) [17][18][23][6] in particular, is still a recent field of the multiagent system paradigm, and its development will allow the application of agent technology in new and more complex application domains. The contribution from agent based computing to the field of computer simulation mediated by ABS (Agent Based Simulation) provides benefits such as methods for evaluation and visualization of multi agent systems, or training future users of the system [7]. Many new technical systems are distributed systems that involve complex interaction between humans and machines, which notably reduces their usability. The properties of ABS make it especially suitable for simulating this kind of system. The idea is to model the behaviour of human users in terms of software agents.

J. Casillas et al. (Eds.): Management Intelligent Systems, AISC 171, pp. 15–24.
springerlink.com   © Springer-Verlag Berlin Heidelberg 2012

However, it is necessary to define new middleware solutions that allow the connection of ABS simulation and visualization software.

There are two ways to visualize Multiagent **System** simulation: agent interaction protocol and agent entity. In the former, a sequence of messages between agents and the constraints on the content of those messages is visualized. The latter method visualizes the entity agent and its iteration with the environment. Most software programs, such as JADE platform [2][15] and Zeus toolkit [5], provide graphical tools that allow the visualization of the messages exchanged between agents.

The toolkits MASON [12], Repast [13][16] and Swarm [19] provide the visualization of the entity agent and its interaction with the environment. Swarm [19] is a library of object-oriented classes that implements the Swarm conceptual framework for agent-based models, and provides many tools for implementing, observing, and conducting experiments on ABS. MASON [12] is a multiagent simulation library core developed in Java. It provides both a model library and an optional suite of visualization tools in 2D and 3D. Repast [16] is a free and opensource agent-based modelling and simulation toolkit. There are other studies like Vizzari et al. [22] that have developed a framework supporting the development of MAS-based simulations based on the Multilayered Multiagent Situated System model provided with a 3D visualization.

This paper describes the results achieved from a multiagent-based middleware for the behavior simulation and visualization of agents. The middleware, called MISIA (*Middleware Infrastructure to Simulate Intelligent Agents*), allows simulation, visualization and analysis of agent behavior [8]. MISIA makes use of technologies for the development of well-known and widely used Multiagent Systems, and combines them so that it is possible to use their capabilities to build highly complex and dynamic systems. Moreover, MISIA presents a reformulation of the FIPA protocol [20] used in JADE [15], achieving several advantages such as independence between the model and visualization components, improvement on the visualization component, which makes it possible to use the concept of "time", essential for simulation and analysis of the behavior of agents, and improvements to user capabilities, including the addition of several tools such as message visualization, 2D (and 3D agents), analysis behavioral, statistics, etc [8].

The main contribution of this paper is to present the modules that form the core of the platform and allow: the integration of both technologies, the construction of Virtual Organizations of agents, and the connection to external modules to represent the entities of the agents (in particular with a three-dimension).

The article is structured as follows: Section 2 introduces a description of the modules within the middleware specifically adapted to the simulation of MAS (MISIA). Section 3 shows some experimental results and, finally, conclusions and future works are provided in Section 4.

2 MISIA Characteristics

MISIA [8] is a framework for agent simulation, the result of the symbiosis of two well-known agent platforms [15][16]. Its purpose is to achieve a complete end tool where developers can build open and dynamic systems, analyze, simulate and visualize the behavior of a MAS design. The platforms used are: (i) JADE [15], a free environment for the development of multi-agent systems (MAS) that simplifies the implementation of MAS through a middleware that complies with FIPA specifications and has graphical tools for debugging, and (ii) Repast Simphony [16][13][14], which is also open source and is intended to the field of Agent Based Simulation (ABS).

MISIA is primarily used for the design and implementation of models for social simulation. The reasons for carrying out a simulation of this type can be either explanatory (for processes carried out in social phenomena) or predictive (providing information on evolutionary, complex and adaptive processes of the system, and enabling the study of emergent behavior of agents in such systems).

A social system can be defined as a collection of autonomous individuals, with perception of the environment and possession of their own beliefs and goals (agents). These individuals can interact with each other and communicate, either directly or through the environment, and moreover, can evolve over time. All of these factors are offered by MISIA: interaction and communication are performed through message passing between agents, in compliance with FIPA standards, as a result of the collaboration of JADE; and time is provided by Repast Simphony, where the unit of time is the *tick*.

In short, this study provides a detailed analysis of simultaneous agent interactions without a predetermined order in any simulation of agents. The objective is to obtain predictive systems models that can produce unexpected behavior and interactions not covered.

The block diagram below shows the architecture of this platform, consisting of three main modules which will be explained in detail in later subsections.

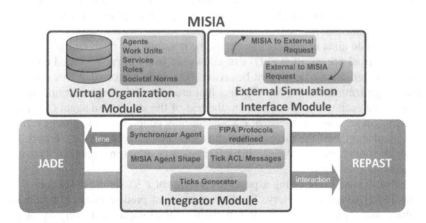

Fig. 1 Block diagram of MISIA

2.1 Integrator Module

For the combination of JADE and Repast Symphony, various components located in this module, which can be considered the core of the platform, were implemented. The main objectvive of the Module Integrator is to provide developers with an API (Application Programming Interface) containing the main features of JADE and Repast adapted to the new platform:

- ACL messages adapted to the notion of time, that is, with information (on *tickticks*) of message delays and information at the moment that messages are sent and received;
- a "time generator" that propagates the *tickticks* generated by Repast to the rest of the platform with the aid of a *synchronizer agent*. This agent informs when a *tick* is passed to a platform agents through specific ACL messages, thus acting as a clock for all agents;
- An agent template that is derived from the template provided by JADE and includes capacities for time management. These capabilities are: an inherent behavior that can process messages received during *tick*, and process messages with time information automatically;
- and finally, a module in charge of redefinint the FIPA protocols available on JADE for use with the notion of time [8].

2.2 Virtual Organization Module

Living together in society fosters the occurrence of unpredictable events that are not possible in isolated environments [3] [1]. Complex adaptive systems have the ability to simulate agents in their social evolution. In these simulations, the agents interact based on partial information, to achieve their individual goals, influenced by the behavior of other agents. Agents somehow compete to adapt their behavior to that of other agents, which leads to the emergence of global characteristics in the simulated model.

This module aims to give the developer the possibility of designing a society of agents, a virtual organization (VO) that allows the study of pre-specified characteristics and prediction of emergent behaviors. The main factors that characterize a virtual organization are the set of agents that are part of the organization, and the services that each of these can offer to the rest of the society, an agent communication language, a set of roles that each agent may adopt, a set of rules governing the organization, and the different subunits into which the agents are grouped as part of the organization [17]. All this information is stored in a database provided by the MISIA platform.

One of the most interesting aspects to consider of a VO is the definition of the norms [4][9] governing society. There are a series of predetermined norms in the database that are common to all VO including, for example, that every agent that exists in the organization must adopt a certain role at all times, or that an agent should always belong to a subunit of the virtual organization. However, the syntax of the norms that a developer can design for their model is limited to the particular case study. The reason for this decision is to avoid restricting the type of rules that can be defined with a predefined syntax, leaving open the possibility of increasing the range of possibilities for creating norms.

Another important aspect to consider is the regulation of access to the VO. Access to the database of the VO is governed by an interface implemented as a set of services provided by the *OVMisiaAgent*. Thus, it is possible to have access control information and the handling thereof from the VO. The main problem that a dynamic virtual organization faces is the action to be taken when an external agent appears and wants to join the organization. The benevolence of agents can be assumed, but real cooperation with other agents is not guaranteed. That is why the OVMisiaAgent exists, as it can enforce strict control required in virtual organization.

In order to avoid having only one agent in charge of the entire database, there is a pre-established norm in the VO: an agent with specific authorization can directly access the VO database, and is also able to generate authorizations for other agents of the VO. This is done to avoid the problems that may arise from having a single agent in charge of this functionality (such as the formation of bottlenecks due to multiple requests at successive times, or the loss of access to database if the single agent were to fall). The *OVMisiaAgent* agent is allowed to authorize other agents to distribute the charge among more entities. This simple scheme provides an optimization in the management of the OV. It is possible to create agents dynamically for this role by taking into account the number of agents that constitute the organization, and the frequency of requests to the organization's database.

2.3 External Simulation Interface Module

One of the reasons that social simulation gains importance when declaring itself a means of modeling social phenomena and a tool for social research, is that the representation of the procedures used in the simulation model are versatile. This means that the resemblance achieved in relation to the processes of the real world is quite acceptable compared to other kinds of models, such as those of a natural or even mathematical language [11]. The process of representation can become as complex as desired, making it easy to specify many details of the real process in the simulated model.

This module emphasizes this characteristic as its main goal. The ABS platforms offer many advantages and possibilities regarding the modeling and simulation of a real environment, but it sometimes needs greater versatility. Moreover, because of the actual power of graphic applications, any real situation can be modeled in as much detail as needed. As an example, the behavior of a group of predators and prey can be modeled in an ABS platform to observe the behavior of both species as a function of the number and arrangement of the animals. However, even though ABS platforms such as Repast Simphony provide 3D simulations, they are generally focused on the visualization of results and the synthesizing of huge amounts of data in 3D representations. These types of simulations are normally based on statistical factors which determine the probability of a predator getting its prey from a certain distance, for example. But the real number of variables which intervene in this decision is higher, such as the condition of the ground and its influence on the ability to reach the prey, the number and type of obstacles standing in the way, or the abilities of each species. ABS platforms are unable to handle this kind of detail, but other types of applications can model these situations in 3D whenever they arise, which means they are able to design a real life simulation.

External Simulation Interface Module is an optional interface which allows the connection between the platform agents and an external application with this type of simulation. The goal is to have agent representations in the external application, so that they can achieve clear communication with the respective platform agents.

There are four main actions which can be made concerning the agents: creation, erasure, modification and interaction. The interface offers the possibility of creating or removing an agent, both from the external application or the platform, so that results are correctly updated both in the platform or the external application. Concerning agent modification, it is possible to change a characteristic parameter of an agent, for example if the prey trips over an obstacle and its speed decreases. Finally, interaction among agents refers to defining a connection between two agents, which is normally the request of a service by one agent to another agent offering it. All these changes may have an influence on the representation of the external application, which is something to take into account. Moreover, the response of an agent to a request of its service might be influenced by the environment of the external application in which it is located at that moment.

The platform offers two kinds of interfaces for this purpose, which can be used simultaneously:

- MISIA to External Request Interface. The purpose of this interface is to ensure that the changes made in the agent platform are updated in the representation of the external application.
- External to MISIA Request Interface. Its function is to update the agent platform with all the changes made in the external application. In this case, when MISIA receives a request it is first necessary to notify the agent or agents which deal with the new information of this change in order to then update it. For this purpose, a specific agent is created to proceed with the notification of this information by means of ACL messages.

Communication between applications is carried out by means of TCP sockets: port 8898 for the first interface and port 8900 for the second interface. The basic unit of information exchanged in both directions is represented as shown in the following figure:

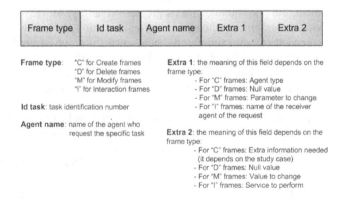

Fig. 2 Basic unit of information exchange

3 Experimental Results

In order to test the MISIA agent platform, a Multiagent System was developed to simulate an office environment and study the problems of accessibility experienced by people with disabilities in performing different jobs. The MAS is designed as a Virtual Organization similar to reality, that is, all workers, jobs and interaction elements such as architectural barriers are modeled as agents, and these are grouped into departments according to their availability and their occupation (Human Resources Department, Quality Department, Production Department with the Costumer Service and Mail sub-departments). This application, which is held in MISIA, connects to another application developed with the Unity 3D engine [21] that simulates the events of the office in 3D. It also includes other features like the ability to create and delete agents, or to configure different architectural barriers from the 3D simulation application. The following figure illustrates an example of using the system and shows the interaction of agents in JADE, in REPAST and 3D application.

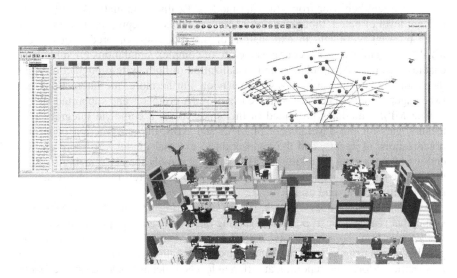

Fig. 3 Application for the study of the employment of disabled people

The main purpose of this application is to search for the optimal working conditions of the employees in the office, allowing for greater efficiency. For this purpose, several simulations of the tasks that workers have to perform, and a 3D simulation application will represent and determine the degree of success of employees in their work. There may frequently occur unusual cases in the simulations, such as a person who needs a wheelchair and cannot access the top floor of the office because the elevator is broken and no ramps are enabled; or a worker who takes a long time to perform certain tasks because the floor contains a step and accessing the destination may require a longer detour. For added versatility in the simulations, and because the application was not dependent on the plan and the disposition of the office, the system uses algorithm A * [10] to search for the shortest path. Thus represented agents are able to find its optimal path for tasks that have to do.

The case study was modeled as a MAS, making it possible to study, at a low-level, all interactions that agents have with their environment, and to then analyze and visualize the results in Repast in order to, after many simulations, predict results. Thus, given an initial configuration for the VO agents, it is possible to predict what the optimal disposition for the work environment is. The three-dimensional simulation of the office environment here is a great incentive to make the visualization more versatile and accurate, and to provide a much more interactive interface for users of simulation application.

4 Conclusions and Future Works

One of the most important features of social simulations is that they can easily observe emergent behavior from studying models. Realistic simulations, with a significant level of detail, though complex, are best suited to represent processes that study or want to obtain an explanation of the processes, or predict outcomes. MISIA is a platform that encourages the use of complex simulations for study, and enables the analysis, simulation and visualization of both system interactions and the results obtained in a multi-agent behavior. Furthermore, the interactions between agents are well defined with the use of FIPA protocols and supported VO.

The MISIA modules presented in this study allow a complete integration of technologies for the development of MAS and ABS, the construction of virtual organizations of agents, and the connection to external modules to represent the entities of the agents. In addition, MISIA provides tools for customizing the display of the simulation, such as simulation speed, pausing, resuming, etc., which is more intuitive for an end user.

As possible future lines of work, we are considering first of all, the design of a syntax that is as generic as possible regarding the specification of the norms of a virtual organization. To achieve this goal, the syntax cannot be too restricting for the possible norms that can be defined, but must instead allow these particular norms to be specified. Moreover, reporting at different levels of detail would be an interesting aspect to consider, as it would be an incentive to understandsystem processes and to improve the study can be performed on the results of the simulations.

Acknowledgments. This work has been supported by the Spanish JCyL project SA225A11-2.

References

[1] Acevedo, D.L.: Aprendizaje y comportamiento social y emergente en sociedades artificiales, Universidad Nacional de Colombia, Seminario de investigación, Maestría en Ingeniería, pp.1–9 (2007)

[2] Bellifemine, F., Caire, G., Poggi, A., Rimassa, G.: Jade a white paper. EXP in Search of Innovation 3(3), 6–19 (2003)

[3] Conte, R., Paolucci, M.: Intelligent social learning. Artificial Society and Social Simulation 4(1), 1–23 (2001)

[4] Castelfranchi, C., Conte, R.: The treatment of norms in modelling rational agents: Cognitive issues. In: Proceedings of Model Age 1995, General Meeting of Esprit Working Group, vol. 8319, INRIA, France (1995)

[5] Collis, J.C., Ndumu, D.T., Nwana, H.S., Lee, L.C.: The zeus agent building tool-kit. BT Technol Journal 16(3) (1998)

[6] Corchado, E., Pellicer, M.A., Borrajo, M.L.: A MLHL Based Method to an Agent-Based Architecture. International Journal of Computer Mathematics 86(10,11), 1760–1768 (2008)

[7] Davidsson, P.: Multi Agent Based Simulation: Beyond Social Simulation. In: Moss, S., Davidsson, P. (eds.) MABS 2000. LNCS (LNAI), vol. 1979, pp. 97–107. Springer, Heidelberg (2001)

[8] García, E., Rodríguez, S., Martín, B., Zato, C., Pérez, B.: MISIA: Middleware Infrastructure to Simulate Intelligent Agents. In: Abraham, A., Corchado, J.M., González, S.R., De Paz Santana, J.F. (eds.) International Symposium on Distributed Computing and Artificial Intelligence. AISC, vol. 91, pp. 107–116. Springer, Heidelberg (2011)

[9] Hales, D.: Group reputation supports beneficent norms. The Journal of Artificial Societies and Social Simulation (JASSS) 5(4) (2002)

[10] Hart, P.E., Nilsson, N.J., Raphael, B.: A Formal Basis for the Heuristic Determination of Minimum Cost Paths. IEEE Transactions on Systems Science and Cybernetics 4(2), 100–107 (1968), ISSN: 0536-1567, doi:10.1109/TSSC.1968.300136

[11] Lozares, C.: La simulación social,?'una nueva manera de investigar en ciencia social? Papers: revista de sociología (Ejemplar dedicado a: Visions alternatives sobre la societat i la realitat social) 72, 165–188 (2004) ISSN 0210-2862

[12] Luke, S., Cioffi-Revilla, C., Panait, L., Sullivan, K.,, M.: A new multiagent simulation toolkit. In: Proceedings of the 2004 Swarm Fest Workshop (2004)

[13] North, M.J., Howe, T.R., Collier, N.T., Vos, J.R.: The repast symphony runtime system. In: Proceedings of the Agent 2005 Conference on Generative Social Processes (2005)

[14] North, M.J., Collier Nicholson, T., Vos Jerry, R.: Experiences Creating Three Implementations of the Repast Agent Modeling Toolkit. ACM Transactions on Modeling and Computer Simulation 16(1), 1–25 (2006)

[15] JADE, Java Agent Development Platform, http://JADE.tilab.com

[16] Repast, http://repast.sourceforge.net/repast_3/index.html

[17] Rodríguez, S., de Paz, Y., Bajo, J., Corchado, J.M.: Social-based Planning Model for Multiagent Systems. Expert Systems with Applications 38(38), 13005–13023 (2011), doi:10.1016/j.eswa.2011.04.101

[18] Rodríguez, S., Pérez-Lancho, B., Bajo, J., Zato, C., Corchado, J.M.: Self-adaptive Coordination for Organizations of Agents in Information Fusion Environments. In: Corchado, E., Graña Romay, M., Manhaes Savio, A. (eds.) HAIS 2010. LNCS (LNAI), vol. 6077, pp. 444–451. Springer, Heidelberg (2010)

[19] Swarm, http://www.swarm.org

[20] Foundation for Inteligent Physical Agents."FIPA Agent Management Specification". Disponible en,
http://www.fipa.org/specs/fipa00001/SC00001L.html

[21] Unity 3D Engine, http://unity3d.com/

[22] Vizzari, G., Pizzi, G., da Silva, F.S.C.: A framework for execution and visualization of situated agents based virtual environments. In: Workshop dagli Oggetti agli Agenti, pp. 22–25 (2007)

[23] Zambonelli, F., Jennings, N.R., Wooldridge, M.: Developing Multiagent Systems: The Gaia Methodology. ACM Transactions on Software Engineering and Methodology 12, 317–370 (2003)

Hybrid Genetic-Fuzzy System Modeling Application in Innovation Management[*]

Kemal Kilic[1] and Jorge Casillas[2]

[1] FENS, Sabanci University, Tuzla, Istanbul, Turkey
[2] Dept. Computer Science and Artificial Intelligence, Research Center on Information and Communication Technologies (CITIC-UGR), University of Granada, Spain

Abstract. In this research a three staged hybrid genetic-fuzzy systems modeling methodology is developed and applied to an empirical data set in order to determine the hidden fuzzy if-then rules. The empirical data was collected in an earlier study in order to establish the relations among human capital, organizational support and innovativeness. The results demonstrate that the model based on the fuzzy if-then rules outperforms more traditional techniques. Furthermore, the proposed methodology is a valuable tool for successful knowledge management.

Keywords: Knowledge Management, Innovation.

1 Motivation

Knowledge Management (KM) focuses to the generation, sharing, maintenance, refinement and application of knowledge in organizations. The earliest applications of KM are mostly limited to various hardware solutions for data storage and processing. On the other hand, with the advancement of information technologies and information system tools terabytes of data is being gathered by the firms from all sorts of processes and transactions. However, *data* has limited value (if any) unless the patterns hidden in it (*information*) are brought to surface and transformed to capability to act (*knowledge*). Therefore, concepts such as artificial intelligence, decision support systems and data mining enhance the KM capabilities, which is invaluable for the sustainable competitiveness of companies.

Operations Research and Management Science (OR/MS) literature mostly focuses on operational level decision making problems and seems to neglect the strategic level problems due to the vagueness and/or lack of the mathematical formulation of such problems. However, strategic level decisions have significant influence on the competitiveness of the companies. Therefore there is a tremendous need for tools that can be utilized by the senior managers, which guides their

---

[*] This work has been supported in part by the *Andalusian Government* (Spain) under grant P07-TIC-3185 and in part by The Scientific and Technological Research Council of Turkey under grant TUBITAK -2219.

J. Casillas et al. (Eds.): Management Intelligent Systems, AISC 171, pp. 25–34.
springerlink.com   © Springer-Verlag Berlin Heidelberg 2012

strategic level decisions based on knowledge. The first step of developing such tools is generating the relevant knowledge. The upper level managers' knowledge on strategic issues is mostly based on experience and common sense. Therefore, the OR/MS researchers should focus more to strategic level issues and generate scientifically validated knowledge based on objective data and methodologies.

This paper aims to contribute to reduce the existing gap in the literature. For this purpose, with the help of artificial intelligence tools, empirical data collected regarding to a strategic level decision making problem is analyzed. Therefore the research is part of *knowledge generation* process, particularly generation of *explicit knowledge* for a strategic level decision making problem based on objective data. On the other hand, the main research problem, i.e., the determination of the role of human capital and organizational support in the firm level innovation process, is also closely related with KM; hence, the resulting knowledge from the analysis *itself* is highly relevant to the KM literature.

The paper also aims to contribute to the literature in terms of the methodology that is utilized in the data mining phase of such research. Generally speaking there are two different methodological perspectives in data mining process. These are namely the *predictive induction* and the *descriptive induction*. The predictive induction is conducted with the purpose of classification and mostly utilizes supervised learning techniques. On the other hand, the main objective of the descriptive induction is identification of *interesting patterns* and mostly utilizes unsupervised learning techniques (Casillas and Martínez-López, 2009).

Both of the perspectives are sound and valid. Therefore, the *most appropriate* perspective that should be utilized in the analysis depends totally on the application itself. In this research, the main objective is generation of explicit knowledge that can be utilized by the senior managers. Basically, the developed methodology should yield "valuable" information for the decision makers. Valuable information implies that it should be understandable by the managers (so that it can easily be transformed to knowledge), it should address various different subgroups so that the resulting knowledge spans a wide range of the universe and at the same time the information should represent a model that *fits* to the empirical data (i.e., have an acceptable predictive accuracy). Hence, the application on hand, requires an approach which synthesis the two perspectives in order to employ the best from the both approaches. Therefore, in this research, a novel methodology which resembles both *descriptive* (Martínez-López and Casillas, 2009) and *predictive* (Casillas and Martínez-López, 2009) induction is developed.

Among various different techniques in the fields of data mining and machine learning, Fuzzy System Modeling (FSM) emerges as a promising tool that can be utilized in the analysis. Hence, a hybrid genetic-fuzzy system modeling methodology which utilizes the multiobjective genetic fuzzy system approach (Casillas and Martínez-López, 2009; Martínez-López and Casillas, 2009) is developed. Note that, the interpretability of the fuzzy if-then rules, which are usually characterized in natural language, is what sets apart FSM among other candidates and makes it the most relevant choice in this paper. On the other hand, by incorporating the *predictive accuracy* as one of the objective functions during the rule selection process, its ability to represent the empirical data accurately is also addressed.

To sum up, in this paper a novel hybrid genetic-fuzzy system modeling methodology is developed which utilizes multiobjective genetic fuzzy systems applied to a strategic level decision making problem namely innovation management in order to generate knowledge that might be useful for the senior managers. The developed framework is not limited to the particular research problem but is also applicable to various strategic level management problems as well.

In the next section, relevant literature regarding to the organizational support, human capital and innovation performance as well as the problem statement will be presented. In Section 3, the methodology that will be utilized in the research will be provided. Later a brief summary of the data set will be provided in Section 4. The details of the experimental analysis and the results will be covered in Section 5. The paper will be concluded with some concluding results.

2 Organizational Support, Human Capital and Innovativeness

Innovation can be described as the value adding changes in business processes, services, products, marketing or the ways that the works are organized in a company. Schumpeter (1934) differentiated between five different types of innovation: new products, new methods of production, new sources of supply, the exploitation of new markets, and new ways to organize business. Literature on innovation management demonstrates that companies should manage their innovation performance carefully in order to stay competitive. On the other hand, the *innovativeness* of the company refers to the capability of making innovations or the degree of success of the innovation management performance.

Various factors are shown to be influencing the firm level innovativeness. Among these *organizational support* enhances the innovativeness particularly at the individual employee level. Organizational support can be shaped by some managerial arrangements, such as *work discretion, rewarding systems, management support* for generation of new ideas, allocation of *time availability*, and *tolerance for failures* in creative undertakings and risky innovation projects (Kuratko et al., 1990; Hornsby et al., 2002; Alpkan et al., 2010).

Work discretion, that is to say the ability to take initiative in decision making is demonstrated to enhance the innovativeness and overall performance of the companies(e.g. Alpkan et al., 2007). On the other hand, high level of trust among the employees in the reward system of their company also positively influences their commitment to innovativeness (Morrison and Robinson, 1997). Management support, i.e., the encouragement of new idea generation and development, positively influence a firms' entrepreneurial behavior and enhance potential intrapreneurs' willingness to innovate (Stevenson and Jarillo, 1990). Free time availability is demonstrated as another critical factor for the employers both daily routines and intrapreneurial ideas and activities (e.g., Fry, 1987) since most of the enthusiastic intrapreneurs make their pioneering steps to actualize their idealized projects in their spare times (Ende et al., 2003). Besides, if the employees feel free from any punishment, adverse criticism, or loss of support in case of failure of their projects or ideas, then their commitment to innovative attempts will be increased (e.g., Morrison and Robinson, 1997; Chandler et al., 2000).

Intellectual capital is the total stocks of all kinds of intangible assets, knowledge, capabilities, and relationships, etc, at employee level and organization level within a company. It is examined in the literature under three subgroups; namely, human, social and organizational capital. The *human capital* is the sum of knowledge and skills that can be improved especially by education and work experience of the employees of an organization (Dakhli and De Clercq, 2004). Hall and Mairesse (2006) states that a great deal of the knowledge created by firm activities is embedded in the human capital. Also, Cohen and Levinthal (1989) suggested that the human capital of a firm is crucial in terms of the innovativeness, due to its ability to obtain and make use of the outcomes of other firms' R&D activities. Human capital also enhances the organizational competencies of the firms by increasing the returns from the innovations and reducing the risks (Hayton 2005). Hence, the human capital not only has direct effect on innovativeness but is also a precious resource that may act as a moderator in the relationship of organizational support and innovativeness (Alpkan et al., 2010).

Alpkan et al. (2010) utilizes an empirical data set in order to analyze the relationship between the organizational support factors, human capital and innovativeness as well as the hypothesized moderator role of human capital. By using multiple linear regression analysis they managed to show that certain dimensions of organizational support factors such as the management support and tolerance for failures have positive influence on innovativeness but the relation of the other three organizational support factors (i.e., time availability, work discretion and reward system) and innovativeness was not supported. On the other hand, human capital is also demonstrated to have positive influence on innovativeness however regarding to the moderation role of human capital, only limited knowledge was attained. The reason for this was based on the limitations associated with the capabilities of the multiple linear regression analysis. Therefore, the knowledge attained as the result of the research was quite limited and left space for further analysis. One of the goals of this paper is utilizing the fuzzy system modeling approach in the same data set in order to attain further valuable knowledge for the decision makers.

3 Hybrid Genetic-Fuzzy System Modeling

The hybrid genetic-fuzzy system modeling methodology that is utilized in the research consists of three stages. The first stage is the identification of a large set of individual fuzzy if-then rules that might be of interest, based on their *confidence* (i.e, a measure of accuracy of the rule) and *support* (i.e., a measure of representativeness of the rule). Hence the problem dealt in this stage is basically a multi objective (accuracy vs. representativeness) decision making problem in which a set of fuzzy if-then rules are obtained as the result. These fuzzy if-then rules should have better scores from the other candidates that were not selected through out the process, in terms of the two conflicting objectives, i.e., their confidence and their support. Therefore, for this stage the multi objective genetic fuzzy system proposed in (Martínez-López and Casillas, 2009) is utilized.

Briefly speaking the genetic algorithm utilizes a gene pool which consists of chromosomes that represents a fuzzy if-then rule regarding to the relationship among the organizational support, human capital and innovativeness. In the model, the innovativeness is set to be the *consequence* of the fuzzy if-then rules and the human capital and organizational support factors (five of them) are set to be the antecedents of the fuzzy if-then rules. Note that, two fuzzy sets are assumed to represent each one of the antecedents (low - L and high - H). On the other hand five crisp values are assumed to represent the consequence in the if-then rules (due to the fact that the data set collected utilized five-point Likert scale as we will discuss in the next section). Therefore the size (i.e., width) of the chromosomes in the gene pool is equal to 17 ($2 + 5 \cdot 2 + 5$; L and H for human capital + L and H for five organizational support factors + 1..5 as the crisp output score for the antecedent, that is to say innovativeness).

The basic genetic operators such as selection, crossover and mutation are utilized as suggested in Martínez-López and Casillas (2009) in the algorithm. Furthermore, the support and the confidence of the fuzzy if-then rules are also determined as suggested in Martínez-López and Casillas (2009). On the other hand, the fitness function scores of the chromosomes are determined based on their domination rank and crowding distance values obtained from their support and confidence.

The second stage of the methodology is merely the elimination of the inconsistent rules among the set of fuzzy if-then rules that are obtained after the first stage. Note that the resulting fuzzy rule set after the first stage might include rules that are redundant and/or contradicting with others in the rule set. An example of an inconsistent rule is a rule which has antecedents that are subsumed by another rule but the output of the subsumed rule is different from the other one. In such cases the rules that has higher confidence is preserved in the set of fuzzy if-then rules and the other one is eliminated. Hence at the end of the second stage, the resulting set of fuzzy rules of stage one reduces in size and only includes consistent fuzzy if-then rules in order to be utilized in the final stage.

The third stage of the methodology is basically, selection of the best set of fuzzy rules among the set of *consistent* fuzzy rules obtained as the result of stage two. This process is also a multi criteria decision making problem since there are two objectives, namely maximization of the *prediction accuracy* (i.e., minimization of the root-mean-square error, RMSE) and minimization of *the number of rules being used* in the set. The prediction accuracy is important since it is kind of a measure of the *goodness-of-fit* of the model to the empirical data. On the other hand the number of rules being used should also be minimized simultaneously in order to assess the knowledge hidden in the data set (i.e., in order to enhance the interpretability and descriptiveness of the attained fuzzy if-then rules). Therefore, for the third stage, the algorithm that is based on the multi objective fuzzy-genetic system which was utilized in the first stage is adopted to the particular problem.

This time the gene pool consists of chromosomes (with size equal to the *cardinality* of the consistent fuzzy rule set which is obtained after the second phase) that represents a candidate set of fuzzy rules (1 represents that the rule is used and 0 represents it is not part of the final rule base). The fitness function is also similar to the fitness function used in stage one, i.e., the domination rank and the crowding distance values obtained from the *prediction accuracy* scores and the *number of rules being used* in the particular chromosome. Again the selection, crossover and mutation genetic operators are utilized as described in Martínez-López and Casillas (2009). Note that the prediction accuracy of each candidate set of fuzzy if-then rules are assessed by means of the training RMSE. The RMSE calculation is conducted by the methodology that was suggested by Casillas and Martínez-López (2009) for the predictive approach.

4 Data Collection

A questionnaire is developed for the empirical survey (Alpkan et al., 2010). In order to measure the human capital five criteria are constructed which were inspired from Subramaniam and Youndt (2005). Similarly organizational support measures were also adapted from several criteria in the Operations Management literature based on previous studies of Kuratko et al. (1990) and Hornsby et al. (2002). On the other hand the innovative performance is measured by means of a scale consisting of the items adapted from the earlier studies of Antoncic and Hisrich (2001), Neely and Hii (1998) and Hagedoorn and Cloodt (2003). All items were measured on a five point Likert scale as suggested in the literature. After the questionnaire was developed, the initial survey draft was discussed with various firms' executives and it was pre-tested through 10 pilot interviews to ensure that the wording, format and sequencing of questions are appropriate. Data was collected over a 7-month period using a self-administered questionnaire distributed to firms' upper level managers operating in manufacturing sectors in the Northern Marmara region in Turkey. A sample of 1,672 manufacturing firms was obtained by selecting randomly from various databases. Afterwards, the questionnaire was applied through a hybrid system of mail surveys and face-to-face interviews. Out of the sample of 1,672 firms, 184 complete responses were obtained resulting in 11% return rate. The data was later controlled with t-test procedure for non-respondent bias and no significant difference ($p \leq 0.05$) was found between the interview and mailing data sets' responses both in terms of the questionnaire items and constructs. Moreover, the issue of Common Method Variance was also attended.

Exploratory factor analysis (EFA) with varimax rotation and confirmatory factor analysis (CFA) to explore and confirm the latent factor structure of the innovative performance, human capital and organizational support factors' scales was conducted. The factor analyses (EFA and CFA) revealed that the hypothesized seven factors were sufficiently valid and reliable (with Cronbach's Alpha value ranging from 0.72 and 0.92 for the constructs). The seven constructs to be used in the analysis were namely, Human Capital, Performance Based Reward System, Management Support for Idea Generation, Tolerance for Risk Taking, Work Discretion, Allocation of Free Time (the latter five factors constitute the components of Organizational Support) and Innovativeness. The former six factors are treated as the inputs (antecedents of the fuzzy if-then rule) in the model where as the Innovativeness is treated as the output (the consequent).

5 Experimental Analysis and Results

A typical GA based algorithm requires at least four parameters to be tuned, namely, *Gene Pool Size* (GPS), *Number of Iterations* (NoI), *Mutation Probability* (MP) and *Crossover Probability* (CP). Since the multi objective genetic fuzzy systems algorithm is employed at two different stages 2·4=8 parameters were considered. For each one of the parameters low and high values are assigned (after some test runs) and an experimental analysis is conducted for parameter tuning. The resulting parameter values used in the analysis is GPS-1=100, NoI-1 = 20, MP-1 = 0.2, CP-1 = 0.6, GPS-2 = 100, NoI-2=40, MP-2=0.2 and CP-2 = 0.9.

The final obtained model consists of seven fuzzy if-then rules and yields a training RMSE equal to 0.388. A similar analysis with a multiple linear regression is conducted and calculated the RMSE in the same manner. The RMSE of the training experiments for the multiple linear regression was determined as 0.455 which suggests that the hybrid genetic-fuzzy system algorithm models the relations in the data better than the MLR (as expected).

Since the Gene Pool Size of the first stage for the tuned parameter set was equal to 100, there were 100 individual fuzzy if-then rules at the end of the first stage. Figure 1, depicts the confidence vs. support degrees of the resulting 100 fuzzy if then rules. Note that from the figure, one can realize how the two objectives, namely the domination rank vs. the crowding distance results.

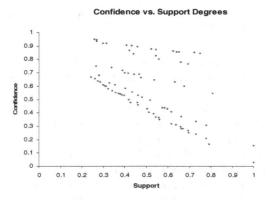

Fig. 1 Confidence vs. Support degrees of the resulting fuzzy if-then rules after the first stage

Among the 100 fuzzy if-then rules, 59 of them were determined to be either subsumed or contradicting with other rules, hence were eliminated during the second stage. Therefore, 41 relevant fuzzy if then rules were used in the third stage of the application. The multi objective genetic–fuzzy system algorithm utilized in the third stage, determined a set of seven rules which yields both good predictive accuracy (so that the empirical data is represented better with the model) and low number of rules (in order to enhance the descriptiveness of the fuzzy sets).

The corresponding seven fuzzy if-then rules that are obtained as the result of the third stage of the methodology and the associated confidence (*Conf.*) and the support (*Sup.*) levels of each fuzzy if-then rules are depicted in Figure 2. Note that Management Support (MS), Tolerance for Risk Taking (RT), Work Discretion (WD), Reward System (RW), Time Availability (TA) and Human Capital (HC) are the antecedents of the fuzzy if-then rules and are represented with two fuzzy sets, i.e., Low (L) and High (H). On the other hand, the consequence is the Innovativeness (I) and represented with five singleton results in which Very Low (VL) refers to a value of 1 and Very High (VH) refers to a value of 5 and the rest accordingly (i.e., Low (L), Medium (M) and High (H), 2, 3 and 4 respectively).

	MS	RT	WD	RW	TA	HC	I	Conf.	Sup.
Rule 1	H	L	H				VH	0.67	0.48
Rule 2	H						VH	0.54	0.81
Rule 3		L	H			L	M	0.87	0.42
Rule 4	L		H		H		M	0.92	0.30
Rule 5		L	L	L		L	VL	0.57	0.34
Rule 6		H	L	H		H	H	0.90	0.46
Rule 7	H	H	L	H			VH	0.69	0.47

Fig. 2 The resulting seven fuzzy if-then rules and associated confidence and support degrees. Training RMSE = 0.388.

The resulting fuzzy if-then rules were parallel with the results of Alpkan et al. (2010) in the sense that these rules also indicated that the Management Support, Tolerance for Risk Taking and Human Capital was positively influencing the innovativeness. On the other hand, the resulting fuzzy rules also demonstrates that the Work Discretion in fact negatively influencing the innovativeness which was not a result of Alpkan et al. but has support in the literature. Note that the Alpkan et al. analysis were merely based on MLR analysis and the utilized hybrid genetic-fuzzy system modeling methodology had a better *goodness-of-fit* (in terms of the prediction accuracy) to the empirical data. Furthermore the resulting fuzzy if-then rules also reveal the lack of the moderator role of human capital on the relation with the organizational support and innovativeness which was hypothesized but couldn't be demonstrated in Alpkan et al. study as well. Particularly Rule 6 demonstrates that whenever the organizational support is high, having human capital high as well not necessarily boosts the innovativeness. Therefore, indicating to a possible *substitute* relation between the human capital and organizational support rather than a *complementary* relation which might suggest synergy among the concepts.

The rules with higher confidence degrees (such as the rules 3, 4 and 7) apparently bear more accurate explicit knowledge in the context. On the other hand, the rules with relatively higher support degrees relate to knowledge on the combinations that are more commonly observed. However, the significance of the proposed methodology lies in the fact that, other rules might be even more interesting instead. In this example Rule 6 (as described above) as well as Rule 5 and Rule 1 reveals highly interesting knowledge for the senior managers.

6 Concluding Remarks

The three staged hybrid genetic-fuzzy system modeling methodology that was applied to a strategic level decision making problem in the context of innovation management demonstrated the strength of the descriptive fuzzy if-then rules in terms of explicit knowledge generation. Furthermore, the better prediction accuracy also hints the ability of fuzzy system modeling to model highly complex and nonlinear systems. The resulting fuzzy if-then rules were highly interesting and revealing and enhance the understanding of the complex interrelations in the problem. Therefore, the developed methodology might be a valuable tool and serve as an engine of a decision support system which might allow the upper level decision makers to make more informed and better decisions.

References

Alpkan, L., Yilmaz, C., Kaya, N.: Market Orientation and Planning Flexibility in SMEs: Performance Implications and an Empirical Investigation. International Small Business Journal 25(2), 152–172 (2007)

Alpkan, L., Bulut, C., Gunday, G., Ulusoy, G., Kilic, K.: Organizational Support for Intrapreneurship and its Interaction with Human Capital to Enhance Innovative Performance. Management Decision 48(5), 732–755 (2010)

Antoncic, B., Hisrich, R.D.: Intrapreneurship: Construct refinement and cross-cultural validation. Journal of Business Venturing 16, 495–527 (2001)

Casillas, J., Martínez-López, F.J.: Mining uncertain data with multiobjective genetic fuzzy systems to be applied in consumer behaviour modeling. Expert Systems with Applications 36, 1645–1659 (2009)

Chandler, G.N., Keller, C., Lyon, D.W.: Unraveling the determinants and consequences of an innovation-supportive organizational culture. Entrepreneurship: Theory and Practice 25(1), 59–76 (2000)

Cohen, W.M., Levinthal, D.A.: Innovation and learning, the two faces of R&D. Economic Journal 99, 569–596 (1989)

Dakhli, M., De Clercq, D.: Human capital, social capital, and innovation: A multicountry study. Entrepreneurship & Regional Development 16, 107–128 (2004)

Ende, J.V.D., Wijnberg, N., Vogels, R., Kerstens, M.: Organizing innovative projects to interact with market dynamics: A coevolutionary approach. European Management Journal 21(3), 273–284 (2003)

Fry, A.S.: The post it note: An intrapreneurial success. SAM Advanced Management Journal 52(3), 4–9 (1987)

Hagedoorn, J., Cloodt, M.: Measuring innovative performance: Is there an advantage in using multiple indicators? Research Policy 32, 1365–1379 (2003)

Hall, B.H., Mairesse, J.: Empirical studies of innovation in the knowledge driven economy. United Nations University Working Paper, 28 (2006)

Hayton, J.C., Zahra, S.A.: Venture team human capital and absorptive capacity in high technology new ventures. International Journal of Technology Management 31(3-4), 256–274 (2005)

Hornsby, J.S., Kuratko, D.F., Zahra, S.A.: Middle managers' perception of the internal environment for corporate entrepreneurship: Assessing a measurement scale. Journal of Business Venturing 17, 253–273 (2002)

Kuratko, D.F., Montagno, R.V., Hornsby, J.S.: Developing an intrapreneurial assessment instrument for an effective corporate entrepreneurship. Strategic Management Journal 11(5), 49–58 (1990)

Martínez-López, F.J., Casillas, J.: Marketing Intelligent Systems for consumer behaviour modelling by a descriptive induction approach based on Genetic Fuzzy Systems. Industrial Marketing Management 38, 714–731 (2009)

Morrison, E.W., Robinson, S.L.: When employees feel betrayed: A model of how psychological contract violation develops. Academy of Management Review 22(1), 226–256 (1997)

Neely, A., Hii, J.: Innovation and Business Performance: A Literature Review. The Judge Institute of Management Studies - University of Cambridge, Cambridge (1998)

Schumpeter, J.A.: The Theory of Economic Development. An Inquiry into Profits, Capital, Credit, Interest, and The Business Cycle. Harvard University Press, Cambridge (1934)

Stevenson, H.H., Jarillo, C.J.: A paradigm of entrepreneurship: Entrepreneurial management. Strategic Management Journal 11(5), 17–27 (1990)

Subramaniam, M., Youndt, M.A.: The influence of intellectual capital on the types of innovative capabilities. Academy of Management Journal 48(3), 450–463 (2005)

Applications for Non-Profit/Public Sector Organizations

Using Data Mining and Vehicular Networks to Estimate the Severity of Traffic Accidents

Manuel Fogue[1], Piedad Garrido[1], Francisco J. Martinez[1], Juan-Carlos Cano[2],
Carlos T. Calafate[2], and Pietro Manzoni[2]

[1] University of Zaragoza, Spain
{m.fogue,piedad,f.martinez}@unizar.es
[2] Universitat Politècnica de València, Spain
{jucano,calafate,pmanzoni}@disca.upv.es

Abstract. New communication technologies integrated into modern vehicles offer an opportunity for better assistance to people injured in traffic accidents. To improve the overall rescue process, a fast and accurate estimation of the severity of the accident represents a key point to help the emergency services to better determine the amount of required resources. This paper proposes a novel intelligent system which is able to automatically estimate the severity of traffic accidents based on the concept of data mining and knowledge inference. Our system considers the most relevant variables that can characterize the severity of the accidents (variables such as the vehicle speed, the type of vehicles involved, and the airbag status). Results show that data mining classification algorithms, combined with an adequate selection of relevant features and a prior division of collisions based on the impact direction, allows generating estimation models able to predict the severity of new accidents.

1 Introduction

During the last decades, the total number of vehicles in our roads has experienced a remarkable growth, making traffic density higher and increasing the drivers' attention requirements. The immediate effect of this situation is the dramatic increase of traffic accidents on the road, representing a serious problem in most countries. As an example, 2,478 people died on Spanish roads in 2010, which means one death for every 18,551 inhabitants [5].

To reduce the number of road fatalities, vehicular networks will play an increasing role in the *Intelligent Transportation Systems* (ITS) area. Most ITS applications such as road safety, fleet management, and navigation, will rely on data exchanged between vehicles and the roadside infrastructure (V2I) or even directly between vehicles (V2V) [12]. The integration of sensing capabilities on-board of vehicles, along with peer-to-peer mobile communication among vehicles, are expected to provide significant improvements in terms of safety in the near future.

Prior to achieving the zero accident objective on the long term, a fast and efficient rescue operation during the hour following a traffic accident (the so-called *Golden*

J. Casillas et al. (Eds.): Management Intelligent Systems, AISC 171, pp. 37–46.
springerlink.com   © Springer-Verlag Berlin Heidelberg 2012

Hour) significantly increases the probability of survival of the injured, and reduces the injury severity. Hence, to maximize the benefits of communication systems between vehicles, the infrastructure should be supported by intelligent systems capable of estimating the severity of accidents, and automatically deploying the actions required, thereby reducing the time needed to assist injured passengers.

In this paper, we take advantage of the use of vehicular networks to collect precise information about road accidents which is used to estimate the severity of collisions. We propose an estimation model based on data mining classification algorithms, trained using historical data about previous accidents. Our proposal does not focus on reducing the number of accidents, but on improving post-collision assistance.

The rest of the paper is organized as follows: Section 2 reviews the related work on data mining for accident severity estimation. Section 3 presents the architecture of our proposed automatic system to improve accident assistance. Sections 4, 5, and 6 provide details of our Knowledge Discovery in Databases (KDD) model adapted to the traffic accidents domain. Finally, Section 7 concludes this paper.

2   Previous Approaches on Traffic Accidents Data Mining

Despite the interest that may arise from understanding the influence that crucial factors will have on road accidents, the number of research works on this topic available the literature is not particularly large. In addition, most attempts to carry out a data mining process related to traffic accidents only considered data coming from a single city or a very small area, making results not so representative.

Several works are based on data obtained from the Traffic Office of Ethiopia, since this country presents one of the largest number of accidents per capita. Beshah et al. [1] used data from 18,288 accidents around Addis Ababa as the basic data set. This study uses Naïve-Bayes, decision trees, and k-nearest neighbors (KNN) algorithms to classify the data using a cross-validation methodology, with accuracy values close to 80%. However, the authors only provided estimations for the whole accident, not for single occupants. Data from Ethiopia was also used to build regression tree models for accident classification in [18], but the feature selection process was not shown, and again only estimations about the whole accident were provided.

The area of South Korea was also used to develop classification models based on artificial neural networks, decision trees, and logistic regression [17]. The data set involved 11,564 accidents, and the authors concluded that the different classification algorithms obtain similar results in terms of accuracy. The use of protection devices, such as the seat belt, and the airbag, was the most relevant factor detected to classify the severity of accidents. This work was later extended [16] using multiple models combined with a prior assignation of instances through clustering, attaching a different classification model to each cluster, which produced a better class assignment.

More recently, Chong et al. [3] selected data from all over the United States, obtained during the 1995-2000 period, to propose a set of models based on artificial neural networks, decision trees, and Support Vector Machines (SVMs). All the

Fig. 1 Architecture of our proposed system for automatic accident notification and assistance using vehicular networks

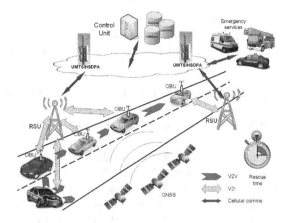

classification models presented similar accuracy results, and they were highly effective at recognizing fatal injuries.

From previous works, we find difficult to combine their results with vehicular networks, since existing works about estimating the severity of road accidents have not been used to improve the assistance to injured passengers.

3   Our Proposal

Our approach collects crucial information when an accident occurs using on-board sensors, and then, based on this information, it will directly estimate the accident severity by comparing the obtained data with information coming from previous accidents stored in a database. This information can be used, for example, to determine the most suitable set of resources in a rescue operation.

3.1   Vehicular Architecture Including V2V and V2I

Figure 1 presents the overview of the vehicular architecture used to develop our system. Our proposed system combines both V2V and V2I communication to efficiently notify an accident situation to the Control Unit (CU). The vehicles should incorporate an On-Board unit (OBU) responsible for: (i) detecting when there has been a potentially dangerous impact for the occupants, (ii) collecting available information from sensors in the vehicle, and (iii) communicating the situation to a CU that will address the handling of the warning notification. Among other features, the CU should integrate mechanisms to estimate the severity of the accident and the injuries of passengers. Notice that our proposal focuses on reducing the consequences of an accident (post-collision assistance), not on reducing the number of accidents. To this end, the Control Unit will be in charge of generating a preliminary and automatic assessment of the damage to the vehicle and its occupants, based on the information received from the involved vehicles to adapt the rescue resources.

A prototype of this architecture was previously built and validated [8]. However, we still need to build accurate severity estimation models to achieve automatic accident assistance, which will be obtained through a KDD process.

3.2 Estimating Accidents Severity Using a KDD-Based Approach

The Knowledge Discovery in Databases, also known as KDD [7], can be defined as the nontrivial process of identifying valid, novel, potentially useful, and understandable patterns from existing data. After the acquisition of initial data, a series of phases are performed during this process: (i) Selection, (ii) Preprocessing, (iii) Transformation, (iv) Data Mining, and (v) Interpretation/Evaluation.

Previous proposals similar to ours do not develop a complete KDD process. In fact, the only phase of the KDD process that has received widespread attention is data mining. Although data mining is a very important phase, the results obtained when omitting the previous phases may lose accuracy; therefore, we propose to develop a complete KDD process. The different phases will be performed using the open-source Weka collection of machine learning algorithms [9]. Weka is open source software issued under the GNU General Public License, which contains tools for data pre-processing, classification, regression, clustering, association rules, and visualization.

We deal with road accidents in two dimensions: damage on the vehicle (indicating the possibility of traffic problems or the need of cranes in the area of the accident), and the passenger injuries. We use the estimations obtained with our system about the damage on the vehicle to help in the prediction of the occupants' injuries as well.

Finally, our system will benefit from additional knowledge to improve its accuracy, grouping accidents according to their degree of similarity. We can use the criteria used in numerous studies about accidents, including some tests such as the Euro NCAP, in which crashes are divided and analyzed separately depending on the main direction of the impact registered due to the collision. The following sections contain the results of applying the different phases of the KDD process.

4 Data Acquisition, Selection, and Preprocessing Phases

Developing a useful algorithm to estimate accident severity needs historical data to ensure that the criteria used are suitable and realistic. The *National Highway Traffic Safety Administration* (NHTSA) maintains a database with information about road accidents which began operating in 1988: the General Estimates System (GES) [13]. The data for this database is obtained from a sample of Police Accident Reports (PARs) collected all over the USA, and it is made public as electronic data sets.

Using the data contained in the GES database, we classify the damage in vehicles in three different categories: (i) *minor* (the vehicle can be driven safely after the accident), (ii) *moderate* (the vehicle shows defects that make it dangerous to be driven), and (iii) *severe* (the vehicle cannot be driven at all, and needs to be towed). Focusing on passenger injuries, we will also use three different classes to determine their severity level: (i) *no injury* (unharmed passenger), (ii) *non-incapacitating injury* (the person has minor injuries that does not make him lose consciousness, or prevent him from walking), and (iii) *incapacitating or fatal injury* (the occupants' wounds impede them from moving, or they are fatal).

After preprocessing the selected GES data and removing incomplete instances, our data sets consist of 14,227 full instances of accident reports (5,604 front crashes, 4,551 side crashes, and 4,072 rear-end crashes).

5 Transformation Phase

This phase consists of developing a reduction and projection of the data to find relevant features that represent the characteristics of the data depending on the objective. We selected a potential subset of variables which could be obtained from the on-board sensors of the vehicle (airbag status, speed, presence of trailer, etc.) or auxiliary devices such as the GPS with integrated maps (road profile, speed limit, etc.). However, when dealing with passengers, there are specific personal characteristics which might help at improving the prediction accuracy. We added two of these personal variables to our data: the age and the sex of passengers. In a near future, this information could be obtained automatically from smartphones or other personal devices, which would be accessed by the on-board system to retrieve this information after an accident using technologies like Bluetooth.

The Weka framework provides a wide variety of feature selection algorithms from which we selected that are commonly used:

- Correlation-based Feature Selection (*CfsSubsetEval*) [11]: This filtering algorithm looks for a subset of attributes highly correlated with the accident severity class, but not correlated with each other.
- Information Gain Selection (*InfoGainAttributeEval*): This metric aims at verifying the entropy change when introducing knowledge about the values of a variable, in order to determine the degree of reduction of uncertainty.
- Wrapper Technique (*WrapperSubsetEval*) [10]: This method is based on a search through all the space of possible variable subsets, aimed at finding the state that presents the highest score determined by a guiding heuristic such as the accuracy. A learning scheme is then required to calculate the score for any given subset.

We determined the optimal variable subset with the three different schemes, and we chose for our final subset those variables selected by, at least, two of the previous algorithms. All the tested variables and the results of the feature selection process appear in Table 1. The top part of the table contains variables about the vehicle involved in the accident, and hence also applicable to the occupants of that vehicle. The bottom part shows variables only applicable to individual occupants. We compared the results of the process when using the whole data set available (*Full*

Table 1 Most relevant variables for vehicle damage and passenger injury estimation

Attribute	Vehicle damage				Passenger Injury			
	Full Set	Front	Side	Rear-end	Full Set	Front	Side	Rear-end
Body Type	✓		✓	✓	✓	✓	✓	✓
Light Condition	✓	✓		✓				✓
Model Year				✓				
Point of Impact	✓		✓		✓			
Road Align[a]						✓		
Road Profile[b]		✓				✓		
Rollover	✓	✓	✓	✓	✓		✓	✓
Speed	✓	✓	✓	✓	✓	✓	✓	
Speed limit	✓	✓	✓	✓	✓		✓	✓
Surface condition				✓				
Trailer			✓	✓	✓		✓	✓
Vehicle role[c]		✓	✓					✓
Weather			✓	✓				
Airbag					✓	✓	✓	✓
Age								
Restraint system					✓	✓	✓	
Seat position							✓	
Sex							✓	
Veh. damage estim.					✓	✓	✓	✓

[a] Roadway alignment just prior to the vehicle's critical precrash event (straight, curve)

[b] Roadway profile just prior to the vehicle's critical precrash event (level, grade, hillcrest, sag)

[c] Determines whether the vehicle was the striking or the struck one

Set), and dividing these data into three subsets depending on the direction of the impact.

As shown, we find noticeable differences between the sets determined for the full set of accident data, and for each of the divisions according to the direction of the impact. The most relevant attributes are the body type, the occurrence of rollover, the speed, the speed limit, the presence of a trailer, the airbag status, and the estimation of vehicle damage. When we divide the instances, new significant values appear, like the road profile for front accidents, and the light condition for rear-end crashes, which were not detected when using the full data set.

For accidents in which the vehicle impacts another element of the road, the attribute that provides a better approximation to determine the accident severity is the speed of the vehicle before the impact. Nevertheless, if you want to estimate the damage when the vehicle is struck by another one, the type of vehicle and the speed limit are more important variables than the speed itself, as can be deduced by studying side and rear-end accidents. The speed limit in the area of the accident is a good estimation factor of the speed of other vehicles, and usually accidents in highways are more severe than those in residential areas, where the speed limit is low.

6   Data Mining, and Interpretation/Evaluation Phases

The most adequate data mining task for our interests is classification. Each instance in our data set contains a record indicating its class membership, while the rest of the attributes are used to predict the class of new instances. We selected three classification algorithms provided by Weka to evaluate which one provides us with the best results:

- Decision trees: *J48* is an open source Java implementation of the Quinlan's C4.5 algorithm [15] included in Weka. It builds decision trees by selecting, at each node of the tree, the attribute that most effectively separates the sample set into subsets with majority of one class value.
- Bayesian networks: Bayesian networks allow modeling a phenomenon with a set of random variables and the dependency relationships between them through a directed acyclic graph. We will use the *BayesNet* implementation with the K2 algorithm [4] to find the graph that better represents the set of dependence or independence in the data.
- Support Vector Machines: SVMs are a set of supervised learning algorithms based on a set of hyperplanes in a high dimensional space, generated by a kernel function, able to separate instances from different classes. We will use the SMO algorithm [14] to train the SVMs from the GES database.

These algorithms present different parameters that must be tuned to maximize the accuracy of the built models: the prune level in decision trees, the number of parents in Bayesian networks, and the kernel function in the SVM case. We carried out several tests to obtain the sets of values that produce the best performance in the selected metrics. To measure the effectiveness of the classification, we can use the True Positive Rate (TP Rate), i.e., the percentage of instances correctly classified. Despite those classifiers which focus their attention on the most frequent class achieve a good value on this metric, their utility may be low as they are not able to differentiate among the existing classes properly.

To cope with the deficiencies of the TP Rate metric, we will also use the area under the ROC (Receiver Operating Characteristics) curve, abbreviated as the AUC [6]. The possible values of this metric vary between 0.5 (for random classifiers with low efficacy) and 1.0 (for a perfect classifier), and it is computed using both the True Positive Rate and the False Positive (FP) Rate. Hence, the AUC metric presents some desirable features when compared to the overall accuracy [2]. It is invariant to *a priori* class probabilities, while giving an indication of the amount of work done by a classification scheme and providing low scores to both random and "one class only" classifiers.

Figures 2 and 3 show the results of the selected algorithms for both the TP Rate and the AUC metrics. Results were obtained by using 10-fold cross validation, which reduces the dependence of the result from the classification process in terms of the partition made for training and validation.

When estimating the damages in vehicles, the three algorithms showed similar performance using the TP Rate metric (although *SMO* is slightly worst in all cases),

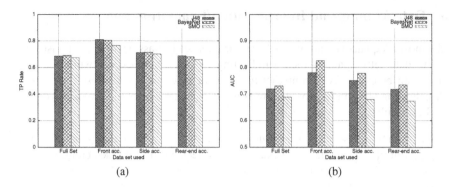

Fig. 2 Comparison of different data mining classification algorithms in the estimation of the damage on the vehicle due to the accident: (a) using the TP Rate metric, and (b) using the AUC metric

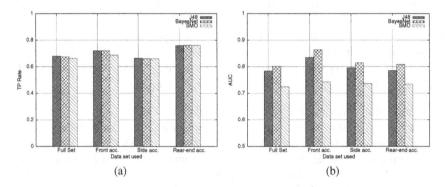

Fig. 3 Comparison of different data mining classification algorithms in the estimation of the injuries of the passengers in the vehicle: (a) using the TP Rate metric, and (b) using the AUC metric

with an overall accuracy about 70 or 80%. However, there are noticeable differences between the schemes under the AUC metric, showing a clear advantage for the *BayesNet* algorithm. This means that Bayesian networks are more robust when facing doubtful cases, and they are not so focused in the majority class. When we divide the accidents depending on the direction of the impact, we obtain a relevant increase on the accuracy for both metrics, showing average results much higher than those achieved with the full data set. Rear-end impacts were the most difficult to estimate, since there was a high number of instances where the car itself was struck by another vehicle, making it harder to estimate the damage without knowing all the details of the other vehicle.

If we estimate the injuries on the passengers, we observe a very similar trend in the results. All the algorithms are very close in terms of the TP Rate metric (even closer than estimating the damage in the vehicle), and again the overall accuracy ranges from 70% to 80%. The differences between algorithms increase when we select the AUC metric, and *BayesNet* outperforms the other algorithms as well. Dividing the accident data into subsets also improves considerably the results under both metrics. However, when studying rear-end collisions, we obtain high values of TP Rate, but lower results of AUC. This effect is due to the distribution of classes in this subset, since there are very few cases where passengers suffered from very severe injuries. This situation might improve if additional information about the colliding vehicle was provided, which will be possible using vehicular networks.

7 Conclusions

Up to now, most of the existing works focusing on data mining in traffic accidents were based on data sets where very limited preprocessing and transformation processes were performed. After a careful selection of relevant attributes, we showed that the vehicle speed is a crucial factor in front crashes, but the type of vehicle involved and the speed of the striking vehicle become more important than speed itself in side and rear-end collisions. The status of the airbag is also very useful in the estimation, since situations where it was not necessary to deploy the airbag are rarely associated with serious injuries on the passengers.

The studied classification algorithms do not show remarkable differences in terms of percentage of instances correctly classified, with accuracies in the range 70-80%. However, since the AUC metric takes into account both the true and false positive rates, we found that Bayesian networks were able to outperform techniques such as decision trees and support vector machines. Dividing the accidents depending on the type of impact allows noticeably increasing the accuracy of the system, especially for front crashes where the vehicle itself is usually the striking one. The accuracy estimation of the severity of side and rear-end crashes could be significantly increased if we provide data regarding other vehicles involved in the collision, which will be possible using inter-vehicular communication technologies.

Our proposed architecture allows receiving all the relevant information about a road accident in a reduced time interval. Estimation the severity of a road accident is invaluable for the emergency services to improve the assistance after a collision. The number of fatalities and the cost of deploying emergency teams due to road accidents could be significantly decreased, since the assistance would be adapted to each individual accident and the effectiveness of available resource usage would be noticeably increased.

Acknowledgements. This work was partially supported by the *Ministerio de Ciencia e Innovación*, Spain, under Grant TIN2011-27543-C03-01, and by the *Diputación General de Aragón*, under Grant "subvenciones destinadas a la formación y contratación de personal investigador".

References

1. Beshah, T., Hill, S.: Mining Road Traffic Accident Data to Improve Safety: Role of Road-Related Factors on Accident Severity in Ethiopia. In: Proceedings of AAAI Artificial Intelligence for Development (AI-D 2010), Stanford, CA, USA (March 2010)
2. Bradley, A.P.: The use of the area under the ROC curve in the evaluation of machine learning algorithms. Pattern Recognition 30, 1145–1159 (1997)
3. Chong, M., Abraham, A., Paprzycki, M.: Traffic accident analysis using machine learning paradigms. Informatica 29, 89–98 (2005)
4. Cooper, G.F., Herskovits, E.: A bayesian method for the induction of probabilistic networks from data. Machine Learning 9, 309–347 (1992)
5. Dirección General de Tráfico (DGT). The main statistics of road accidents. Spain (2010), http://www.dgt.es/portal/es/seguridad_vial/estadistica
6. Fawcett, T.: ROC Graphs: Notes and Practical Considerations for Researchers. Technical report, HP Labs (2004)
7. Fayyad, U., Piatetsky-Shapiro, G., Smyth, P.: The KDD process for extracting useful knowledge from volumes of data. Communications of the ACM 39, 27–34 (1996)
8. Fogue, M., Garrido, P., Martinez, F.J., Cano, J.-C., Calafate, C.T., Manzoni, P., Sanchez, M.: Prototyping an automatic notification scheme for traffic accidents in vehicular networks. In: 2011 IFIP Wireless Days (WD), pp. 1–5 (2011)
9. Hall, M., Frank, E., Holmes, G., Pfahringer, B., Reutemann, P., Witten, I.H.: The WEKA data mining software: an update. SIGKDD Explorations 11, 10–18 (2009)
10. Kohavi, R., John, G.H.: Wrappers for feature subset selection. Artificial Intelligence - Special Issue on Relevance 97, 273–324 (1997)
11. Hall, M.: Correlation-based feature selection for machine learning. PhD thesis, Department of Computer Science, University of Waikato, Hamilton, New Zealand (2008)
12. Martinez, F.J., Cano, J.-C., Calafate, C.T., Manzoni, P., Barrios, J.M.: Assessing the feasibility of a VANET driver warning system. In: Proceedings of the 4th ACM Workshop on Performance Monitoring and Measurement of Heterogeneous Wireless and Wired Networks, PM2HW2N 2009, Tenerife, Spain, pp. 39–45. ACM (2009)
13. National Highway Traffic Safety Administration (NHTSA). FTP Site for the General Estimates System, GES (2012), ftp://ftp.nhtsa.dot.gov/GES/
14. Platt, J.C.: Fast training of support vector machines using sequential minimal optimization, pp. 185–208. MIT Press, Cambridge (1999)
15. Ross Quinlan, J.: C4.5: programs for machine learning. Morgan Kaufmann Publishers Inc., San Francisco (1993)
16. Sohn, S.Y., Lee, S.H.: Data fusion, ensemble and clustering to improve the classification accuracy for the severity of road traffic accidents in Korea. Safety Science 41(1), 1–14 (2003)
17. Sohn, S.Y., Shin, H.: Pattern recognition for road traffic accident severity in Korea. Ergonomics 44(1), 107–117 (2001)
18. Tesema, T., Abraham, A., Grosan, C.: Rule Mining and Classification of Road Accidents Using Adaptive Regression Trees. International Journal of Simulation Systems, Science & Technology 6(10-11), 80–94 (2005)

ContextCare: Autonomous Video Surveillance System Using Multi-camera and Smartphones

Gonzalo Blázquez Gil, Alvaro Luis Bustamante,
Antonio Berlanga, and José M. Molina

Group of Applied Artificial Intelligence, University Carlos III of Madrid, Colmenarejo,
Spain
{gbgil,aluis}@inf.uc3m.es, {aberlan,molina}@ia.uc3m.es

Abstract. In the future, Ambient Intelligence (AmI) technology could assist people autonomously and interpret their intentions. Current technology can already be used to recognize the presence of a person in a private or public space and trigger an automatic response or reaction depending on the user activity. This work describes ContextCare, an extension for an video surveillance system in a health care scenarios based on activity recognition using sensor smartphones. Both systems are coordinated using ECA parading.

1 Introduction

According to the current financial crisis situation around the world, reduced budgets is a real challenge for governments. In this case, to freeze or reduce the healthcare budget is a priority while at the same time the service improves its quality. Thereby, AmI applications aim to contribute to reduce costs and offer a better and more efficient services. Anytime and anywhere assistance requires several underlying mechanisms and tools [8]: ranging from wireless-enabled monitoring, location systems that permit us to identify where they are located in case of needed.

There are many potential uses for AmI in Business management scenarios [4], however, ContextCare architecture is focused on human resources monitoring. Especially, hospitals and nursing homes scenarios where it possible to increase the efficiency of their services by monitoring patient's health, progress, and routines through the analysis of their activities, decreasing budgets.

Activity recognition is traditionally carried out through video systems like those described in [3] and [7]. However, video activity recognition systems present few problems like: huge computational cost, speedy bandwidth to transmit information, and complex techniques to interpret video data. Recent researches in activity recognition shows that Micro-Electro-Mechanical Systems (MEMS) is becoming another way to face with activity recognition problem [1, 5].

Normally, activity recognition researches based on MEMS are intrusive, for that reason, actual researches try to collect unobtrusively MEMS data using smartphones

J. Casillas et al. (Eds.): Management Intelligent Systems, AISC 171, pp. 47–56.
springerlink.com © Springer-Verlag Berlin Heidelberg 2012

[6]. Smartphones may obtain and process physical phenomena from embedded sensors (accelerometer, gyroscope, compass, magnetometer, proximity sensor, light sensor, GPS, etc.) and transmit this information to remote locations without any human intervention.

Taking into account these advantages, it may be possible to consider a smartphone like a non-intrusive device to obtain people activities, since smartphones experience almost the same physical forces, temperature and noise that the person who carries them out. So, if you track mobile phone actions you are tracking person actions [11].

The huge potential of the smartphones devices has motivated us to design a framework that can intelligently capture different sensory data from a smartphone in real-time in order to coordinate a video surveillance system. ContextCare is capable to react and adapt video surveillance systems functionalities depending on the human resources context actions. Smartphone are not considered a mere communication device, otherwise it is introduced into an autonomous platform which allows to monitor efficiently patients with a video surveillance system.

ContextCare architecture rely on the ECA (Event-Condition-Action) paradigm which is composed by a defined set of reactive rules working over an event-driven architecture [10].

In this paper is presented the ContextCare architecture as an event-driven Semi-supervised video surveillance system that involves the use of smartphones and visual sensors.

2   Related Works

ContextCare architecture is an monitoring system that englobes two different architectures: inContexto [6] which monitors user activity and location and video surveillance system [2] which uses inContexto information to focus on an anomalous situation reducing the time spend by security personnel in front of a screen.

2.1   InContexto: Architecture Definition

inContexto is a multimodal architecture which infers actions from users who carries out a smartphone (Figure 1).

A low-level sensing module continuously collects relevant information about the user activities. Thanks to Android OS that provides background processing is possible to run services without human control since the presented architecture is developed to obtain physical actions in a non-intrusive way. Data collection level gathers single raw data from smartphone sensors (Accelerometer, Gyroscope, GPS and so on) in order process and transform in features. Pre-processing is often required to improve performance, removing noise and redundancy in measurements.

Features extraction level which is also implemented in the smartphone, involves the extraction of symbolic features from smartphone sensor data. Hence, the

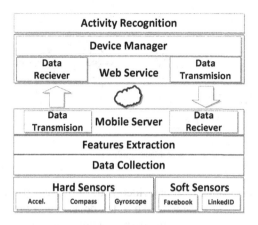

Fig. 1 inContexto architecture overview

objective of feature extraction is to represent an activity with the main characteristics of a data segment. In [5] has been defined the best features to infer user activity (Signal mean and variance, correlation between axes and signal energy).

Activity recognition module uses selected features to infer what activity is the user engaged in. In this component is implemented a J48 classification tree which was off-line trained.

2.2 Video Surveillance Architecture

The Video Surveillance architecture is basically a system that allows the control of PTZ cameras by local or remote processes. In this case, this architecture mainly resides in the *Sensor Manager*, which is the responsible of attending control flows (allowing the positioning of PTZ cameras), and streaming video sequences to the different terminals. This component is used by ContextCare for monitoring patients, allowing an easy management of the underlying cameras.

The *Sensor Manager* used by ContextCare is the responsible of video acquisition, compression, and transmission, as well as to handle the communication protocols to perform the different movements in the cameras. The internal organization of this component is briefly outlined in figure 2.

There are two main functions that are controlled by the *Sensor Manager* for each video camera. The first one is related with the control of the camera and its movements. As we can see in figure 2, there is a PTZ server which allows other processes to interact with the camera, so this controller can provide a standard interface to control homogeneously any underlying device. It has defined some high-level operating primitives like *goTo X Y*, *zoom amount*, etc. These high-level primitives are exposed as a non-connection oriented UDP Server, with a simple request-response protocol in the client-server computing mode.

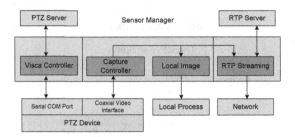

Fig. 2 Sensor manager for local/remote camera control

The second main feature is about video acquisition and transmission. Generally, the access to limited resources like video devices is a problem if share the information between different terminals or systems is a requirement. In order to solve this issue we have defined two different strategies depending on the video destination. For local processes running on the same computer it is created a shared region of non-paged memory which can be accessed to retrieve the latest video frames. On the other hand, for remote processes, it is used a JPEG2000 [9] compression and real-time streaming system [2] with the aim of provide frames with the minimum delay. This allows to transmit real-time video sequences to remote processes like operators, agents, backup systems, mobile phones, etc.

3 Proposed Architecture: ContextCare

The main functionality of the ContextCare Surveillance Platform (See figure 3) is to connect seamlessly patients and the health professionals, reporting to the latter alerts and health measurements obtained from the patients. Besides, ContextCare is designed to manage video surveillance flow and determine which emergency member is the most suitable to look after the patient taking into account patient health context. Next subsection introduces the characteristics of the scenario where ContextCare architecture may be deployed and its architectural particularities.

3.1 ContextCare Scenario Model

Our interaction scenario considers a space populated with video cameras responsible of monitoring people. As we previously said, video surveillance systems are hardly tedious for security personnel due to within this scenario, there are plenty of people to be tracked.

The architecture is implemented following a event-driven model where patient context (event) triggers actions depending on a given condition (ECA rules). An ECA rules is divided in three different parts: the *event* is the signal that triggers a set of rules; the *condition* which if is satisfied makes the execution of the rule to continue and finally, the *action* defines the execution flow of a process.

ECA Rules are created by experts (security personnel or emergency services) and represent patient emergency situations. ContextCare architecture contains three different applications to deliver the global functionality:

- *Video surveillance system:* has been developed in C++ modules (as described in the previous works section) and it was evaluated in previous works [2]. Summarizing, the frame-rate obtained is the same as the video sensor provides (25 FPS).
- *Patient smartphone application:* The technological platform in the current prototype is an Android smartphone. Patient Smartphone app mainly consist in a monitoring service implementing inContexto architecture which collect patient sentinel data and establish user activity and location. Besides, is also able to reason with the raw sensor data to identify higher level information including patient activities or even though beat rate, temperature, etc.
- *Security mobile terminal (SMT):* This application provides the necessary tools to access patients information. When a event triggers a rule, SMT receive a message which contains: patient profile, video tracking about patient situation, patient location and the event that was happened. Moreover, it is also provided, which allows the emergency services to evaluate the most recent medical details obtained from sensors, perform new measurements, and communicate with the caretakers.

Consider a patient in a nursing home wearing a smartphone with the normal sensors and blood pressure sensor connected. inContexto architecture captures every sensor data and generates patient context information. This information or event may be dispatched a rule (ECA rules) depending on the ECA rules condition. For example, an unusual blood pressure level triggers one rule and provokes a set of actions to be followed, sending a message to his surveillance system or even his family members, physician, emergency services, friends or colleagues.

3.2 Rule Manager Component

Rule Manager component is probably the most important component of the ContextCare architecture. it involves many IT disciplines like database access, knowledge reasoning, video camera resource management and ECA rules evaluator. ECA model allows bidirectional communication between users (patients and security personnel) and video surveillance system in order to make possible to control and coordinate human resources and video cameras.

Rule Manager is a centralized server where the whole surveillance system is managed. Rule manager component's inputs are ECA rules provided by surveillance personnel which has the form: ON *event* IF *condition(s)* DO *action(s)*. Besides, user context activity information from inContexto architecture which follows this structure: *(userID, action, location)*. On the contrary, rule manager outputs are guidelines (ECA rules actions) to carry out by the flow control component, coordinating cameras and also inform to the surveillance staff that something wrong is happening.

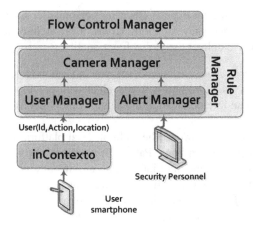

Fig. 3 ContextCare architecture: Rule Manager Component, video surveillance system and inContexto

When ContextCare receives a new user context action from inContexto architecture, ECA rules is constantly evaluated in order to detect configured events, executing the associated actions if the conditions are fulfilled. In that case, first of all a message sequence is activated to notify from the users to Security personnel that something is happening. Later, ContextCare architecture decides which camera is the most propitious to track the situation. Finally, the system starts to monitoring autonomously the involved person.

3.2.1 User Manager

User Manager receives and stores every user context action and location generated by inContexto architecture. This component manages inContexto User State Message (USM) and also stores into a database with the purpose of providing this information to camera manager which will check if any alert is active. inContexto USM is composed by the following attributes:

- *User id* which consist in a string which identifies every user, in this case smartphone direction IP. The Id permits to determines if the performed action is allowed to this person or not.
- *Action*, this field contains a concrete action or sensor value (depending on the embedded sensors on the system). Smartphone normally provides inertial sensors, however, inContexto allows to connect other sensors such as blood pressure, heart beat sensor, and so on.
- *Location* this field depicts the place where the user is in that moment. The suitability of each method depends on whether the location is outdoors or indoors and also the technology used. This field necessarily contains an absolute position

like GPS coordinates. It could be filled by a symbolic position like corridor, room number, the nearest access point or wherever.

3.2.2   Alert Manager

Alert manager aims to communicate ContextCare architecture with human resources staff (Security or emergency personnel) terminals. The set of ECA rules are configured by an expert and define those events a flow control manager should be aware of.

As we explain previously, ECA Rules are composed by by three fields (events, conditions, actions) and they are described below:

- *Events*: describes a situation (user activity or location in this case) to which the rule may be able to respond. Events can be essentially divided into two categories: (i) primitive events which correspond to elementary occurrences and (ii) composite events that are composed for more than one primitive events.
- *Conditions*: specifies the conditions to trigger the ECA rule. Once the result of the condition evaluation is true, the condition is satisfied and the action field is executed.
- *Actions*: describes the task the rule considers relevant to the event and the condition. Actions field indicates the subsequent activities if the condition is satisfied.

Alert manager component generates two different responses, the first one is ECA rules created from the human resources terminal to Camera Manager and the second one is the alert protocol when an rules is triggered in the Camera Manager Component.

3.2.3   Camera Manager

Camera Manager component contains ECA Rules engine which is responsible of generate alerts and guidelines to manage the video surveillance system. These rules involves to evaluate 'online' conditions, i.e., those which require to access an external resource (user context information, video camera control), therefore it is mandatory a real time response.

Mainly, there are two ways to control video flow. The first one is manually via the main terminal. Human Resources moves manually the cameras looking for situation of interest. The second way to control video flow is automatically via actions defined by ECA rules manager. Hence, it operates video cameras three different actors but each one play different roles:

- *Human Resources* monitoring the video streamed by different cameras and controlling their orientation manually.
- *Mobile human resources* walking around the monitored area. This person has a smartphone where receives alerts with security problems.

- *Users* are the principal actor. They also have a smartphone which is used to track their actions. When a non-usual action takes place the mobile phone will launch an alert to Rule manager.

Camera Manager gathers user context actions and ECA rules to build a unified view of the scene. Besides, it creates a goal which represent the overall objective of the video surveillance system. For example, a goal may be *track user which ID is 10001* and Camera Manager send user information (location and action) to the control flow manager which will be decided which camera is the most suitable to track this person.

Summarizing, Camera Manager decides according to the environment situation and the ECA rules which goal is a priority, sorting a list of goals actions for the Control flow Manager and creating the alert to the Security personnel.

4   ContextCare Architecture Evaluation: Study Case

ContextCare architecture has been created to monitor autonomously dependent people (Elderly people, children, etc.) and give health care support to hospital patients. Taking into account that this architecture is an autonomous monitor system, depending on the ECA Rules depicted by the human resources, it is mandatory a real-time performance to efficiently manage the information given by sensors.

In order to better assist the evaluation process the ContextCare system deployment will take place firstly in the University Carlos III of Madrid for pre-evaluation. The working scenario consists in a visual sensor network which contains six Pant-Tilt-Zoom (PTZ) control video devices around two rooms, three in the corridor and the other ones in the main room.

The scenario is a public place, so every person who enter to the track zone was under system control, however, there are just four people (playing user role) connected to ContextCare Application. In this cases, inContexto was configured to recognize five different actions (Lying, Standing, Walking, Running and Falling down).

In this case, an expert defines the ECA rules which describes what is considered a problem for each patient. Thereby, the ContextCare system may reach different rules like *'ON userContex:fall_down IF user:X DO follow'* or *'ON userContexto:walking IF user=Y DO track'*). Every single action is detected by inContexto architecture and send it to ContextCare application where it is checked in the rule-based engine if it actives a rule.

First figure 4 depicts video surveillance system tracking. There are two persons sitting inside the room (bottom left camera), one of them stands up and start to walk. inContexto generates two different USM as follows:

- *(userID:10001,action:Stand,location:corridor)*
- *(userID:10001,action:Walking,location:(2,1))*

First alert shows a symbolic location *corridor* and the second one an absolute position according to video surveillance coordinates. After that, the user fall down and the smartphone generates another USM which dispatches the next rule in the camera

Fig. 4 ContextCare tracking a person who fell down on the floor. Figure a shows the monitored environment and Figure b shows a screem alert over the person

Manager Component. Alert manager rule engine checks the actual situation and the next ECA rule is dispatched.

- *'ON userContex:fall_down IF user:10001 DO follow AND Inform SP'*

Finally, alert system inform to camera manager component where is the user (Figure 4 shows the person on the floor inside a red rectangle) and also it creates the message to inform every security personnel.

5 Conclusions

AmI technology is developing fast and will promote a new generation on business management with some characteristics in the area of context awareness, anticipatory behavior and video surveillance. The presented architecture improves multicamera tracking applications performance, selecting the most suitable camera for any situation assessment during video tracking analysis. Using this new approach, the time the security personnel spends in front of the screen is reduced, taking this time for other tasks.

ContexteCare architecture is well-suited in many human resources surveillance situations, for example the prevention of labor risks, management of human resources. However, we think that eHealth context is the most suitable situation.

Considered future works extending the development of the Activity Recognition system with more complex activities or even user emotional state, according to voice, face gestures, or typing patterns.

Acknowledgements. This work was supported in part by Projects CICYT TIN2011-28620-C02-01, CICYT TEC2011-28626-C02-02, CAM CONTEXTS (S2009/TIC-1485) and DPS2008-07029-C02-02.

References

1. Avci, A., Bosch, S., Marin-Perianu, M., Marin-Perianu, R., Havinga, P.: Activity recognition using inertial sensing for healthcare, wellbeing and sports applications: A survey. In: 23rd International Conference on Architecture of Computing Systems (ARCS), pp. 1–10. VDE (2010)
2. Bustamante, A.L., Molina, J.M., Patricio, M.A.: Multi-camera Control and Video Transmission Architecture for Distributed Systems. In: Molina, J.M., Corredera, J.R.C., Pérez, M.F.C., Ortega-García, J., Barbolla, A.M.B. (eds.) User-Centric Technologies and Applications. AISC, vol. 94, pp. 37–45. Springer, Heidelberg (2011)
3. Cilla, R., Patricio, M., Berlanga, A., Molina, J.: A probabilistic, discriminative and distributed system for the recognition of human actions from multiple views. Neurocomputing (2011)
4. Cook, D., Augusto, J., Jakkula, V.: Ambient intelligence: Technologies, applications, and opportunities. Pervasive and Mobile Computing 5(4), 277–298 (2009)
5. Gil, G.B., de Jesús, A.B., Lopéz, J.M.M.: Comparing Features Extraction Techniques Using J48 for Activity Recognition on Mobile Phones. In: Omatu, S., Paz Santana, J.F., González, S.R., Molina, J.M., Bernardos, A.M., Rodríguez, J.M.C. (eds.) Distributed Computing and Artificial Intelligence. AISC, vol. 151, pp. 141–150. Springer, Heidelberg (2012)
6. Gil, G., Berlanga, A., Molina, J.M.: incontexto: A fusion architecture to obtain mobile context. In: 2011 Proceedings of the 14th International Conference on Information Fusion (FUSION), pp. 1–8. IEEE (2011)
7. Gómez-Romero, J., Serrano, M., Patricio, M., García, J., Molina, J.: Context-based scene recognition from visual data in smart homes: an information fusion approach. Personal and Ubiquitous Computing, 1–23 (2011)
8. López, G., Custodio, V., Moreno, J.: Lobin: E-textile and wireless-sensor-network-based platform for healthcare monitoring in future hospital environments. IEEE Transactions on Information Technology in Biomedicine 14(6), 1446–1458 (2010)
9. Luis, A., Patricio, M.: Scalable streaming of jpeg 2000 live video using rtp over udp. In: International Symposium on Distributed Computing and Artificial Intelligence (DCAI 2008), pp. 574–581. Springer (2009)
10. Michelson, B.: Event-driven architecture overview. Patricia Seybold Group (2006)
11. Want, R.: You are your cell phone. IEEE Pervasive Computing 7(2), 2–4 (2008)

Clustering of Fuzzy Cognitive Maps for Travel Behavior Analysis

Lusine Mkrtchyan[1], Maikel León[2], Benoît Depaire[3], Da Ruan[1],
and Koen Vanhoof[3]

[1] Belgian Nuclear Research Centre SCK CEN, Boeretang 200, Mol, Belgium
    {lmkrtchy, druan}@sckcen.be
[2] Center of Studies on Informatics, Central University of Las Villas, Santa Clara, Cuba
    mle@uclv.edu.cu
[3] Transportation Research Institute, Hasselt University, Diepenbeek, Belgium
    {koen.vanhoof,benoit.depaire}@uhasselt.be

Abstract. The increasing of public transportation or bike use became an important issue in addressing economic, energy and environmental challenges. With this regard one of the most main tasks is to find and analyze the factors influencing car dependency and the attitudes of people in terms of preferred transport mode. In this paper Fuzzy Cognitive Maps (FCM) are explored to show how travelers make decisions based on their knowledge of different transport modes properties. The results of this study will help transportation policy decision makers in better understanding of people's needs and actualizing different policy formulations and implementations.

1 Introduction

Travel behavior studies are important for decreasing travel-related energy consumption, and in the transportation planning travel demand forecast is one of the most important instruments to evaluate various policy measures to influence travel supply and demand.

The benefits of bike and public transportation use are multi fold. The use of bicycles gives many benefits such as health, safety, environmental, quality of life as well as economic benefits. The use of public transportation will decrease the traffic considerably, and will contribute to road safety with less accidents. The economic benefits are obvious: the use of bikes and public transportation helps increasing the capacity of roads with lower costs as well as decreasing the parking related costs.

Individuals' travel selections can be considered as actual decision problems, generating a mental representation of the decision situation. To this end, the development of the mental map concept seems to benefit from the knowledge provided by individual tracking technologies.

Records regarding individuals' decision making processes can be used as input to generate mental models. Such models treat each individual as an agent with

J. Casillas et al. (Eds.): Management Intelligent Systems, AISC 171, pp. 57–66.
springerlink.com   © Springer-Verlag Berlin Heidelberg 2012

mental qualities, such as viewpoints, objectives, predilections, and inclinations. Among various Artificial Intelligence techniques, Fuzzy Cognitive Maps (FCMs) are applied in this study to simulate individuals' decision making processes.

The research on Cognitive Maps (CMs) and travel focuses primarily, in fact almost exclusively on some route choice. In contrast, the other steps such as trip generation (how many trips?), trip distribution (where to go?) have been given far less attention by cognitive mapping researchers [1].

In [2] the authors suggest CMs modeling to extract the mental representation of individuals in the planning of trips, related to daily travels. [14] has centered on the location, possible destinations, and feasible alternatives for any travel mode choice; the authors claim that the CMs of people who mostly walk and use public transit may vary systematically from those who are mostly chauffeured in private vehicles, or who usually drive themselves. At an individual level it is important to realize that the relationship between travel decisions and the spatial characteristics of the environment is established through the individuals' perception and cognition of environment [3].

This study has been made in the city of Hasselt in Belgium. The city has around 73 000 habitants, with a traffic junction of important circulation arteries from all directions. Hasselt made public transport by bus zero-fare from July 1st of 1997, and bus use was said to be as much as "13 times higher" by 2010, being the first city in the world that had entirely zero-fare bus services on the whole of its territory.

We explored the clustering of FCMs as a descriptive tool to analyse different groups of users and understand the main features of choosing a transport mode. The results of FCMs clustering will serve as a guide for transportation policy decision makers for future plans.

In this paper we first we give background information about FCMs main concepts, then we discuss a new algorithm to cluster the FCMs, and finally, we discuss the results of applying FCMs clustering to traveler's behaviors analysis, ending the study with concluding remarks about the limitations and possible future extensions of the proposed algorithm.

2 Modelling with Fuzzy Cognitive Maps

Cognitive Maps were first introduced by Axelrod [4] who focused on the policy domain. Since then, many researchers have used CMs in various fields where the problems are ill structured or not well-defined.

Cognitive Maps have two types of elements: *concepts* and *causal beliefs*. The former are variables while the latter are relationships between variables. Causal relationships can be either positive or negative, as specified by a '+', respectively a '-', sign on the arrow connecting two variables. The variable that causes a change is called a cause variable and the one that undergoes the effect of the change is called an effect variable. If the relationship is positive, an increase or decrease of the *cause variable* causes the *effect variable* to change in the same direction (e.g., an increase in the cause variable causes increase of the effect variable). In the case of a negative relationship, the change of the effect variable is in the opposite direction (e.g., an increase in the cause variable causes decrease of the effect variable).

However, CMs, whatever their types, are not easy to define and the magnitude of the effect is difficult to express in numbers. Usually CMs are constructed by gathering information from experts/users and generally, they are more likely to express themselves in qualitative rather than quantitative terms.

To this end, it may be more appropriate to use FCMs, suggested by Kosko [5]. Actually, FCMs are weighted CMs where the weights are associated with fuzzy sets. The degree of the relationship between concepts in an FCM is either a linguistic term, such as: often, extremely, some, etc.; or a number in [-1; 1] associated with fuzzy sets.

Some other notions that we use in the next sections are introduced below. For a given node C_j, the *centrality* value $CEN(C_i)$ is decided as follows: $CEN(C_i) = IN(C_i) + OUT(C_i)$, where $IN(C_i)$ is the column sum of absolute values of a variable in the adjacency matrix and shows the cumulative strength of variables entering the node, and $OUT(C_i)$ is the row sum of absolute values of a variables and shows the cumulative strengths of connections exiting the node.

The conceptual centrality represents the importance of the node for the causal flow in the CM. A node C_i is a *transmitter* if $IN(C_i) = 0$ and $OUT(C_i) > 0$, and is called a *receiver* if $IN(C_i) > 0$ and $OUT(C_i) = 0$. The total number of receiver nodes in a map is considered a complexity index.

Among several ways of developing CMs and FCMs, the most common methods are extracting knowledge from questionnaires, extracting knowledge from written texts, conducting interviews, or drawing maps from data. Note that these methods can be used also in combinations, such as questionnaires with interviews.

The following design is used in our approach:

- Personal information about individual: useful for demographic analyses.
- Cognitive subsets: the interaction in the decision making process such as *situation-benefit-attribute* variables.
- Expert criteria: all cognitive subsets are captured by using artificial scenarios. Situational variables are assigned with random states, and the respondent specifies the utility of those conditions in terms of using bus, car or bike.
- Causal influences among variables: the experts evaluate the causal relations among the variables they had selected.
- Benefits importance: experts assign an importance level to all benefit variables.

3 Clustering of Fuzzy Cognitive Maps

3.1 The Distance Matrix of FCMs

While there are many examples of the use of CMs or FCMs in different application fields, there are only few attempts for CMs clustering. Similar studies of finding the similarity between CMs are done in [6] and in [7]. The first study is restricted only on a similarity algorithm without further use of it for cluster analysis. The second study extends the similarity measurement of the first approach and, in addition, cluster analysis is provided. However there are some drawbacks of this study. For cluster analysis Ward linkage method is applied which commonly uses the squared Euclidean distance for the similarity matrix calculation. In addition, the number of optimum clusters is not discussed; neither any validation method is applied for cluster analysis. An interesting study of FCMs clustering is done in [8]. However, while the clustering based on a map structure is comprehensive enough to be used for different application domains, the clustering based on the map content is rather a case specific and is restricted by principal component analysis. Moreover, neither the weights of the links nor the links' signs are considered: thus the functionalities of CMs are not fully discussed in cluster analysis. Another study of clustering is done in [9]. This approach differs from previous ones as it clusters not FCMs into different groups but the nodes of an FCM. A similar approach of clustering the nodes into hierarchical structure is proposed in [10]. These last two studies are less interest for us because first we already have structured representation of the nodes in our dataset, and, second, we are more interested in FCMs clustering (not nodes clustering). In [7] the authors suggest an improvement of FCMs similarity measurement algorithm described in [6] mentioning that the algorithm does not consider the missing values properly as well as it lacks of generalizability. We will take into account only the comment about generalizability, and will adjust the algorithm to be applicable in our study. Notice that by generalizability, we mean that in [6] the number of linguistic terms is fixed and the comparison formula cannot handle different number of linguistic terms. More specifically, the study fixed the number of linguistic terms to 7, assigning maximum strength to 3, minimum strength to -3, and consequently the similarity measurement algorithm cannot be applied for the cases with more/less linguistic terms. Therefore, for our task, we use the following distance ratio (DR), expressed in 1 and 2.

$$DR = \frac{\sum_{i=1}^{p} \sum_{1=0}^{p} (a_{ij}^* - b_{ij}^*)}{2p_c^2 + 2p_c(p_{u_A} + p_{u_B}) + p_{u_A}^2 + p_{u_B}^2 - (2\alpha p_c + p_{u_A} + p_{u_B})} \tag{1}$$

where

$$m_{ij}^* = 1(i), \text{ if } m_{ij} \neq 0 \text{ and } i \text{ or } j \notin P_c m_{ij}(ii), \text{otherwise} \tag{2}$$

In (1), a_{ij} and b_{ij} are the adjacency matrices of the first and second map respectively, p is the total number of possible nodes, p_c is the set of common nodes for both maps, p_c is the number of such nodes, p_{u_A}/p_{u_B} is the number of unique nodes for user u_A/u_B respectively. In (2) m_{ij} is the value of the i-th row and j-th column in the zero augmented adjacency matrix.

3.2 Hierarchical Clustering of FCMs

Cluster analysis is an unsupervised learning method to examine the dataset by dividing it into groups so that the similarity within the clusters and dissimilarity between different clusters are maximized. In the previous subsection we presented the distance ratio algorithm to find the distance between two FCMs. There are two widely used clustering techniques based on similarity or distance measurements: the hierarchical approach and the partitional approach (e.g., K-means). There are also some hybrid approaches discussed in the literature. In this study we explore only hierarchical clustering algorithms as they suit the best our task.

Hierarchical clustering algorithms produce a nested series of partitions for merging or splitting clusters based on the similarity. The most widely used linkage methods are single, complete and ward linkage methods [11]. The Ward method is not efficient for our study as we do not use the Euclidian distance for similarity measurement. To cluster FCMs the single-linkage method takes the distance between two clusters as the minimum of the distances between all pairs of maps from the two clusters. On the other hand, the complete-linkage algorithm calculates the distance between two clusters as the maximum of all pairwise distances between maps in the two clusters. Single linkage suffers from a chaining effect producing elongated clusters.

The study in [11] shows that the complete linkage produces more compact and more useful hierarchies in many applications than the single-linkage algorithm. Besides to decide which algorithm best suits our data, we calculated the Cophenetic Coefficient (CC) for single, complete, weighted, average, ward and centroid linkage methods. Note that the CC shows how strong the linking of maps in the cluster tree has a correlation with the distances between the maps in the distance vector. This coefficient usually is used to compare different linkage methods. The closer the CC to one the more accurately the clustering solution reflects the data. For our dataset, the results of CC calculation show that the complete and centroid methods are the best for our data as they gave maximum values for CC. However, the centroid method also suffers from chaining effect, and the best option that we used for our cluster analysis is the complete linkage.

In this section we discuss the optimum number of the clusters and we propose the adjustment of well-known *Davies − BouldinIndex* (DBI). The results are twofold: first we find the optimum number of clusters; second, we use the concept of *centralmap* to analyze the clusters separately. There are several cluster validity indexes that lead to the decision of the optimum number of clusters. We will explore two of them only, namely the *SilhouetteIndex* (SI) and DBI [12], [13]. We compare the results of SI and we propose an extension of DBI index to be used for FCMs. For each sample in a cluster the SI assigns a confidence value showing how good the sample has been classified. To find the optimum number of clusters the average value of SI is calculated for different number of clusters, and the one with maximum value is taken as the optimum number. As SI, DBI also aims to find the optimum number of clusters. DBI is defined as follows:

$$DBI = \frac{1}{c} \sum_{i=1}^{c} max_{i \neq j} \left\{ \frac{\Delta(X_i) + \Delta(X_j)}{\delta(X_i, X_j)} \right\} \qquad (3)$$

where $\Delta(X_i)/\Delta(X_j)$ is the average distance of the samples in the i-th/j-th cluster to the center of the cluster, $\delta(Xi, Xj)$ is the distance between the centers of the i-th and j-th clusters. The cluster configuration minimizing DBI is taken as an optimum number of clusters. As we are clustering FCMs, we adjust DBI to be applied for FCMs. We propose the concept of a central map: hence we first derive a central map for each cluster, and then calculate the distance of each map in a cluster from its central map.

Note that all maps in our dataset have the same hierarchical structure, some nodes are only transmitters (situation nodes), some are only receivers, and some others are both transmitters and receivers. A node is included in the central map if it exists in more than half of maps in a cluster. The weights of the links are calculated as the average value of the weights from all maps that contain both nodes that comprise the link. Once we have all central maps of all clusters we calculate BDI for different number of clusters to find the optimum number. Note that once we identify the number of clusters, we use central maps for further analysis of each cluster.

3.3   Cluster Estimation and Validation

In this section we explain the results of cluster analysis of 221 FCMs implemented in Matlab. The dendrogram in Figure 1 illustrates the arrangement of the clusters produced by 1 and 2 for the distance matrix and by complete linkage for inter-cluster distance. To find the optimum number of clusters we calculated the SI for 10 clusters as shown in Table 1. Moreover, we derived also DBI for again 10 clusters first finding the central maps for each cluster configuration.

Table 1 SI for different clustering configurations

Clusters Number	S	S_1	S_2	S_3	S_4	S_4	S_5	S_6	S_7	S_8	S_9	S_{10}
2	0.354	0.342	0.367									
3	0.389	0.449	0.376	0.342								
4	0.555	0.535	0.392	0.449	0.472							
5	0.612	0.768	0.659	0.435	0.648	0.552						
6	0.829	0.848	0.734	0.868	0.735	0.948	0.842					
7	0.707	0.789	0.633	0.748	0.634	0.868	0.535	0.742				
8	0.524	0.652	0.518	0.489	0.473	0.571	0.456	0.678	0.357			
9	0.529	0.491	0.632	0.352	0.418	0.691	0.548	0.468	0.435	0.342		
10	0.419	0.461	0.515	0.591	0.422	0.252	0.318	0.389	0.448	0.368	0.435	

The optimum number of clusters with both SI and DBI is found equal to six (see Figure 1) with 15, 53, 33, 23, 59 and 38 maps in each cluster respectively. After finding the optimum number of clusters, we first analyze "situation-benefit-attribute"

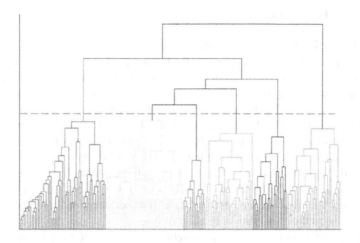

Fig. 1 The clustering results interpretation with a dendrogram

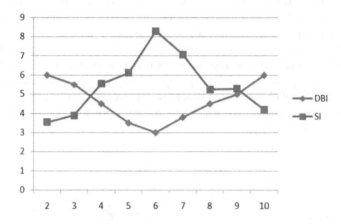

Fig. 2 SI and DBI values for different number of clusters

sets for each cluster; afterwards, we analyze the demographic features of the clusters. For that we use central maps that we derived for DBI: thus, we have six maps representing each cluster. For demographic analysis we take into consideration the users' age, gender, income, household size, education level, occupation, parking type (paid or free), bike ownership, number of owned cars and having a bus card. The results of chi-square test indicate that some of these variables namely the income, occupation, bus card, parking availability and number of cars are dependent variables.

3.4 Clustered Groups

As already stated in the previous section, we obtained six different clusters. Figure 3 (the width of the relations corresponds to the frequency of links evaluation by group members) shows two central maps of two clusters out of six clusters. For simplicity of the map visualization we omit the three decision nodes and the final utility node.

In the first group we have young travelers (less than 30 years old), with medium income (from 2000 to 4000 Euros) having more than two cars and a small household size (living alone or in two). The main benefits combining the travelers in this group is normally their desire to be free while traveling, to have convenience, paying attention on the required time and effort of a specific travel mode. The most important attributes for them are the travel time as well as the flexibility and independence. In addition, they take into account also the travel time, the treatment of bags; mental effort needed and the possibility of direct traveling.

In the second group we have mostly retired people (more than 60 years old), having only one car, with household more than two persons and who mainly use paid parking. This group has similar preferences as the previous one; in addition members of this group pay more attention on physical comfort, easiness of parking, physical effort, and the situation precipitation, the shelter availability as well as the reliability of a transport mode under the chosen situation variables.

In the third cluster we have the travelers who are older than 40 years. They are either employed or retired, with high income (more than 4000 Euros), using mainly paid parking. The preferences of this group are very similar to the second group with the difference of choosing as a benefit being healthy, convenience and paying no attention on reliability. Actually, this group and the previous one are in the same cluster after maps' learning.

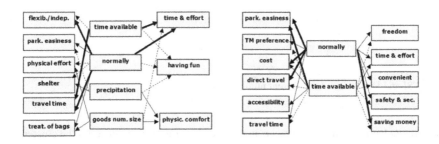

Fig. 3 Central maps of two different clusters

The fourth group is different from previous ones (shown in left of Figure 3). In this group mainly are students or young employers with medium or high income, and they mainly use free parking. There is one benefit variable unique to this group: having fun. Another difference with respect to the other groups is the set of situation variables. Namely, this group considers the situations related with precipitation, number and size of the goods purchased, as well as the available time.

The fifth group includes the travelers with low education, high income preferring not to provide information about their income or declaring themselves as unemployed. This group has a unique benefit variable that is the assurance and certainty related with their decision. The unique situation that the group considers and other groups do not, is the availability of parking and some attributes that the group considers as important are the reliability, accessibility, travel time, transport mode preference, etc.

The last group is the only group that gives importance to safety and security, saving money, and the traveling cost (shown in right of Figure 3). Here travelers have low education, low income (less than 2000 Euros), mainly students and unemployed or retired with small household.

4 Conclusions

In this study we proposed FCMs as a modeling tool to analyze the behavior of complex systems, where it is very difficult to describe the entire system by a precise mathematical model. Consequently, it is easier and more practical to represent it in a graphical way showing the causal relationships between concepts.

As a case study we discussed the travel behavior and we analyzed the main concepts that affect on the travelers choice of a specific transport mode such as car, bike or bus. We explored the clustering algorithms for FCMs to offer policymakers a framework and real data to deal with, in order to study and simulate individual behavior and produce important knowledge to use in the development of city infrastructure and demographic planning.

Some of the suggestions from this study are related with several improvements which according to the travelers will increase their will to use public transportation more frequently. Namely, the travel time, the easiness of parking, the direct travel, the treatment of bags, the independence, the shelter/staying dry, direct travel, the physical effort were important (with different strength) situation variables for almost all groups: thus improving those factors will increase the use of public transportation for all kinds of travelers. Note that, the cost which is an important decision factor in many life-situations, was mentioned as important only by one group of travelers. Mainly saving money is not the main concern while choosing a certain travel mode but other external factors which can be improved by policy makers.

With an increasing concern of air pollution, fuel and car dependency, these kinds of studies will contribute to the development of a better infrastructure for city transportation facilities.

As a future work of this study, we intend to work on the distance measurement of the CMs to make it simpler and taking into account also the complexity of the map. It would be interesting to compare the current results applied the improved distance formula.

References

1. Taco, P., Yamashita, Y., Souza, N., Dantas, A.: Trip generation model: A new conception using remote sensing and geographic information systems (2000)
2. Leon, M.: Cognitive maps in transport behavior. In: Proceedings of MICAI Mexican International Conference on Artificial Intelligence. IEEE Computer Society Press (2009)
3. Janssens, D., Hannes, E., Wets, G.: Tracking down the effects of travel demand policies. Urbanism on Track. Research in Urbanism Series (2008)
4. Axelrod, R.: Structure of Decision: the Cognitive Maps of Political Elites, Princeton (1976)
5. Kosko, B.: Fuzzy cognitive maps. International Journal on Man-Machine 24(1), 65–75 (1996)
6. Langfield-Smith, K., Wirth, A.: Measuring differences between cognitive maps. The Journal of the Operational Research Society 43(12), 1135–1150 (1992)
7. Markoczy, L., Goldberg, J.: A method for eliciting and comparing causal maps. Journal of Management (2), 305–333 (1995)
8. Ortolani, L., McRoberts, N., Dendoncker, N., Rounsevell, M.: Analysis of Farmers' Concepts of Environmental Management Measures: An Application of Cognitive Maps and Cluster Analysis in Pursuit of Modelling Agents' Behaviour. In: Glykas, M. (ed.) Fuzzy Cognitive Maps. STUDFUZZ, vol. 247, pp. 363–381. Springer, Heidelberg (2010)
9. Alizadeh, S., Ghazanfari, M., Fathian, M.: Using data mining for learning and clustering fcm. International Journal of Computational Intelligence 4(2) (2008)
10. Eden, C.: Analyzing cognitive maps to help structure issues or problems. European Journal of Operational Research 159(3), 673–686 (2004)
11. Kianmehr, K., Alshalalfa, M., Alhajj, R.: Fuzzy clustering-based discretization for gene expression classification. Knowledge and Information Systems 24(3), 441–465 (2010)
12. Davies, D., Bouldin, D.: A cluster separation measure. IEEE Transactions on Pattern Analysis and Machine Intelligence (2), 224–227 (1979)
13. Bolshakova, N., Azuaje, F.: Cluster validation techniques for genome expression data. Signal processing 83(4), 825–833 (2003)
14. Dijst, M.: Spatial policy and passenger transportation. Journal of Housing and the Built Environment 12, 91–111 (1997)

Production and Operations Management

Smart Objects System: A Generic System for Enhancing Operational Control

Gerben G. Meyer[1], W.H. (Wilrik) Mook[2], and Men-Shen Tsai[1]

[1] National Taipei University of Technology, Taiwan
   gerben.meyer@gmail.com, mstsai@ntut.edu.tw
[2] University of Groningen, The Netherlands
   w.h.mook@rug.nl

Abstract. Many companies are making considerable investments in tracking technology, such as GPS and RFID. Although tracking technology captures vast amounts of information about the ongoing operations, companies struggle to effectively apply this captured information for enhancing their operational control. In order to contribute in solving this problem, this paper presents a generic system for enhancing operational control, which applies the captured information in a more effective way. The proposed system is based on the approach of intelligent products. The intelligent products represent physical objects, and are capable of autonomously performing some of the repetitive tasks required for operational control. The usefulness of the system is demonstrated by presenting the results of several applications of the system.

1 Introduction

Many companies are making considerable investments in tracking technology, such as GPS and RFID, which is capable of capturing vast amounts of real-time information about the on-going operations (see e.g. [1, 24]). These investments are typically made with the aim to enhance the control of the ongoing operations, as the captured information can be applied for detecting unexpected events as well as for directing the decision making aimed at mitigating the consequences of these unexpected events [3]. Examples of unexpected events include delays in transportation and deliveries, failing resources, and products which are missing or malfunctioning. Hence, operational control is required for determining whether the operations are being performed as intended, and to make short-term adjustments for mitigating the negative impact of unexpected events on the ongoing operations [25].

Many companies however struggle to effectively apply the captured information for enhancing their operational control. Regarding this issue,

J. Casillas et al. (Eds.): Management Intelligent Systems, AISC 171, pp. 69–78.
springerlink.com   © Springer-Verlag Berlin Heidelberg 2012

Crainic et al. [5] state that *"The information is there. One only needs the appropriate methodology to transform these data into accurate and timely decisions.".* Many authors confirm this observation of Crainic et al., showing that computer-based support systems often fail to provide the required information and functionality to support planners responsible for the control of ongoing operations (see e.g. [2, 6, 11, 13, 23]). Therefore, operational control is still typically a manual task, leaving most of the captured information unused.

The goal of the research as described in this paper is to develop a generic system for enhancing operational control by applying the captured information in a more effective way. A design science approach [8, 21] has been adopted for the development and evaluation of this system. The proposed system as presented in this paper is based on the approach of intelligent products [16]. Intelligent products can represent individual physical objects such as products and resources in the system, and are capable of autonomously performing some of the repetitive tasks required for operational control. Such tasks include analyzing the captured information on a low level of granularity, presenting the captured information in a more comprehensive way, and providing notifications when unexpected events occur. In this way, the proposed system enables companies to enhance their operational control by applying the information captured by tracking technologies in a more effective way.

The development of the system reported here has been supported by extensive research related to the concept of intelligent products, which comprises two main sources of knowledge. The first source is formed by state of the art theories and frameworks in the field of intelligent products (see e.g. [10, 12, 26]). The recent research trends, emerging architectural designs, and novel application domains in this field are surveyed in [16], leading to the formulation of a classification model which has been used as a research framework. The second source of knowledge is based on application oriented experience, which was gained during the iterative development of an intelligent products architecture and the evaluation of that architecture by means of validating various prototypes [14].

The remainder of this paper is structured as follows. Next, a short analysis is presented, leading to the formulation of system requirements. Afterwards, Section 3 presents the proposed generic system for enhancing operational control. In order to demonstrate its usefulness, Section 4 presents the results of several applications of this system. Conclusions are provided in the last section.

2 Requirements

As mentioned in the introduction, operational control is typically still a manual task, leaving most of the captured information about the ongoing operations unused. This is due to the fact that current computer-based support systems often fail to provide the required information and functionality to support planners responsible with this task. Hence, new system requirements are needed, in order to develop a system which is able to contribute in solving this problem.

Tracking technology typically captures vast amounts of information about the ongoing operation. For example, fleet management systems capture every movement of every vehicle in a company's fleet (see e.g. [9]), manufacturing control systems capture every operation performed by a company's manufacturing resources (see e.g. [12]), and product lifecycle management systems capture every action performed on every individual product (see e.g. [7, 22]). In order to apply this captured information in a more effective way, the system should first of all be able to access all this information.

- *Requirement #1*: The system should be able to access all captured information.

Presenting all captured information to the planners will not improve the operational control of the ongoing operations, as the amount of captured information is likely to overwhelm the planners. Hence, the system should only present the information which the planners require for performing operational control.

- *Requirement #2*: The system should only present the information relevant for performing operational control.

In order to effectively perform operational control, it is key that unexpected events are detected in time. An early detection of an unexpected event gives the planners more time to assess the situation at hand and to determine which short-term adjustments are needed for mitigating the consequences of the unexpected event. Therefore, it is important that the system is able to pro-actively notify the planners when unexpected events occur.

- *Requirement #3*: The system should pro-actively provide notifications when unexpected events occur.

For planners, manually finding a suitable control decision for mitigating the consequences of an unexpected event can be troublesome, due to the complexity of the operations under control and the time available for making such a decision. Hence, it is beneficial if the system is able propose potential control decisions, allowing the planners to choose the most appropriate decision.

- *Requirement #4*: The system should discover and propose potential control decisions.

3 System Design

This section describes the Smart Objects System (SOS), the proposed generic system for enhancing operational control. First, the overall system architecture

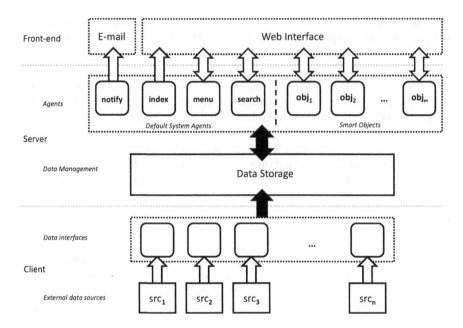

Fig. 1 System architecture

is presented. Afterwards, the structure and the behavior of the proposed system are described.[1]

3.1 Architecture

The overall architecture of SOS is shown in Figure 1. As can be seen from the figure, an application built with SOS needs information about physical objects from one or more external data sources. For each of these external data sources, a data interface needs to be provided, which interprets the external data and converts the structure of this data into a structure compatible with the SOS data storage. For every physical object of which information is stored in the data storage, the server creates an agent which can add intelligence to this physical object. In this way, a so called *smart object* is created. Every agent can execute its own application specific behavior, for example to determine which information will be displayed in the web interface. Besides the agents representing physical objects, several default system agents are always present. The *index*, *menu*, and *search*-agent are responsible for generating the generic parts of the web interface, and the *notify*-agent is responsible for generating e-mail notifications.

---

[1] A more extensive description of the system can be found on the SOS project page:
http://code.google.com/p/smart-objects-system/

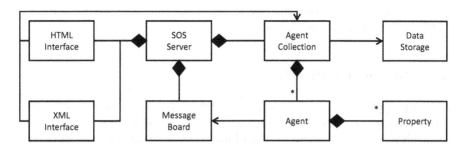

Fig. 2 System structure

3.2 Structure

The basic structure of the system is shown in Figure 2, which shows a simplified UML class diagram containing the most important classes of the system. The purpose of each of these classes will be shortly explained next.

- The *SOSServer* class is the starting point of the system. It starts the HTTP and XML interfaces, prepares the AgentCollection and MessageBoard, and initializes the default system agents.
- The *Agent* class is an abstract class, which acts as a base for all application specific agents. Every agent represents one physical object in the system, of which the information is maintained through a set of properties. Methods for getting and setting these properties, as well as learning its current status and executing application specific behavior are defined in this class.
- A *Property* is a basic data structure which defines a single property of an agent. A number of different property types have been defined, such as text, number, time, and location.
- The *AgentCollection* is used for managing all agents which are currently present in the system, and the *DataStorage* is used for storing all the properties of agents in a database.
- The *MessageBoard* is used for communication between agents.
- The *HTMLInterface* provides the web interface for system users, and the *XMLInterface* provides the interface through which external data about physical objects can be added to the data storage.

3.3 Behavior

The behavior of the agents is largely application specific, and therefore different for every application developed with SOS. This section describes on a more generic level how the behavior of the agents can be used to meet the requirements introduced in Section 2 when developing an application with SOS. This behavior is introduced according to the three levels of intelligence for intelligent products as prescribed by Meyer et al. [16]: information handling, problem detection, and decision support.

3.3.1 Information Handling

When an application is developed with SOS, it is key to determine which physical objects will be represented by agents in the system. Information about these objects captured by tracking technology or other information systems has to be made available to the agents representing them. For this purpose, application specific data interfaces have to be developed, which translate the external data into data understandable by the agents. In that way, *Requirement #1* will be met, and every agent is enabled to collect all available information related to the physical object it is representing.

Presenting all available information to the system users will not enhance operational control. As every agent has to determine which information it will present to the system users (see Figure 1), the agents representing physical objects have to be developed in such a way, that only information important for performing operational control in the specific application is displayed. In that way, *Requirement #2* will be met.

3.3.2 Problem Detection

By means of the web interface, SOS enables the system users to train the agents, as the system users can inform them whether their current situation is problematic or not. The trained agent will generate a training instance based on the provided status and the information available on the physical object it is representing. This training instance will be stored in the data storage and will be shared among other agents representing the same type of physical object. By using a machine learning classifier [20], the agents are enabled to continuously determine their status. In order to meet *Requirement #3*, the system users will be directly notified by means of an e-mail message when the status of an agent suddenly becomes problematic. Moreover, when requested by a system user, the agents will also provide this information through the web interface.

3.3.3 Decision Support

Although no generic behavior for agents is provided on this level of intelligence, every agent developed for SOS can perform application specific behavior needed for decision support. Agents can for example communicate and negotiate with each other by using the message board, in order to discover potential control decisions. By developing such application specific agent behavior, *Requirement #4* can be met, as one of the applications presented next demonstrates.

4 Applications

Two applications of SOS have been developed, in order to demonstrate the usefulness of the proposed system. These applications will be shortly discussed next.

4.1 Transportation

One application developed with SOS is an application for enhancing operational control of a medium-sized road freight transportation company. The planners of this company struggle to detect unexpected events in time, as the amount of information collected by the tracking technology in place is too high to be analyzed manually. Despite the available tracking technology, the planners are only informed about unexpected events through conventional methods, such as phone calls with customers and on-route truck drivers, often resulting in problems being detected too late.

Within the application developed with SOS, every truck of the company as well as every pallet to be transported by the company is represented by an agent. This application is focused on the behavior of these agents on the level of information handling and problem detection, as the main goal of the application is to assist the planners in handling the available information and detecting unexpected events in time. Hence, on the level of information handling, every agent collects all information available in the existing information systems of the company related to the physical object it represents and provides the planners with a comprehensive overview of the plan it is involved in combined with the actual progress of this plan. On the level of problem detection, pallet agents are trained to have a problematic status when their expected delay is one hour or more, and truck agents are trained to have a problematic status when they are transporting one or more pallets with a problematic status. The web interface of the application enables the planners to directly determine which agents have a problematic status. Moreover, e-mail messages are send automatically to the planners to pro-actively inform them when the status of an agent suddenly becomes problematic.

The developed system has been used for conducting a pilot study at the aforementioned transportation company. This study showed that the system was able to correctly detect problems caused by unexpected events. The planners were also correctly notified about problems which were not yet observed by themselves, but nevertheless required immediate control decisions. Moreover, the pilot study showed that the information presented by the agents enabled a better understanding of the unexpected events including their impact on the ongoing operations. More details on the developed application as well as the performed evaluation can be found in [15] and [17].

4.2 Production

Another application developed with SOS is an application for performing operational control of a manufacturer of personal computers within the Trading Agent Competition Supply Chain Management (TAC SCM) simulated supply chain [4]. Within the TAC SCM simulated supply chain, every manufacturer has to procure customer orders, buy computer components, assemble computers, and ship the computers to the customers. For this purpose, the manufacturer has a computer factory containing an assembly cell capable of assembling any type of computer, and a warehouse that stores both components and assembled computers.

Within the TAC SCM manufacturer developed with SOS, every product ordered by a customer is represented by an agent. Such an agent is responsible for the complete processing of one final product. Hence, an agent is responsible for acquiring the components required for assembling the product, acquiring the required production capacity, as well as arranging the shipment of the assembled product to the customer who ordered it. The developed application is focused on the behavior of these agents on the level of decision making, as information handling and problem detection are less important in a simulated environment with no involvement of human planners. Hence, on the level of decision making, every product agent negotiates with other agents to plan and control the distribution of components, production capacity, and shipping capacity. For each of these tasks, an auctioning approach is used, in which the agents can bid on the required components and resources.

The results of the conducted simulations showed that the developed TAC SCM manufacturer is very robust in terms of handling unexpected events. Although the manufacturer had to deal with late supplier deliveries as well as broken components, it was still able to finish nearly all of the requested products in time, without increasing the component inventory "safety stock". More details on the developed application as well as the simulation results can be found in [18] and [19].

5 Conclusions

Many companies struggle to effectively apply the vast amounts of information captured by tracking technology for enhancing their operational control. Therefore, a generic system for enhancing operational control has been developed, with the goal to apply the captured information in a more effective way. By adopting the intelligent products approach, the proposed system is capable of autonomously performing some of the repetitive tasks required for operational control. The presented applications have demonstrated that the proposed system contributes to enhancing operational control by applying the captured information in a more effective way.

Two managerial implications can be gleaned from the research as presented in this paper. Firstly, investments of companies in tracking technology are not likely to directly enhance operational control. Secondly, a system based on intelligent products as presented in this paper can be applied to overcome this problem. Due to the generic nature of this system, other companies facing similar problems when applying captured information for enhancing operational control are likely to benefit from the system as well. Accordingly, future work will be focused on developing additional applications for companies facing similar problems. These applications can contribute to confirming as well as generalizing the results and insights as presented in this paper.

Acknowledgements. The authors would like to express their gratitude to Gijs Roest for his assistance in developing the Smart Objects System. This research is supported by the National Energy Project of the National Science Council, Taiwan, under grant number NSC 101-3113-P-006-020.

References

1. Angeles, R.: Rfid technologies: Supply-chain applications and implementation issues. Information Systems Management 22(1), 51–65 (2005)
2. Budihardjo, A.: Planners in action: roadmap for success: an empirical study on the relationship between job decision latitude, responsiveness and planning effectiveness in road transport companies. PhD thesis, University of Groningen, Groningen, The Netherlands (2002)
3. Buijs, P., Szirbik, N.B., Meyer, G.G., Wortmann, J.C.: Situation awareness for improved operational control in cross docking: An illustrative case study. In: Proceedings of the 14th IFAC Symposium on Information Control Problems in Manufacturing (2012)
4. Collins, J., Arunachalam, R., Sadeh, N., Eriksson, J., Finne, N., Janson, S.: The supply chain management game for the 2007 trading agent competition. Technical Report CMU-ISRI-07-10, Carnegie Mellon University, Pittsburgh, Pennsylvania, USA (2006)
5. Crainic, T.G., Gendreau, M., Potvin, J.-Y.: Intelligent freight-transportation systems: Assessment and the contribution of operations research. Transportation Research Part C: Emerging Technologies 17(6), 541–557 (2009)
6. de Snoo, C., van Wezel, W., Jorna, R.J.: An empirical investigation of scheduling performance criteria. Journal of Operations Management 29(3), 181–193 (2011)
7. Främling, K., Ala-Risku, T., Kärkkäinen, M., Holmström, J.: Agent-based model for managing composite product information. Computers in Industry 57(1), 72–81 (2006)
8. Hevner, A.R., March, S.T., Park, J., Ram, S.: Design science in information systems research. MIS Quarterly 28, 75–105 (2004)

9. Ichoua, S., Gendreau, M., Potvin, J.-Y.: Planned Route Optimization for Real-Time Vehicle Routing. In: Dynamic Fleet Management: Concepts, Systems, Algorithms & Case Studies, pp. 1–18. Springer (2007)
10. Kärkkäinen, M., Holmström, J., Främling, K., Artto, K.: Intelligent products - a step towards a more effective project delivery chain. Computers in Industry 50(2), 141–151 (2003)
11. MacCarthy, B.L., Wilson, J.R.: Influencing Industrial Practice in Planning, Scheduling and Control. In: Human performance in planning and scheduling, pp. 451–461. Taylor & Francis (2001)
12. McFarlane, D., Sarma, S., Chirn, J.L., Wong, C.Y., Ashton, K.: Auto ID systems and intelligent manufacturing control. Engineering Applications of Artificial Intelligence 16(4), 365–376 (2003)
13. McKay, K.N., Wiers, V.C.S.: Practical production control: a survival guide for planners and schedulers. J. Ross Publishing (2004)
14. Meyer, G.G.: Effective Monitoring and Control with Intelligent Products. PhD thesis, University of Groningen (2011)
15. Meyer, G.G.: System Prototype for Transportation. In: Effective Monitoring and Control with Intelligent Products, pp. 111–139. University of Groningen (2011)
16. Meyer, G.G., Främling, K., Holmström, J.: Intelligent products: A survey. Computers in Industry 60(3), 137–148 (2009)
17. Meyer, G.G., Roest, G.B., Szirbik, N.B.: Intelligent products for monitoring and control of road-based logistics. In: Proceedings of the 2010 IEEE International Conference on Management and Service Science, Wuhan, China (August 2010)
18. Meyer, G.G., Wortmann, J.C(H.): Robust Planning and Control Using Intelligent Products. In: David, E., Gerding, E., Sarne, D., Shehory, O. (eds.) AMEC 2009. LNBIP, vol. 59, pp. 163–177. Springer, Heidelberg (2010)
19. Meyer, G.G., Wortmann, J.C., Szirbik, N.B.: Production monitoring and control with intelligent products. International Journal of Production Research 49(5), 1303–1317 (2011)
20. Mitchell, T.M.: Machine learning. McGraw-Hill (1997)
21. Peffers, K., Tuunanen, T., Rothenberger, M.A., Chatterjee, S.: A design science research methodology for information systems research. Journal of Management Information Systems 24(3), 45–77 (2007)
22. Rönkkö, M., Kärkkäinen, M., Holmström, J.: Benefits of an item-centric enterprise-data model in logistics services: A case study. Computers in Industry 58(8-9), 814–822 (2007)
23. Rushton, A., Oxley, J., Croucher, P.: The handbook of logistics and distribution management. Kogan Page (2000)
24. Schumacher, J., Feurstein, K. (eds.): Proceedings of the 3rd European Conference on ICT for Transport Logistics, Bremen, Germany (November 2010)
25. Slack, N., Chambers, S., Johnston, R.: Operations management. Pearson Education (2004)
26. Valckenaers, P., Saint Germain, B., Verstraete, P., Van Belle, J., Hadeli, K., Van Brussel, H.: Intelligent products: Agere versus essere. Computers in Industry 60(3), 217–228 (2009)

Distributed Cognition Learning in Collaborative Civil Engineering Projects Management

Jaume Domínguez Faus and Francisco Grimaldo

[1] Centre for 3D GeoInformation, Aalborg University. 9220 Aalborg, Denmark
`jaume@land.aau.dk`
[2] Departament d'Informàtica, Universitat de València, Av. Universitat, s/n, 46100, Burjassot, Spain
`francisco.grimaldo@uv.es`

Abstract. Due to the diversity and complexity of its projects, the Civil Engineering domain has historically encompassed very heterogeneous disciplines. From the beginning, any Civil Infrastructure project is systematically divided into smaller subprojects in order to reduce or isolate the overall complexity. However, as a parallel design work, these subdesigns may experience divergences which often lead to design conflicts when they are merged back to the global design. If a high-quality design is desired, these conflicts need to be detected and solved. We present a Multiagent system able to manage these design conflicts by detecting them, by assisting the engineers in the negotiation of solutions, and finally by learning how to solve future similar problems. The advantage of the system is that what is learned is not one individual's knowledge but the project's global distributed cognition.

Keywords: Multi-agent systems, Civil Engineering Projects, Semantic validation, Negotiation, Machine learning.

1 Introduction and Related Work

Civil Engineering projects are a collaborative work. The integral development of any civil infrastructure demands expertise in many different disciplines. Since developing a project as a whole is an overwhelming job, the works to be done are systematically classified and assigned to teams of experts holding the required skills for each of them. Thus, for instance, structure engineers take responsibility for designing the skeleton of the infrastructure; sewerage engineers design the management of waste waters and drainage; signaling is done by traffic experts, etc.

When working on the project, each team develops its part separately and periodically they all align their work by merging all their "subdesigns". Then, engineers evaluate the progress in order to find and solve the problems the project contains. Nowadays, this task is performed manually by interpreting the project's documentations and, when possible, by navigating a 3D computer model created by combining all the subdesigns. However, this process can fail in detecting all the issues since

J. Casillas et al. (Eds.): Management Intelligent Systems, AISC 171, pp. 79–88.

documentation may become difficult to be analyzed and some information may be lost when constructing the global model. As a result, if a problem remains unde- tected during the design time, it causes potentially large loses in resources after- wards. Some authors have estimated the average costs caused by these undetected problems in 5-10% of the total project budget [4]. Thus, given the size of Civil Engineering projects, there is significant potential for improvement.

To improve interoperability between the different stakeholders, the industry has traditionally defined new data exchange formats and it has extended the existing ones . Whereas probably the most common format is DWG from AutoDesk, which is still the *de facto* standard, there are others, e.g. CityGML or IFC [7], that also enjoy a relative success. The latter even include extension mechanisms for "user- defined" data structures as a way to consider particular needs. However, they can hardly guarantee that those user-defined structures are understood by outsiders, given that consumer applications need to implement mechanisms to support them. This shows that the interoperability has not been completely reached yet. On the other hand, Building Information Model (BIM) servers [4] offer a higher degree of interoperability by storing CAD models together with management spreadsheets and other documents in a centralized way and, thus, easing the project management.

Even though these efforts facilitate the control of the project, the detection of problems and their resolution still require human interaction. Traditional approaches have shown difficultes in tackling the challenges derived from distributed collab- orative work but literature suggests that Multi-agent Systems (MAS) can better equipped for facing these kind of problems. MAS have been already used in Civil Infrastructure research in different situations. An example of controlling construc- tion equipment is the system described by Zhang, Hammad, and Bahnassi where sets of cranes synchronize their movements to transport materials from one loca- tion to another in the construction area without crashing [12]. It is also possible to find examples of MAS that focus on the distinct phases a typical project consists of. Thus, Denk and Schnellenbach describe an agent-based tendering procedure that covers the initial phase when the project is published and interested compa- nies bid for it [13]; Udeaja and Tah [14] focus on the construction material sup- ply chain that is managed collaboratively. Within the Construction phase the main focus has been put on the formalization of expert knowledge and the negotiation mechanisms to solve unexpected situations. Peña-Mora and Wang [10] presented a Game Theory approach to solve various known conflict situations between the Ar- chitect/Designer/Constructor settings when each agent competes for reducing the impact of unexpected situations in the Construction area in its own interest. More recently Shen et al. studied the applicability of cognitive maps MAS in collaborative- competitive working in construction projects where these maps are used to parame- terize the agent's beliefs within the MAS negotiation [16]. We propose using MAS to assist in the detection of problems in the Design phase, where the definition of a problem depends on how an expert sees the world according to her perspective. Problems are solved attending to the project's global benefit by means of negotiat- ing alternatives to it. Finally, the MAS uses humans' decisions as a way to capture the project's *distributed congition* that is exploited afterwards to train the system so that it can suggest its own alternatives from the knowledge it has gained.

This work aims at developing an intelligent system devoted to aid engineers in managing design conflicts in civil infrastructure projects. The proposed system is able to detect these conflicts, to assist engineers in the negotiation of solutions, and finally to learn how to solve future similar problems. It follows a distributed multi-agent approach incorporating well-known artificial intelligence techniques in order to tackle the problem in a flexible and extensible way, thus providing the necessary level of abstraction to be applied to different projects within this domain. Hence, this ad hoc intelligent system contributes to increase strategic intelligence (i.e., knowledge management + business intelligence + competitive intelligence) of companies whose projects join teams with heterogeneous expertise working collaboratively.

Section 2 introduces the main components of the system: Semantic Conflict Detection, Distributed Conflict Solving, and the Learning Subsystem. A deep description of the Semantic Conflict Detection and Distributed Conflict Solving components is out of the scope of this paper since it can be found in [5]. However, they are are briefly introduced in sections 2.1 and 2.2 for the sake of readability. The Learning Subsystem is described in section 2.3. Finally, in section 3 we describe a use case in which the system was applied in order to illustrate its usefulness.

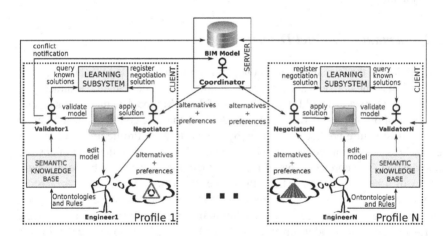

Fig. 1 Overview of the system

2   The Multi-Agent System

Our system follows a distributed architecture approach allowing the engineers of different expertises to design, through their client interfaces, a common Building Information Model (BIM) that is stored at the server (see Figure 1). Notice that the meaning (the semantics) of an object in the model, and in particular what it implies, varies from who is looking at it. For instance, while a gas conduction could be seen as a mere tube-shaped obstacle by an electrician engineer, it means much more for the expert that is designing it. The former may only need to ensure that her designs are not putting anything in the location already used by the gas conduction to avoid object overlapping. On the other hand, the latter will have other concerns like, e.g., whether the material of the pipe is compatible, in terms of safety, with the distance of such an electric installation. Hence, it is not only the object's shape what is important, but also what the object *means* (i.e. its semantics) and, therefore, what it implies. We refer to a semantic conflict when a problem of this type is detected. In other words, a situation that may be correct from a single engineer's point of view but where different interests collide when it is considered globally. In our system, we define the semantics that each engineer cares about by means of OWL [1] ontologies and SWRL [9] rules. They are used by each engineer's Validator agent for detecting and inferring problematic situations. When a conflict is detected, a solution for it can be negotiated. The Validator agent that detected it notifies a Coordinator agent residing at the BIM server. The Coordinator then conducts the negotiation with all the stakeholders' Negotiators which, upon an agreement on the solution for the conflict, apply it and register it in the Learning Subsystem for further analysis in order to suggest solutions for future similar situations. Our agent ecosystem uses JADE [2] as the MAS engine. We define a Profile as the set of the human expert engineer, her semantic knowledge base, her Validator and Negotiator agents, and her Learning Subsystem.

2.1   Semantic Conflict Detection

Ontologies are the formalization of concepts from which knowledge is built and that, in contrast to the classical attribute/value approach, provide semantic abstraction to the model. Our system builds the semantics of the BIM model by means of a layered structure of ontologies. A Base Ontology defining the core concepts needed for any civil infrastructure project is provided by default to each Profile and, on top of it, engineers can stack more Profile-specific ontologies to achieve the level of abstraction desired. The core concepts of the Base Ontology are: 1) Feature, the basic object of a model; 2) Geometry, the shape of a Feature; 3) hasGeometry and its inverse isGeometryOf, used to set the relationship between a Feature and a Geometry; 4) Attribute, defining one of the properties of a Feature; 5) hasAttribute and its inverse isAttributeOf, used to set the relationship between a Feature and an Attribute; 6) hasRelationship and its inverse isRelationshipOf, defining a generic relationship between two Features; 7) Problem, to capture individual errors in the project; and 8) Conflict, which is used to mark a subclass of Problem that has to be negotiated in order to be solved. In turn, Profile-specific ontologies extend these concepts with other ones that are of interest for particular fields of expertise such as: a Building or a Road for a building designer. Complementing the ontologies, engineers also provide SWRL rules to allow advanced reasoning. SWRL rules infer more concepts by specifying an antecedent that, when it turns out to be true, it implies that what is expressed in the consequent is also true. For instance, a RoadExhaustedConflict can be detected by a road designer when the building designer places a new building in a parcel where the road connecting to that parcel cannot hold the new population brought by the building. As engineers work in parallel, changes are performed to the model. These changes are monitorized by the Validator agents, that executes the Reasoning Engine when changes to the model are detected. Thus, both the set of ontologies and the rules defined by the engineers are used by the Validator agents to analize the model from each Profile's perspective in order to find `Conflicts`. More details about the semantic conflict detection can be found at [5].

2.2 Conflict Solving Protocol

When conflicts are detected, engineers have the possibility to solve them by means of negotiation. The negotiation follows a Multi-Agent Resource Allocation [3] approach, as a general mechanism that assists the engineers in evaluating the possible alternative solutions and in making socially acceptable decisions. Thus, they are required to express their preferences about the solutions, which are then used to select those that maximize the welfare of the group. As the allocation procedure, we use a ContractNet-like protocol involving two rounds. In the first round, once a conflict has been detected by a Validator agent, it is notified to the Coordinator agent. Then, the Coordinator informs all the Negotiators about the conflict and asks for alternatives to solve it through a Call For Solutions. In turn, Negotiators record the alternatives provided by the engineers and send them back to the Coordinator, who is in charge of collecting all the proposals. The solutions consists of operations that, when applied to the model, avoid the conflict. Invalid or repeated solutions are discarded and the set of remaining solutions is distributed again, thus starting the second round of the protocol. In this phase, engineers express their preferences by giving a score ranging from 0 (lowest) to 10 (highest) to each alternative and their corresponding Negotiators send this information to the Coordinator. Then, the Coordinator performs a winner determination process that leads to the selection of the alternative that maximizes the global welfare. The Nash social welfare function is used, as it ensures that the chosen solution is the most preferred one while also balancing the utility level among the negotiators. Finally, the solution is broadcasted to all the engineers and it is applied to the BIM model. As previously stated, more details about this distributed conflict solving mechanism can be found at [5].

2.3 Learning Subsystem

It seems possible to predict what the result of a negotiation will be if we have a model that is semantically rich enough to contain sufficient information. We used Machine Learning (ML) techniques to capture the distributed cognition that results in the selection of a feasible solution for a given type of conflict. From the advent of Nash's formalization of the bargaining problem [8], the negotiation can be methodically approached. Thus, the negotiation can be seen as a black box that captures all the subtleties (the context, the points of view, etc.) in which we input the problem and a solution is obtained in the output. But it requires that the engineers express their utility functions which are especially elusive to be defined a priori in dynamic and complex environments. That is the case of Civil Infrastructure projects where not even the set of parameters to such functions are easily known. The negotiation protocol we use solves this problem by asking each engineer's preferences on the possible alternatives, resulting in a solution. This way, we can capture the global behavior pattern while the complexity of the system remains at reasonable levels. We can keep a record of the inputs to the problem (i.e. the conflict and the context in which it occurs) and the solution (consisting of operations applied to the model), given by the engineers and use it as a training set from which the agents in the MAS can learn. As a result, after several rounds of negotiation to conflicts showing similarities, the agent can suggest the solution to a newly detected conflict.

In our system, when the Validator agent detects a conflict, it queries the Learning Subsystem (LS) for known solutions to the conflict. The query consists of the conflict and the `Features` involved in it and the relationships among them, that is, the conflict's context. If possible solutions are known, they are presented to the user so she can choose and directly apply the most suitable. If there are no known solutions or the ones proposed by the Validator are not satisfactory then the negotiation described in section 2.2 occurs. After this negotiation, all engineers will agree on a new solution. The Negotiators then apply it to the model and register it into the LS.

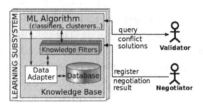

Fig. 2 Learning subsystem

Reich [11] carried out an extensive review where he analyzed experiments with ML in Civil Infrastructure concluding that there is no ML technique that solves all the problems but instead some work better than others depending on the problem. We designed our learning system in a way that it is easily extendable with standard techniques (e.g, ID3, C4.5, K-Means, etc. [15]) as well as with experimental ones.

As already mentioned, a solution consists of operations applied to the model that change its state from a, say, "conflicted status" to a "valid status". Thus, the solutions directly rely on what the supported operations are, and also what and how the parameters for those operations are. In Civil Infrastructure, spatial operations are predominant. Consider, for instance, the operation "move object A 3 meters in direction D". A closer look at that operation shows that it takes five parameters: the object, the distance the object is going to be moved and the direction (which in 3D needs three coordinates). The spatial nature of the operations forces some preprocessing to be done before it becomes a useful solution for the learning. Imagine that this solution was produced as a result of the conflict "the object A is overlapping the object B". If we store the solution as it is, in the eventual event that new similar "object F is overlapping the object G" conflict is detected, upon a request to the LS for a solution to this conflict, it will respond with the "move object A 3 meters in direction D" which is obviously wrong. A way to overcome this problem is to store the solutions in a symbolic way. In other words, instead of "move object A 3 meters towards direction D" we would express it as "move *TheObjectThatIsOverlapping AwayFromTheObjectItOverlapsWith*". And make the corresponding substitutions in each different case of the conflict. The Knowledge Base of the LS is equipped with the Knowledge Filters (see Figure 2) which is an extendable set of filters meant for endowing the solutions with a level of abstraction as necessary. The filters are applied forward and in reverse order for storing and querying respectively.

On the other hand, Machine Learning (ML) algorithms are very sensitive to the way data is stored and to its *noise*. In order to accommodate the data to the algorithm in use, the Data Adapter (DA) is included. The prediction of a solution the algorithm does is based on the conflict and its context. However, while the conflict belongs to a finite set of conflict types (otherwise it could not have been detected), the context where it occurs varies. The DA is in charge of transforming the data in the database to fit the algorithm's needs while no relevant information is lost.

3   Use Case

To test the proposed approach, we have applied our system to a real project consisting in the design of the electricity installation of a power plant carried out by "Vianova Plan og Trafikk AS" in Norway. We prefer using a real use case as there is to our knowledge no standard and suitable benchmark to test our approach against. Even though some CAD benchmarks have been proposed [6], they focus on the size of the dataset as well as on the computation time. Instead, we aim at finding conflicts and learning how to solve them without the need to write a routine on purpose.

In this use case, the design is done by two profiles, one designing the foundation of the installation (Foundation Profile) and the other designing the structure holding the electric cables (Structure Profile). The use case consists of 4592 ontology entities and the reasoning time takes 50 seconds. During validation, conflicts can be detected such as the bolts of a foundation that is used to fix a structure not fitting in the holes the Structure Profile designed for them due to some measurement misunderstanding. This kind of conflict is detected 80 times, one per each bolt in the foundation, through the following SWRL rule:

$$Bolt(?b) \land Hole(?h) \land$$
$$isGeometryOf(?b, ?bg) \land isGeometryOf(?h, ?hg) \land$$
$$isClosest(?bg, ?hg) \land distance(?bg, ?hg, ?d) \land$$
$$isGreaterThan(?d, 0) \rightarrow BoltDoesntFit(?b, ?h)$$

This rule would tag the feature b as a conflict of type `BoltDoesntFit` if, when transversing the model, b is a `Bolt` feature and h is a `Hole` feature such that they are the closest Hole/Bolt pair in the model (given by their geometries) and the distance between them is greather than 0.

The two profiles negotiate to solve the first conflict. In the alternative proposal round, the Foundation profile suggests to move the left leg of the two-legged structure so it fits with the bolts. The leg is represented by a `StructurePart` feature which contains an attributed called `Leg` with value "Right". To make the leg fit in the foundation bolts it needs to move 3 centimeters to the left (X axis) and 1 cm downwards (Y axis), i.e. it needs to move to the distance $d1 = (-0.03, -0.01, 0)$ (alternative $A1$). The Structure profile suggest to move the bolt to the distance $d2 = (0.03, 0.01, 0)$ (alternative $A2$). In the second round, the profiles express their preferences. Structure profile gives a value of 1, and 5 to $A1$ and $A2$ respectively because, say, she knows that the structure parts are already delivered by the supplier and changing them would be very costly while moving the bolts does not seem to be a big deal. The Foundation profile, in turn, gives a value of 7, and 4 because she does not know about the structure parts situation but would prefer others to change while she admits that moving the bolts is a relatively small effort. As a result, $A2$ is selected as a solution. Similar negotiations follow for the rest of conflicts with same results with the only difference that when the bolt is supposed to fit the "Left" leg of the structure, it is moved in the opposite direction, i.e. $d1 = (-0.03, -0.01, 0)$. The results were stored in the LS using only one simple Knowledge Filter that replaces the name of the features involved by symbolizers (#@#, in this case) so they can be more generally applied. Table 1 shows the (simplified) database created from this. We only show the necessary columns on the table due to space restrictions. In this case, the actual table is composed of 15 columns storing other relationships the bolts have and attributes that are not relevant since they take the same values in all the records. We used an ID3 decision tree as the ML algorithm which automatically detects that what defines the direction where the bolt is to be moved is the Leg attribute of the "StructurePart" feature.

Table 1 Simplified database for this conflict. Each row represents a conflict and its negotiated solution (CTF=ConflictedFeatureType, CWF=ConflictedWithFeature, CWFRF=CWF-RelatedFeature, "#@#" is a symbol refering to the feature that is in conflict that is substituted and restored by the Knowledge Filter)

CTF	CWFType	CWFRF1	...	CWFRF1AttrLeg	...	Solution
Bolt	Hole	StructurePart		Right		MoveFeatureCmd(#@#,-0.03,-0.01,0.0)
Bolt	Hole	StructurePart		Left		MoveFeatureCmd(#@#,0.03,0.01,0.0)

Thus, once the system collects enough information, Validator agents can find the learned solutions from the LS that can be applied without further negotiations. Notice, however, that the amount of negotiations required could be reduced by having an extra Knowledge Filter substituting the distance vector within the parameters of the solution by a symbol as explained in section 2.3.

4 Discussion

As computers became more powerful, they allowed to execute bigger and more complex programs. The industry producing software for Civil Engineering creates software packages with newer and more powerful features. Today's engineers can perform more complex tasks with these advanced tools. However, these tools consist of pre-established algorithms with the only possibility of parameterizing them as a means to adapt them to one situation or to another. Furthermore, the tools are meant to be executed once the user decides to invoke them. In other words, they are not designed to make decisions but to execute them.

It is assumed that the only way to improve is creating new algorithms. In our opinion, this approach is, somehow, preventing the creation of software that solves conflicts in distributed design settings. Because of the overwhelming amount of factors converging in a conflict, it seems not possible to write algorithms covering all the issues. Even worse, the necessary information that influences a distributed decision may simply not be available for the computer system. In fact, most of the information is still residing only in the engineers head. It is simply too difficult to input every detail into a system which, in turn, needs to be prepared to store it.

The proposed system takes another approach. By following a semantic knowledge-driven approach, the system is able to detect conflicts and to propose automated solutions, which is of great utility for solving any managerial problem and decisional scenarios. While humans make decisions, the system tracks them and captures the distributed cognition that allows making the right decisions in the particular context the project occurs in. Thus, the system capitalizes on the conflict resolution made by humans. What is a problem in the begining turns out to be a valuable asset that is exploited by the system to learn and to improve. Then, engineers are no longer limited to what software packages allow since it is the software what evolves and adapts to their needs.

The future work will focus on testing the system in more use cases to deeper study how the system performance compares to humans with regard to the quality of the decisions made.

Acknowledgements. This work was supported by Norwegian Research Council, Industrial PhD scheme case no: 195940/I40 through Vianova Systems AS, Norway; the Spanish MICINN, Consolider Programme and Plan E funds, as well as European Commission FEDER funds, under Grants CSD2006-00046 and TIN2009-14475-C04-04. It was also partly supported by Development and Planning Department of the Ålborg University (AAU), the Vice-rectorate for Research of the Universitat de València (UV) under grant UV-INV-AE11-40990.62. Authors also want to thank Dr. Erik Kjems from the AAU for his help without which this paper would not be possible.

References

1. Bao, J., Calvanese, D., et al.: OWL 2 Web Ontology Language Document Overview. In: W3C (2009)
2. Bellifemine, F., Caire, G., Greenwood, D.: Developing Multi-agent Systems with JADE. John Wiley & Sons Ltd. (2007)

3. Chevaleyre, Y., Dunne, P.E., Endriss, U., Lang, J., Lemaitre, M., Maudet, N., Padget, J., et al.: Issues in multiagent resource allocation. Informatica 30, 3–31 (2006)
4. Eastman, C., Teicholz, P., Sacks, R., Liston, K.: BIM Handbook, a guide to Building Information Modeling for Owners, Managers, Designers, Engineers, and Contractors. John Wiley & Sons, Inc., New Jersey (2008)
5. Faus, J.D., Grimaldo, F.: Multiagent system for detecting and solving design-time conflicts in civil infrastructure. Advances in Intelligent and Soft Computing, vol. (157) (2012)
6. Jayanti, S., Kalyanaraman, Y., Iyer, N., Ramani, K.: Developing an engineering shape benchmark for CAD models. Computer-Aided Design 38(9), 939–953 (2006)
7. Kolbe, T.H., Gröger, G., Plümer, L.: Citygml: Interoperable access to 3d city models. In: Geo-Information for Disaster Management, pp. 883–899 (2005)
8. Nash, J.F.: The bargaining problem. Econometrica, 155–162 (1950)
9. O'Connor, M.F., Knublauch, H., Tu, S., Grosof, B., Dean, M., Grosso, W., Musen, M.: Supporting rule system interoperability on the semantic web with SWRL. In: Gil, Y., Motta, E., Benjamins, V.R., Musen, M.A. (eds.) ISWC 2005. LNCS, vol. 3729, pp. 974–986. Springer, Heidelberg (2005)
10. Peña-Mora, F., Wang, C.-Y.: Computer-supported collaborative negotiation methodology. Journal of Computing in Civil Engineering, 64–81 (1998)
11. Reich, Y.: Machine learning techniques for civil engineering problems. Microcomputers in Civil Engineering (1997)
12. Ren, Z., Anumba, C.: Multi-agent systems in construction–state of the art and prospects. Automation in Construction 13, 421–434 (2004)
13. Schnellenbach, M., Denk, H.: An agent-based virtual marketplace for aec-bidding. In: Proceedings of the 9th International EG-ICE Workshop Advances in Intelligent Computing in Engineering, Darmstadt, Germany, pp. 40–48 (2002)
14. Udeaja, C., Tah, J.: Agent-based material supply chain integration in construction. In: Perspectives on Innovation in Architecture, Engineering and Construction. CICE, pp. 377–388. Loughborough University (2001)
15. Witten, I.H., Frank, E., Hall, M.A.: Data Mining Practical Machine Learning Tools and Techniques. Elsevier (2011)
16. Xue, X., Ji, Y., Li, L., Shen, Q.: Cognition driven framework for improving collaborative working in construction projects: Negotiation perspective. Journal of Business Economics and Management (2010)

Designing Lines of Cars That Optimize the Degree of Differentiation vs. Commonality among Models in the Line: A Natural Intelligence Approach

Charalampos Saridakis[1], Stelios Tsafarakis[2], George Baltas[3], and Nikolaos Matsatsinis[2]

[1] Leeds University Business School, University of Leeds, United Kingdom
[2] Department of Production Engineering & Management, Technical University of Crete, Greece
[3] Department of Marketing & Communication, Athens University of Economics & Business, Greece

Abstract. The product life cycle of cars is becoming shorter and carmakers constantly introduce new or revised models in their lines, tailored to their customer needs. At the same time, new car model design decisions may have a substantial effect on the cost and revenue drivers. For example, although a new car model configuration with component commonality may lower manufacturing cost, it also hinders increased revenues that could have been achieved through product differentiation. This paper applies a state of the art, nature inspired approach to design car lines that optimize the degree of differentiation vs commonality among models in the line. Our swarm intelligence mechanism is applied to stated preference data derived from a large-scale conjoint experiment that measures consumer preferences for passenger cars in a sample of 1,164 individuals. Our approach provides interesting insights on how new and existing car models can be combined in a product line and suggests that differentiation among models within a product line elevates customer satisfaction.

Keywords: Car line design, differentiation vs commonality, swarm intelligence, conjoint analysis.

1 Introduction

In technology and capital-intensive industries product lines need to constantly evolve in response to market and technology changes. In the automotive industry the process of designing a line of cars is extremely costly and requires extended investments in R&D, whilst product variety within the line is a critical marketing-mix decision that may determine a firm's survival (Jan and Hsiao, 2004).

Industry practice and research to date suggest that product line design decisions in the car industry range between two options, namely, differentiation and

J. Casillas et al. (Eds.): Management Intelligent Systems, AISC 171, pp. 89–97.

commonality. Differentiation among car models in a line enables the manufacturer to charge price premiums, due to greater product variety, but is criticized for escalating product design, development, and manufacturing costs (Heese and Swaminathan, 2006). Commonality and component sharing among car models in the line has been suggested as a means to lower design and manufacturing cost, but is criticized for hindering price premiums and revenues. A product configuration with commonality may distort the perceived value of the product to consumer when the component sharing among products in the line is visible or is known to the consumer (Robertson and Ulrich, 1998). For example, General Motors was negatively criticized for its look-alike car line-up and Honda lost significant market share for its Acura model which was considered to be nothing more than a Honda Accord. Even the best hidden common components will diminish perceived valuation, especially when shared attributes are highly valued by the consumers (Desai et al., 2001).

Evidently, car manufacturers face a considerable dilemma regarding the balance between differentiation and commonality among models in a line-up. This paper applies a state of the art, nature inspired mechanism that can assist in designing car lines that optimize the degree of differentiation vs commonality among car models in a line. Our approach provides important implications for managers in the automotive and other capital-intensive industries who attempt to reduce manufacturing and design costs, whilst maintaining their ability to charge price premiums through variation in key product characteristics.

Alternative heuristic procedures are available in the literature that could potentially handle such complex optimization problems, including Dynamic Programming (Kohli and Sukumar, 1990), Beam Search (Nair et al., 1995), and Lagrangian Relaxation with Branch and Bound (Belloni et al., 2008). Recently, nature-inspired approaches have been also introduced, including Genetic Algorithms (Steiner and Hruschka, 2003) and Ant Colony Optimization (Albritton and McMullen, 2007). In this paper, we apply Particle Swarm Optimization (PSO), a state of the art optimization algorithm inspired from natural intelligence.

More specifically, the study follows a two step methodology: First, consumer preferences for car attributes have to be determined. To do that, stated preference data are derived from a large conjoint experiment involving preferences for automobiles. In the second stage, the derived measures of individual preferences are utilized to predict the valuation for any new concept car configuration that was not originally assessed by the respondents. In our study, we allow product attributes to vary over a continuum of values and not over a set of predetermined discrete levels. Thus, we apply Particle Swarm Optimization, a state of the art optimization mechanism which has an excellent compatibility with continuous data. The rest of the paper is organized as follows. In the next section, we provide an overview of the conjoint experiment which was carried out to analyze consumer preferences for car attributes. Section 3 provides an overview of the Particle Swarm Optimization algorithm and introduces our approach for the car market. Finally, section 4 discusses the empirical results, whilst a concluding section summarizes the paper.

2 The Conjoint Task: Estimating Heterogeneous Preferences for Car Attributes

In order to decide on the optimal configuration of models in its product line, a car manufacturer must first understand the manner in which consumers evaluate product alternatives. In this direction, a large scale conjoint experiment was carried out to estimate consumer preferences for certain car characteristics. The so-called part-worth model uses dummy variables to estimate part-worths at discrete levels for each attribute. The so-called vector model treats product attributes as linear variables (Green et al., 2001). The vector model is used in this study since product attributes of the optimal derived configurations are allowed to take on any value from a continuous range.

After consultation with car marketers and experts, the following six attributes are selected (levels in brackets): (1) engine horsepower units [75; 100], (2) price in euros [11,000 euros; 15,000 euros], (3) maximum speed measured in km/hr [170 km/hr; 180 km/hr], (4) acceleration measured in seconds required to accelerate from 0 to 100 km/hr [11 sec; 13 sec], (5) fuel consumption measured in litres/100 km [5 lt/100 Km; 6.5 lt/100 Km], (6) the existence of ESP, automatic air-conditioning and alloy wheels in the standard equipment [No; Yes].

As noted above, the vector model estimates a single coefficient for each attribute. We generate a symmetric and orthogonal fractional factorial design. Respondents evaluate 16 concept cars and assign a number between 0 and 100 points to reflect purchase probability. In total, 1,164 individuals participated in the study. Our dataset was treated as a balanced panel, in which we observe a large number of panellists ($N = 1,164$) responding to the same number of stimuli ($T = 16$).

Individual attribute coefficients are estimated by the application of a random coefficients (RC) regression model. Our econometric model allows variation in parameters across respondents and permits heterogeneity of individual preferences (Western, 1998; Beck, 2001; Beck and Katz, 2007). The random coefficients can be considered outcomes of a common mean plus an error term representing a mean deviation for each individual n (Hsiao, 1995). More formally the following model was estimated,

$$U_t^{(n)} = \left(\alpha + \delta^{(n)}\right) + \sum_{ij}\left(\beta_{ij} + \gamma_{ij}^{(n)}\right)x_{ij} + \varepsilon_t^{(n)} \tag{1}$$

where $U_t^{(n)}$ is the utility (evaluation) of product profile t by individual n, α is a common mean intercept, β_{ij} is a common mean attribute coefficient of level j of attribute i, and $\delta^{(n)}$ and $\gamma_{ij}^{(n)}$ are individual deviations from mean intercept α and mean preference parameter β_{ij}. Both $\delta^{(n)}$ and $\gamma_{ij}^{(n)}$ are random variables. Thus, the RC model has a unique set of coefficients (both slope and intercept) for each individual n. Finally, ε_t is the group-wise heteroscedastic error term which allows a different variance for each individual, $\text{var}(\varepsilon_t^{(n)}) = \sigma_n^2$.

3 The Particle Swarm Optimization Algorithm

Particle Swarm Optimization is a method for optimizing continuous, nonlinear functions. It was introduced by Kennedy and Eberhart (1995) and it has roots in two main component methodologies. The first is artificial life and swarming theory, that is, analogues of social behavior found in nature, such as fish schooling and bird flocking. The second is evolutionary computation, genetic algorithms and evolutionary programming in particular. PSO possesses some unique advantages. First, it comprises a very simple concept that can be implemented in a few lines of computer code, second, it requires only primitive mathematical operators, and third, it does not require excessive computer memory and speed.

The algorithm is population based, meaning that it works with a group (swarm) of agents (particles) that collectively move in the d-dimensional real space, where d is the number of the problem's dimensions. The location of each particle in the real space corresponds to a potential solution to the problem and it is represented by a vector $\vec{x}_i \in \Re^d$:

$$\vec{x}_i = (x_{i1}, x_{i2}, ..., x_{id}), i = 1, 2, ..., n, x \in \Re$$

where n is the number of particles in the swarm (population size). The algorithm works as follows. The particles are placed on specific locations of the problem space if there is prior information about potential good solutions; otherwise the particles are placed in a random manner. The performance of each particle in the objective function is evaluated and an iterative process begins. During the process each particle "moves" in the search space by following both its current personal best location (solution) $\vec{p}_i$, as well as the location of the best particle of the entire swarm $\vec{p}_g$, with some random permutations. The rate of the particle's location change is represented by its velocity $\vec{v}_i$. In each algorithm's iteration the location and the velocity of particle i are adjusted for each dimension d using the following functions:

$$v_{id}(t+1) = v_{id}(t) + c_1 * rnd_1 * (pbest_{id} - x_{id}(t)) + c_2 * rnd_2 * (pbest_{gd} - x_{id}(t)) \quad (2)$$

$$x_{id}(t+1) = x_{id}(t) + v_{id}(t+1) \quad (3)$$

where t is the iteration number, rnd_1 and rnd_2 are two random numbers in the range [0, 1], and $pbest_{id}$ and $pbest_{gd}$ are dimension's d values of the $\vec{p}_i$ and $\vec{p}_g$ respectively. The weights of the "cognition" part that simulates the private thinking of the particle itself, and the "social" part, which simulates the collaboration among particles are controlled by the two positive constants c_1 and c_2 (Kennedy, 1997). They are both usually set to 2, which on average makes the two weights to be 1. After all particles have completed their move, their fitness score is evaluated, the values of $\vec{p}_i$ and $\vec{p}_g$ are updated, and the algorithm proceeds to the next iteration. The iterative process terminates when a convergence criterion is met, or after a preselected number of iterations.

3.1 Binary PSO

In order to extend the application of the algorithm to discrete domains Kennedy and Eberhart (1997) developed a binary version of PSO. In this version, the velocity v_{ik} represents the probability of particle's dimension x_{ik} taking the value 1. If for instance, $v_{ik} = 0.6$ then there is a 60% likelihood that $x_{ik} = 1$ and a 40% likelihood that $x_{ik} = 0$. Since the velocity calculated in Equation 2 plays now the role of a probability threshold it should be limited to the range [0, 1]. A sigmoid function is used:

$$s(v_{ik}) = \frac{1}{1 + \exp(-v_{ik})} \tag{4}$$

The particle's location is now updated as follows:
    If $s(v_{ik}) > rnd_3$ then $x_{ik} = 1$

$$\text{else } x_{ik} = 0 \tag{5}$$

where rnd_3 a random number drawn from a uniform distribution in [0, 1].

3.2 A PSO Algorithm for Designing Optimal Lines of Cars

We now turn to deal with our optimization problem (i.e., designing car lines that optimize the degree of differentiation vs commonality among car models in the line). Possible solutions to the problem, i.e., lines of cars, are represented by particles that move in a search space of $d = l*m$ dimensions, where l is the number of cars in the line and m is the number of attributes per car.

As noted above, each car profile consists of six attributes, from which the attributes 1-5 are real numbers and the sixth attribute is a binary variable. The velocity for each attribute is updated using Equation 2. The dimensions of the particle's location are updated using Equations 2 and 3 for attributes 1-5 and Equations 4-5 for the sixth attribute. Hence, if we are looking for a single-car line, then the location of each particle corresponds to a single car profile, and is represented by a vector $\vec{x}_i = (x_{i1}, \ldots, x_{i5}, x_{i6})$, where $x_k \in \Re$ for $k = 1, \ldots, 5$ and $x_6 = 0/1$. Each location's dimension represents the value of the corresponding car attribute. If we are looking for a multiple-car line, then we aggregate the different car profiles into a single particle with a size of $d=m*6$, represented by a vector $\vec{x}_i =(x_{i1},\ldots, x_{ik},\ldots,$

$x_{id})$, where $x_k=0/1$ for $k=j*6$, $j=1,\ldots,m$, and $x_k\in \Re$ otherwise.

To illustrate consider a two-car line that is represented by the particle x=(191.44, 21000.5, 180.54, 8.2, 7.1, 1,\58.78, 12500.8, 130.12, 13.15, 5.63, 0). The first candidate car profile represented by this particle has the following characteristics: engine horsepower is 191.44 horsepower units, price is 21,000.5€, maximum speed is 180.54 km/h, acceleration is 8.2 seconds to reach a speed of 100 km/h, fuel consumption is 7.1 liters per 100 km, and the car includes ESP, automatic air-conditioning and alloy wheels in the standard equipment.

As noted earlier, the algorithm begins with the creation of an initial population $P(0)$ of n particles, that is, $P(0)=\{ x_1(0),\ldots,x_n(0)\}$, where $x_i(0)$, $i=1, \ldots, n$, corresponds to the ith particle of the initial population ($iter=0$). We generate the particles at random, since there is no prior knowledge about potential good solutions that should be included in the initial population. Then, we evaluate each particle according to an objective function, and assign the derived value as the particle's fitness. To calculate the fitness score of a particle that represents a car profile x, we first estimate the utility value of x for each respondent y. The utility value (U) is the sum of the partworths (u) of y that correspond to the values of the attributes that form the x, that is, $U_{yx}= \sum_k u_{yk}$, where $k=1, \ldots, 6$. Next, we aggregate the

utility values of x across the entire sample of respondents to get a degree of the overall customer satisfaction f_x provided by x, that is $f_x=\sum_j U_{xj}$, where

$j=1,\ldots,1164$ is the number of respondents.

When the solution contains more than one car profiles, we assume that a respondent will deterministically select the car that provides him/her with the maximum utility. Hence, in a solution that includes three car profiles, each respondent will be assigned the car that maximizes his/her utility. After each respondent is assigned a single car profile, the utilities of all respondents are aggregated and the overall utility value is assigned as the particle's fitness. The process is then repeated from the evaluation step, until a pre-specified number of iterations are completed.

The proposed mechanism derives optimal lines of cars whilst compensating for look-alike profiles that may have been randomly included in the initial population. More specifically, assuming that there were three look-alike profiles in the initial population, product lines were derived in such a way that the cumulative probability of consumer choosing all three look-alike profiles is lower than the sum of the partial choice probabilities of these three profiles if each of them had been included on its own in the initial population. The algorithm has been implemented using the MATLAB programming platform. Different population sizes, as well as different values for the maximum number of iterations were tested in the three-car line problem. The results indicated that for maximum number of iterations more than 600 there is no gain in performance, while the best performance was achieved for a population size of 60 particles.

4 Results

We run our PSO algorithm to find the optimal solutions for a car line consisting of one, two, or three different car models. Twenty replications are performed in each case. A final population of 60 particles along with their fitness scores is provided in each replication, from which we can choose the best or any other solution with fitness close to the best. In the single-car line as well as the two-car line, the algorithm reaches approximately the same solution in all replications. In the three-car line case the algorithm provides several different solutions throughout the 20 replications, of which we chose the solution with the highest fitness score. Table 1 reports the derived car lines, the utility for each car profile, the line fitness indices (overall portfolio utility) and the percent of customers assigned to each car profile.

Table 1 Empirically derived optimal lines of cars for the different scenarios

Solution	Single-car line	Two-car line		Three-car line		
	1st	1st	2nd	1st	2nd	3rd
Horsepower	200	200	55	200	55	92.35
Price (in euros)	23653	23653	10607	23653	10607	13919
Maximum Speed (in Km/hr)	197.9	197.9	164.3	197.9	164.3	172.4
Acceleration (in sec.)	7.83	7.83	13.44	7.83	13.44	11.27
Fuel consumption (in lt/100km)	7.12	7.12	4.95	7.12	4.95	6.53
Extra equipment (ESP, air-conditioning, alloy wheels)	1 (Yes)	1 (Yes)	0 (No)	1 (Yes)	0 (No)	1 (Yes)
Car utility	35.76	34.95	7.71	34.13	6.35	2.53
Choice share	100%	80.07%	19.93%	78.35%	15.89%	5.76%
Product line fitness index	35	42.66		43.01		

Inspection of Table 1 reveals some interesting patters. First, all the derived car portfolios share some common models, suggesting that such optimization algorithms are necessary in identifying how to combine new and existing car models. Second, models within each car line are sufficiently heterogeneous with respect to some characteristics and more homogenous with respect to some others, suggesting that our approach could be particularly useful in balancing the degree of differentiation vs commonality among models in a car line. We remind our reader that optimal car model configurations are derived based on consumer preferences for car attributes and thus look-alike model configurations within a car line-up reduce consumer's utility and thus result in lower fitness scores. Third, the derived utility levels suggest that variation-differentiation among car models of a product line elevates customer satisfaction. Fourth, the choice shares for the two and three-product portfolios reveal that choice shares differ markedly. Thus the distribution of demand across the elements of a product portfolio is asymmetric.

5 Conclusions

Competition in the car market pressures for low costs and prices, whilst customer demand pressures for high product variety. This situation presents a considerable dilemma for many car manufacturers. Industry practice suggests that although approaches based on component commonality can substantially lower the costs of proliferated car lines, the manufacturer's overall profits may decline as well, due to reduced differentiation among models in the car line. Evidently, car manufacturers face a considerable dilemma regarding the balance between differentiation and commonality among models in a line-up. This paper presents a novel probabilistic approach for designing car lines that optimize the degree of differentiation vs commonality among models in the car line and provides valuable insight into how to combine new and existing car models. The proposed framework has direct and important implications for managers in the automotive and other capital-intensive industries. Component sharing among models in a line results in economies of scale and thereby direct savings in manufacturing and design costs, whilst differentiation in key car characteristics enhances firm's ability to charge price premiums and thereby increases overall profits. Some illustrative car lines are constructed directly from consumer preferences using the Particle Swarm Optimization Algorithm. Our approach was applied to stated preference data derived from a large conjoint experiment involving preferences for automobiles. The results are promising and generally demonstrate that variation within the product line elevates customer satisfaction. It is hoped that the approach presented in this paper might motivate further developments in this key area of marketing.

References

Albritton, M.D., McMullen, P.R.: Optimal Product Design Using a Colony of Virtual Ants. European Journal of Operational Research 176(1), 498–520 (2007)

Beck, N.: Time-series-cross-section data: What have we learned in the past few years? Annual Review of Political Science 4, 271–293 (2001)

Beck, N., Katz, J.: Random coefficient models for time-series–cross-section data: Monte carlo experiments. Political Analysis 15(2), 182–195 (2007)

Belloni, A., Freund, R., Selove, M., Simester, D.: Optimizing Product Line Designs: Efficient Methods and Comparisons. Management Science 54(9), 1544–1552 (2008)

Desai, P., Kekre, S., Radhakrishnan, S., Srinivasan, K.: Product differentiation and commonality in design: Balancing revenues and cost drivers. Management Science 47(1), 37–51 (2001)

Green, P., Krieger, A., Wind, Y.: Thirty years of conjoint analysis: Reflections and prospects. Interfaces 31, S56–S73 (2001)

Heese, H.S., Swaminathan, J.M.: Product Line Design with Component Commonality and Cost-Reduction Effort. Manufacturing and Service Operations Management 8(2), 206–219 (2006)

Hsiao, C.: Analysis of Panel Data. Cambridge University Press, Cambridge (1995)

Jan, T.S., Hsiao, T.: A four-role model of the automotive industry development in developing countries: A case in Taiwan. Journal of the Operational Research Society 55, 1145–1155 (2004)

Kennedy, J.: The Particle Swarm: Social Adaptation of Knowledge. In: Proceedings of the IEEE International Conference on Evolutionary Computation, pp. 303–308. IEEE Service Center, Indianapolis (1997)

Kennedy, J., Eberhart, R.C.: A discrete binary version of the particle swarm algorithm. In: Proceedings of the IEEE International Conference on Systems, Man, and Cybernetics, Orlando, FL, USA, pp. 4104–4108 (1997)

Kennedy, J., Eberhart, R.C.: Particle Swarm Optimization. In: Proceedings of the IEEE International Conference on Neural Networks, pp. 1942–1948. IEEE Service Center, Piscataway (1995)

Kohli, R., Sukumar, R.: Heuristics for product line design using conjoint analysis. Management Science 36(12), 1464–1478 (1990)

Nair, S.K., Thakur, L.S., Wen, K.: Near optimal solutions for product line design and selection: Beam Search heuristics. Management Science 41(5), 767–785 (1995)

Robertson, D., Ulrich, K.: Planning for product platforms. Sloan Management Review 39(4), 19–31 (1998)

Steiner, W., Hruschka, H.: Generic Algorithms for product design: how well do they really work? International Journal of Market Research 45(2), 229–240 (2003)

Western, B.: Causal heterogeneity in comparative research: A bayesian hierarchical modelling approach. American Journal of Political Science 42, 1233–1259 (1998)

E-Business and E-Commerce

Semantic Web Mining for Book Recommendation

Matilde Asjana, Vivian F. López, María Dolores Muñoz,
and María N. Moreno

Dept. of Computing and Automatic, University of Salamanca, Plaza de los Caídos s/n,
37008 Salamanca

Abstract. A current strategy for improving sales as well as customer satisfaction
in the e-commerce field is to provide product recommendation to users. The increasing acceptance of web recommender systems is mainly due to the advances
achieved in the intensive research carried out for several years. However, in spite
of these improvements, recommender systems still present some important drawbacks that prevent from satisfying entirely their users. In this work, a methodology
that combines an association rule mining method with the definition of a domain-
specific ontology is proposed in order make efficient book recommendations.

Keywords: Semantic Web Mining, Recommender Systems, Associative
Classification.

1 Introduction

Endowing Web systems with mechanisms for selective recovery of the information is nowadays a highly demanded requirement. Many Web applications, especially e-commerce systems, already have procedures for personalized recommendation that allow users to find products or services they are interested in.
Recommender systems are becoming indispensable in the e-commerce environment since they constitute a way of increasing customer satisfaction and taking
positions in the competitive market of the electronic business activities. For many
years traditional companies have improved their competitiveness by means of
business intelligence strategies supported by techniques like data mining. Data
mining algorithms find consumers' profiles and purchase patterns in the corporate
databases that can be used for effective marketing and, in general, for business decision making. In the field of the e-commerce these procedures can also be applied
but they have been extended to deal with specific domain problems. In spite of the
advances achieved in this field, the recommendations provided by this type of systems have some important drawbacks, such as low reliability and high response
time. Therefore, it is necessary to research in new recommender methods that join
precision and performance as well as solving other usual problems of these systems. Web mining is one of the techniques providing better results in these two

J. Casillas et al. (Eds.): Management Intelligent Systems, AISC 171, pp. 101–109.
springerlink.com   © Springer-Verlag Berlin Heidelberg 2012

important aspects. Patterns extracted from web data by means of data mining algorithms constitute reliable recommendation models. The most critical drawbacks of recommender system are the following:

- *Sparsity*: Caused by the fact that the number of ratings needed for prediction is greater than the number of the ratings obtained from users.
- *Scalability*: Performance problems presented mainly in memory-based methods where the computation time grows linearly with both the number of customers and the number of products in the site.
- *First-rater problem*: Takes place when new products are introduced. These products, never have been rated, therefore they cannot be recommended.
- *Cold-start problem*: affects to new users, which cannot receive recommendations since they have no evaluations about products.

Sparsity is the most difficult problem to be solved and usually its improvement entails an increase of the computation time. Scalability problems can be avoided by using web mining methods since data mining models are already built at recommender time, thus, the building time has no impact in the user response time.

In order to deal with the last two drawbacks, new strategies using ontology are being tested in the last years. The idea is to enrich the data to be mined with semantic annotations in order to produce more interpretable results and to obtain patterns at different levels of abstraction, which allow to provide recommendation based on product characteristics and users profiles.

In this paper, the referred problems are tackled by means of the proposal of a semantic based web mining method. Recommendations are obtained by means of applying an associative classification data mining algorithm to data annotated with semantic metadata according to a domain-specific ontology.

The rest of the paper is organized as follows: Section 2 includes a brief description of the state of the art of recommender systems, associative classification and semantic web mining. In section 3 the proposed methodology and its application in a specific recommender system is presented. Finally, the conclusions are given in section 4.

2 Background

This section is devoted to present the fundamentals and the main works in the literature related to the methods that our proposal encompasses.

2.1 Recommender Systems

There is a great variety of procedures used for making recommendations in the e-commerce environment. They can be classified into two main categories [Lee et al., 2001]: *collaborative filtering* and *content-based* approach. Nearest neighbor algorithms were originally the basis of the first class of techniques. These algorithms predict product preferences for a user based on the opinions of other users. The opinions can be obtained explicitly from the users as a rating score or by using some implicit measures from purchase records as timing logs [Sarwar et al., 2001]. In the content based approach text documents are recommended by comparing between their contents and user profiles [Lee et al., 2001]. Currently there are two approaches for collaborative filtering, *memory-based* (*user-based*) and *model-based* (*item-based*) algorithms. Memory-based algorithms, also known as nearest-neighbor methods, were the earliest used [Resnick et al., 1994]. They treat all user items by means of statistical techniques in order to find users with similar preferences (neighbors). The prediction of preferences (recommendation) for the active user is based on the neighborhood features. A weighted average of the product ratings of the nearest neighbors is taken for this purpose. The advantage of these algorithms is the quick incorporation of the most recent information, but they have the inconvenience that the search for neighbors in large databases is slow [Schafer et al., 2001]. Model-based collaborative filtering algorithms use data mining techniques in order to develop a model of user ratings, which is used to predict user preferences.

Collaborative filtering, specially the memory-based approach, has some limitations in the e-commerce environment. Rating schemes can only be applied to homogeneous domain information. Besides, sparsity and scalability are serious weaknesses which would lead to poor recommendations [Cho et al., 2002]. Sparsity is due to the number of ratings needed for prediction is greater than the number of the ratings obtained because usually collaborative filtering requires user explicit expression of personal preferences for products. The second limitation is related to performance problems in the search for neighbors in memory-based algorithms. These problems are caused by the necessity of processing large amount of information. The computer time grows linearly with both the number of customers and the number of products in the site. The lesser time required for making recommendations is an important advantage of model-based methods. This is due to the fact that the model is built off-line before the active user goes into the system, but it is applied on-line to recommend products to the active user. Therefore, time spent on building the model has no effects in the user response time since small amount of computations is required when recommendations are requested by the users, contrary to the memory based methods that compute correlation coefficients when user is on-line. Nevertheless, model based methods present the drawback that recent information is not added immediately to the model but a new induction is needed in order to update the model.

2.2 Associative Classification

Associative classification methods obtain a model of association rules that is used for classification. These rules are restricted to those containing only the class attribute in the consequent part. This subset of rules is named class association rules (CARs) [Liu et al., 1998].

Since Agrawal and col. introduced the concept of association between items [Agrawal et al., 1993a][Agrawal et a l., 1993b] and proposed the Apriori algorithm [Agrawal and Srikant, 1994], association rules have been the focus of intensive research. Most of the efforts have been oriented to simplify the rule set and improve the algorithm performance. The number of papers in the literature focused in the use of the rules in classification problems is lesser. A proposal of this category is the CBA (Classification Based on Association) algorithm [Liu et al., 1998] that consists of two parts, a rule generator based on Apriori for finding association rules and a classifier builder based on the discovered rules. CMAR (Classification Based on Multiple Class-Association Rules) [Li et al., 2001] is another two-step method, however CMAR uses a variant of FP-growth instead Apriori. Another group of methods, named integrated methods, build the classifier in a single step. CPAR (Classification Based on Predictive Association Rules) [Yin and Han, 2003] is the most representative algorithm in this category. In [Wang and Wong, 2003], the weight of the evidence and the association detection are combined for flexible prediction with partial information. The main contribution of this method is the possibility of making prediction on any attribute in the database. Moreover, new incomplete observations can be classified. The algorithm uses the weight of the evidence of the attribute association in the new observation in favor of a particular value of the attribute to be predicted. This approach uses all attributes in the observation, however in many domains some attributes have a minimal influence in the classification, so the process can be unnecessary complicated if they are taken in consideration. Our proposal evaluates the importance of the attributes on the classification

2.3 Semantic Web Mining

Semantic Web Mining is an emerging research area that aims at addressing the challenges of current web systems by means of using the Semantic Web and the Web mining in a complementary way. The lack of structure of most of the data in the Web can be only understood by humans; however, the huge amount of information can only be processed efficiently by machines. The semantic Web deals with the first problem by enriching the Web with machine-understandable information, while web mining deals with the second one by automatically extracting the useful knowledge hidden in web data. Semantic web mining aims at improving the results of web mining by exploiting the semantic structures in the web as well as building the semantic web making use of web mining techniques [Stumme et al., 2006].

This work is focused on the first approach where ontology is used to describe explicitly the conceptualization behind the knowledge represented in the data to be mined. An ontology is "an explicit formalization of a shared understanding of a conceptualization". Most of them include a set of concepts hierarchically organized and additional relations between them. This formalization allows integrating heterogeneous information as a preprocessing step of web mining tasks [Liao et al., 2009].

Semantic information obtained from taxonomies can be used to find out patterns at high abstract level. In that way, regularities between categories of products instead of between specific products can be found. Recommender systems can apply these patterns in the recommendation of new products that still have not been rated by the users [Huang and Bian, 2009]. This is a major problem since new products introduced in the catalog cannot be recommended if classical collaborative filtering algorithms are used because the induced models do not include these products. However, when taxonomies are used, new products can be classified into one or more categories and recommendations can be done from models enclosing more general patterns, which relate user profiles with categories of products. In addition, taxonomic abstraction provided by an ontology is often essential for obtaining meaningful results.

In spite of data mining techniques are widely used in recommender systems, their use in combination with semantic techniques is scarce. However, in the last years it is being used increasingly. In [Mossavi et al., 2009] concepts of classical market analysis are extended by means of ontological techniques in order to enhance electronic market. Substitution is one of these concepts which represent the similarity degree between related products [Resnik, 1999]. Usually, techniques for computing the similarity degree are based on finding product properties [Resnik, 1999], [kanappe, 2005], [Ganjifar et al., 2006]. Complement is another useful concept , mainly used for designing marketing strategies, since it provide them and additional value. In some circumstances recommendation of similar products do not give the desired results, but recommendation of complementary products does. Mossavi et al. [Mossavi et al., 2009] apply an ontological technique to determine complement products and the OWL language to model types of products. In this direction, Liao et al. [2009] have updated the marketing concept of branding, which is traditionally used in the business field for making a brand by means of differentiating the brand products from those of the competitors [Baker, 1996]. A brand is "a combination of features (what the product is), customer benefits (what needs and wants the product meets) and values (what the customer associates with the product)". Liao et al [2009] analyze a specific market segment by means of brand spectrums depicted from data relating to consumer purchase behaviors and beverage products. They develop a set of ontologies for describing the integrated consumer behavior and a set of databases related to these ontologies. In further steps two data mining techniques are applied. First, a clustering algorithm is used for segmenting the data according to customer information, lifestyles, and purchase behavior. Then, the relationship among the clusters is analyzed by means of the Apriori algorithm for association rule induction. In [Kim et al., 2011] a new way of enriching user data in recommender systems is proposed. The idea consists on discovering relevant and irrelevant topics for users and generating new tags for building the user model. They use the technique of the nearest neighbors, therefore, though the approach can be effective for users who do not contain enough topics in their user model, it does not address other weaknesses of this kind of collaborative filtering methods.

3   Recommendation Methodology

The procedure proposed in this work is intended for overcoming sparsity, scalability, first rater and cold-start problems, which are frequent drawbacks of recommender systems, already described in the introduction. The methodology combines an associative classification data mining method with the definition of a domain-specific ontology. The study was carried out with data from BookCrossing system.

3.1 BookCrossing Database

This study has been carried out with the data set BookCrossing by Cai-Nicolas Ziegler [Ziegler et al., 2005]. The database used for the study contains demographic information (*age* and *location*) of the users as well as the evaluations they provide about the books. The available book information is the following: *isbn*, *title*, *author* and *publication year*. Given that *book category* it is an important not present property we extracted this attribute from the interface GoogleBooks8.

The sample used for this study was randomly extracted from the BookCrossing database. It consists of 17,150 evaluations (ratings between 1 and 10) about 783 books from 10,557 users. The age attribute was discretized in seven intervals of values and the attribute year was split in intervals of ten years. On the other hand, since the rating attribute is used to decide if a book is going to be recommended to a user, we changed such attribute in order to have only two values: "Not recommended" (scores 1-5) and "Recommended" (scores 6-10).

3.2 Ontology Development

An ontology consists of abstract concepts (classes) and relationships defined to be shared and reusable in a particular application domain. Web data can be considered the instances of ontology entities when they are classified according to a specific ontology. In this way, web mining techniques can be applied to these instances giving more meaningful knowledge to the user [Lim and Sun, 1996].

The application domain of this book system has been represented by means of the ontology showed. Figure 1 shows the concepts defining the ontology, their attributes and their relationships. In that way, user demographic information, book attributes and ratings given to books by users are related in a structure that is used for designing the database where the data to be mined are stored.

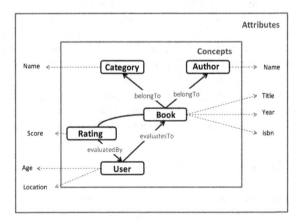

Fig. 1 Ontology for the book system

In this work, we have applied a variant of the Apriori algorithm in order to classify the dataset records in two classes, recommended and not recommended. The algorithm produces a special kind of rules, class association rules, which contain in the consequent part a unique attribute, the class. Therefore, this algorithm belongs to the category of associative classification methods. We have chosen one of these methods with the purpose of dealing with the data sparsity problem because these methods are less sensitive to data sparsity.

Since the main objective of this work is to show how the first-rater and cold-start limitations are solved by means of the semantic, association rules are obtained to an abstract level by means of the procedure described in the section (3.3).

3.3 Recommendation Procedure

The target of the proposed methodology is to predict in an efficient way the user preferences in order to recommend him products he is interested in.

To overcome scalability problems a model based approach was applied. The predictive models are induced off-line (before the entry of the users in the system) but are used on-line (when active user is connected to the system). Therefore, the time spent on building the models does not influence the user response time. This is the main advantage of this approach that avoids problems associated with traditional memory-based techniques.

Predictive models are induced by means of an associative classification method due to the better behavior of these methods in sparse data contexts [Moreno et al., 2009]. Consequently more reliable recommendations can be obtained with a lesser number of ratings. This is the way of dealing with the sparsity problem.

Category=Fiction age=[32-42] author=Gregory Maguire 19 → class=Recommended

Fig. 2 Example of rule of the associative classification model

Table 1 Precision of the rules' model

Rules	Precision		
	Minimum	**Maximum**	**Average**
First 10	0.9912	0.9938	0.9926
First 50	0.9841	0.9938	0.9887
First 100	0.9687	0.9938	0.9829

The class association rule algorithm is applied in two different abstraction levels. Semantic annotations provided to the data following the defined ontology allow inducing patterns at a more abstract level. Figure 2 shows a rule example obtained with the predictive Apriori algorithm at this level of abstraction. The rule relates user attributes (age) and book attributes (category and author) instead of a specific user with a specific book, thus, this kind of patterns can be used for re-commending new products that still have not been rated by the users, avoiding in this way the first-rater problem. In a similar way, new users can receive recom-mendations according his profile, which is defined from properties and relations given by the ontology, thus, cold-start problem is solved. Table 1 shows the preci-sion of the rules' model. We can observe the high precision of the rules in spite of the sparsity of the data.

4 Conclusions

The incorporation of efficient recommendation methods in e-commerce applica-tions contributes significantly to improve clients´ satisfaction and, hence, to increase business benefits. Many web systems are endowed with such personaliza-tion mechanisms, however, most of them do not provide reliable recommenda-tions. Semantic Web Mining is an effective approach to be used in recommender system, especially in the e-commerce field, in order to deal with their main short-comings. In this context, we propose a methodology combining an associative classification method with the definition of a specific domain ontology. The methodology has been validated with data from BookCrossing database. The ap-plication of the proposed data mining algorithm to data, annotated with semantic information according to the defined ontology, provides a high level predictive model formed by rules relating categories of books and user profiles instead of specific books or particular users. This model is used for recommending not rated books or for making recommendation to new users avoiding in this way the first-rater and cold-start problems respectively. On the other hand, the application of the associative classification method to data without semantic information gene-rates a low level model relating users, books and ratings for making the recom-mendations. These models are used in traditional recommendation.

In addition, the use of associative classification methods provides more accu-rate models in sparse data contexts such as the ones managed usually in recom-mender systems. Moreover, the induction of the models off-line allows to prevent scalability problems.

References

Agrawal, R., Srikant, R.: Fast algorithms for mining association rules in large databases. In: Proc. of 20th Int. Conference on Very Large Databases, Santiago de Chile, pp. 487–489 (1994)

Agrawal, R., Imielinski, T., Swami, A.: A. Database mining. A performance perspective. IEEE Trans. Knowledge and Data Engineering 5(6), 914–925 (1993a)

Agrawal, R., Imielinski, T., Swami, A.: Mining associations between sets of items in large databases. In: Proc. of ACM SIGMOD Int. Conference on Management of Data, Wa-shinton, DC, pp. 207–216 (1993b)

Baker, M.: Marketing-An introductory text. Macmillan Press, London (1996)

Cho, H.C., Kim, J.K., Kim, S.H.: A Personalized Recommender System Based on Web Usage Mining and Decision Tree Induction. Expert Systems with Applications 23, 329–342 (2002)

Economides, N., Brian, V.: Pricing of Complements and networks effects. NET Institute Working Paper, NYU Working Paper No. EC-05-12, NYU Working Paper No. 2451/26105

Ganjifar, Y., Abolhasani, H., et al.: A similarity measure for OWL-S annotated web services. Web Intelligence, 621–624 (2006)

Huang, Y., Bian, L.: A Bayesian network and analytic hierarchy process based personalized recommendations for tourist attraction over the Internet. Expert Systems with Applications 36, 933–943 (2009)

Kanappe, R.: Measures of semantic similarity and relatedness for use in ontology-based information retrieval. Thesis of Doctor. Denmark, Roskilde University (2005)

Kim, H.N., Alkhaldi, A., El Saddik, A., Jo, G.S.: Collaborative user modeling with user-generated tags for social recommender systems. Expert Systems with Applications 38, 8488–8496 (2011)

Lee, C., Kim, Y.H., Rhee, P.K.: Web Personalization Expert with Combining collaborative Filtering and association Rule Mining Technique. Expert Systems with Applications 21, 131–137 (2001)

Li, W., Han, J., Pei, J.: CMAR. Accurate and efficient classification based on multiple class-association rules. In: Proc. of the IEEE International Conference on Data Mining (ICDM 2001), California, pp. 369–376 (2001)

Liao, S., Ho, H., Yang, F.: Ontology-based data mining approach implemented on exploring product and brand spectrum. Expert Systems with Applications 36, 11730–11744 (2009)

Lim, E., Sun, A.: Web Mining – the Ontology Approach. In: Proc. of International Advanced Digital Library Conference, IADLC 2005 (2006)

Liu, B., Hsu, W., Ma, Y.: Integration classification and association rule mining. In: Proc. of 4th Int. Conference on Knowledge Discovery and Data Mining, pp. 80–86 (1998)

Moosavi, S., Nematbakhsh, M., Farsani, H.K.: A semantic complement to enhance electronic market. Expert Systems with Applications 36, 5768–5774 (2009)

Moreno, M.N., Pinho, J., Segrera, S.Y., López, V.: Mining Quantitative Class-association Rules for Software Size Estimation. In: Proc of. IEEE 24th International Symposium on Computer and Information Sciences (ISCIS 2009), pp. 199–204. IEEE, Northern Cyprus (2009)

Resnick, P., Iacovou, N., Suchack, M., Bergstrom, P., Riedl, J.: Grouplens: An open architecture for collaborative filtering of netnews. In: Proc. of ACM CSW 1994 Conference on Computer, Supported Cooperative Work, pp. 175–186 (1994)

Resnik, P.: Semantic similarity in a taxonomy: An information based measure and its application to problems of ambiguity in natural language. Journal of Artificial Intelligence, 94–130 (1999)

Sarwar, B., Karypis, G., Konstan, J., Riedl, J.: Item-based Collaborative Filtering Recommendation Algorithm. In: Proceedings of the Tenth International World Wide Web Conference, pp. 285–295 (2001)

Schafer, J.B., Konstant, J.A., Riedl, J.: E-Commerce Recommendation Applications. Data Mining and Knowledge Discovery 5, 115–153 (2001)

Stumme, G., Hotho, A., Berendt, B.: Semantic Web Mining. State of the art and future direction. Journal of Web Semantics 4, 124–143 (2006)

Wang, Y., Wong, A.K.C.: From association to classification. Inference using weight of evidence. IEEE Transactions on Knowledge and Data Engineering 15, 764–767 (2003)

Yin, X., Han, J.: CPAR. Classification based on predictive association rules. In: SIAM International Conference on Data Mining (SDM 2003), pp. 331–335 (2003)

Ziegler, C., McNee, S.M., Konstan, J., Lausen, G.: Improving Recommendation Lists Through Topic Diversification. In: Proceedings of the 14th International World Wide Web Conference, Chiba, Japan, pp. 22–32 (2005)

An Automated Approach to Product Taxonomy Mapping in E-Commerce

Lennart Nederstigt, Damir Vandic, and Flavius Frasincar

Erasmus University Rotterdam, Econometric Institute
len_nederstigt@xs4all.nl, {vandic,frasincar}@ese.eur.nl

Abstract. Due to the ever-growing amount of information available on Web shops, it has become increasingly difficult to get an overview of Web-based product information. There are clear indications that better search capabilities, such as the exploitation of annotated data, are needed to keep online shopping transparent for the user. For example, annotations can help present information from multiple sources in a uniform manner. This paper proposes an algorithm that can autonomously map heterogeneous product taxonomies for Web shop data integration purposes. The proposed approach uses word sense disambiguation techniques, approximate lexical matching, and a mechanism that deals with composite categories. Our algorithm's performance on three real-life datasets was compared favourably against two other state-of-the-art taxonomy mapping algorithms. The experiments show that our algorithm performs at least twice as good compared to the other algorithms w.r.t. precision and F-measure.

Keywords: e-commerce, taxonomy mapping, word sense disambiguation.

1 Introduction

The interchange of information has become much easier with the advent of the Web. This ease of sharing information has lead to a worldwide surge in activity on the Web. According to a recent study [13], the Web is doubling in size roughly every five years. The ever-growing amount of information stored on the Web poses several problems. Due to the vast amount of Web sites, it is becoming increasingly difficult to get a proper overview of all the relevant Web information. While traditional search engines help to index the Web, they do not understand the actual information on Web pages. This is due to the fact that most Web pages are geared towards human-readability, rather than machine-understandability. Differently than machines, humans are able to extract the meaning of words from the context of Web pages, but a machine cannot do so. This is particularly a problem when searching using words that can have multiple senses, like 'keyboard', which can either be a computer device or musical instrument. The search engine will include every page that contains the search term in the search results, regardless of whether it is actually relevant or not.

J. Casillas et al. (Eds.): Management Intelligent Systems, AISC 171, pp. 111–120.
springerlink.com   © Springer-Verlag Berlin Heidelberg 2012

A Web domain in which this search problem manifests itself is e-commerce. There is virtually an endless amount of products available and just as many Web shops from which you could order them. There are clear indications that better search functionalities are needed in order to keep online shopping transparent for the user. A study on online shopping in the USA, performed in [4], indicates that more than half of the respondents had encountered difficulties while shopping online. Information was often found to be lacking, contradicting, or overloading the user. This emphasises the need for aggregating the information found in those Web shops and presenting it in a uniform way.

While existing price comparison sites, such as [12], already show aggregated information, they are often restricted in their use. Most price comparison sites only include regional price information and therefore compare only a limited amount of Web shops [14]. Furthermore, in order to take part in the comparison, Web shops often have to take the initiative and provide their data in a specific format that is defined by each price comparison site. This can be laborious because there is no standardised semantic format for exchanging information. In other words, sharing product information on the Web requires a significant amount of manual work.

A solution to the search problems encountered on the Web would be to annotate the data found on Web pages using standardized ontologies. In this way the data becomes fully understandable for computers as well. For e-commerce, there is already a standard ontology emerging, called GoodRelations [3]. Unfortunately, not that many Web pages have included a semantic description for their content so far. Furthermore, even when a semantic description is available, not every Web page might use the same ontology. That is why there is a need for algorithms that are able to map product ontologies to each other in a (semi-)automatic way.

In this paper we propose an algorithm for mapping product taxonomies. Taxonomies are the backbone of an ontology, as they contain the type-of relations. Our algorithm is based on the approach presented by Park & Kim [11]. The proposed algorithm can autonomously map heterogeneous product taxonomies from multiple sources to each other. Similar to the Park & Kim algorithm, our algorithm employs word sense disambiguation techniques to find the correct sense of a term using the semantic lexicon WordNet [9]. Differently than the Park & Kim algorithm, our algorithm considers for lexical matching various lexical similarity measures, like the Jaccard index and the Levenshtein distance. Our proposed algorithm also exploits the hierarchical structure of taxonomies by taking into account the distance between each candidate path and already existing mappings. In addition, a different similarity aggregation function is used to make the algorithm more robust against outliers.

2   Related Work

The algorithm we propose in this paper is based on the approach presented in [11], where the focus is more on product taxonomy mapping in particular, rather than ontology mapping. Due to this focus, it manages to achieve a higher recall than more general approaches for ontologies when mapping product taxonomies [11].

These approaches only map the classes when the similarity between these is very high. However, the design of product taxonomies is a subjective task, which makes their mapping a loosely-defined domain. Furthermore, the mapping of product taxonomies is aimed at reducing search failures when shopping online. Thus, in order not to lose potentially user-desired products from the Web shop presentation, it is better to map more product classes, even when classes are not very similar.

While the algorithm presented in [11] is suited for product taxonomy mapping, we found aspects that can be significantly improved. For instance, the algorithm does not consider the existence of composite categories, which are categories that consist of multiple concepts, e.g., 'Movies, Music & Games'. Mapping these categories often fails, because the word sense disambiguation process is not applicable for these categories. This is due to the fact that the algorithm is unable to find the senses of the complete name in WordNet. Furthermore, the algorithm has difficulties disambiguating categories with short paths to the root, because of the lack of information content. This could be improved by also considering children and siblings of a category node when disambiguating. Another drawback of the algorithm is its bias towards mapping to short paths, which are sometimes proven to be too general.

Another approach for taxonomy mapping is Anchor-PROMPT, available in the PROMPT Suite [10]. This algorithm provides a (semi-) automatic ontology mapping process. As the performance of Anchor-PROMPT largely depends on the accuracy of the initial mappings that are provided (a requirement of the algorithm), it is not suitable for fully automatic ontology mapping [10]. Because the algorithm uses relatively simple lexical matching techniques to find the initial mappings, it would be better to manually create the initial mappings instead. This can become an issue when having to map many large product taxonomies. We have chosen to include this algorithm in our evaluation in order to investigate how a general ontology mapping algorithm performs in the context of product taxonomy mapping.

There are several other approaches that, despite the fact that their focus is on ontology mapping in general, are interesting to mention. The authors of [8] propose an algorithm to semi-automatically map schemas, using an iterative fixpoint computation, which they dubbed *similarity flooding*. In [6], the authors propose a semi-automatic ontology mapping tool, called Lexicon-based Ontology Mapping (LOM). The Quick Ontology Mapping (QOM) approach, presented in [2], is designed as a trade-off between the quality of a mapping and the speed with which a mapping can be performed. In [1], the authors present a system of Combination of Matching Algorithms (COMA++), capable of performing both schema and ontology mapping. Cupid [7] is a general-purpose schema matching algorithm. It is a hybrid matcher that exploits both lexical and semantic similarities between elements.

3   Algorithm

Before going into the algorithm details, we first present a high level overview of the algorithm, indicating the differences between our approach and the approach of Park & Kim. Our algorithm requires two inputs. The first input is a category and its path in the source taxonomy. The second input is the target taxonomy of categories to which the source category has to be mapped. Our algorithm starts with the pre-processing of the category name. It splits the name on ampersands, commas, and the word 'and', which results in a set containing multiple terms, called the *split term set*. This step is performed to enhance the word sense disambiguation process for categories that consist of multiple concepts, which are called *composite categories*.

The first major process (the same as in the Park & Kim approach) is the word sense disambiguation process. This process tries to determine the correct sense of the term in the leaf node of the source taxonomy by finding the term in WordNet (a semantic lexicon) [9]. The correct sense of the term can be found by comparing the hyponyms of each sense found in WordNet with all the ancestor nodes in the path of the source taxonomy. A path is a list of nodes from the root to the current node. Our algorithm repeats this process for each term in the *split term set*. The result of this process is the *extended term set*, which contains the original term and also its synonyms (if the algorithm was able to determine the correct sense of the term). Because our algorithm splits the original category name, we define the *extended split term set* as the set of all extended term sets (one for each split term).

Using the extended split term set that is obtained from the word sense disam-biguation process, the algorithm analyses the target taxonomy and determines which paths are considered to be candidate paths for the mapping of the path from the source taxonomy. It does that by searching for paths that end with leafs that contain at least half of the terms in the extended (split) term set.

In order to determine which candidate path is the best path to map to, both algo-rithms compute the *co-occurrence* and *order-consistency* similarities for each path. The *co-occurrence* expresses the level of overlap between the source taxonomy path and one of the candidate target paths, while disregarding the hierarchy. The order-consistency is the ratio of common nodes (nodes that occur in both the source path and the candidate path) that appear in the candidate path according to the hierarchi-cal order in the source path. Our algorithm adds a third measure, the *parent mapping distance*, which is the normalised distance in the target taxonomy between a candi-date path and the path to which the parent in the source path was mapped to.

Using the similarity measures obtained in the previous step, the algorithms deter-mine the best path to map the source path to. While Park & Kim use the arithmetic mean of the co-occurrence and order-consistency to obtain the overall similarity, our algorithm uses the harmonic mean (including the parent mapping distance). The path with the highest overall similarity is selected as the path to map to, assuming that the overall similarity is higher than a configurable threshold. If it fails to reach this threshold, or if no candidate paths were found, the algorithm of Park & Kim will not map the source path for a given input category. In this situation, our algorithm will map the source path to the same path in the target taxonomy to which its parent is mapped. If the parent of the source path was not mapped, our algorithm will also not map the source path.

Word Sense Disambiguation. As discussed previously, our algorithm splits com-posite categories and the split term set contains the individual terms that result from this process. This means that rather than using the entire category term for the word sense disambiguation process, like the algorithm of Park & Kim does, it will perform this process separately for each term in the split term set. Other than that, the implementation of this part of the algorithm does not differ from the implementation used by Park & Kim. Both algorithms enhance their ability to perform a correct mapping by first trying to determine the correct sense of a category term from the source taxonomy. This is useful, because it helps to identify semantically similar categories from different taxonomies, even when they are not lexically similar. For instance, if the path from the source taxonomy is 'Computers/Notebook', we can deduce that the correct sense would be a laptop in this case, rather than a notepad. We could then include the word 'laptop' in the search terms that are used for identifying candidate paths in the target taxonomy. This might yield better candidate paths than only searching using the term 'notebook'.

In order to find the meaning that fits most closely to the source category that needs to be mapped, the algorithm identifies matches between an upper category, i.e., an ancestor of the current node from the source taxonomy, and a sense hierarchy obtained from WordNet. This is done by finding the set of matching lemmas between an upper category in the source taxonomy and a sense hierarchy defined by hypernym relations. By comparing each upper category from the source taxonomy with all sense hierarchy nodes that are in the set of matching lemmas, we can measure how well each upper category of the source taxonomy fits to each sense hierarchy. As the information content per node in a sense hierarchy increases when a node is closer to the leaf, we aim to find the match with the shortest distance to the sense hierarchy leaf. The similarity score increases when this distance is shorter, it is defined as:

$$\text{hyperProximity}(t, S) = \begin{cases} \frac{1}{\min\limits_{x \in C}(\text{dist}(x, \ell))} & \text{if } C \neq \emptyset \\ 0 & \text{if } C = \emptyset \end{cases}$$

where t is an upper category to be matched, C is the set of matching lemmas, and ℓ is the leaf of the sense hierarchy S. The dist() function computes the distance between each matching lemma x (of a synonym of a hypernym) and the leaf node ℓ in the sense hierarchy. The distance is given by the number of edges that are being traversed when navigating from the node with the matching lemma to the leaf node in the sense hierarchy.

After we have determined the hyperproximity between each upper category from a source path and a particular sense hierarchy from WordNet, we compute the overall similarity between an entire source category path and a sense hierarchy of the (split) category term. This is done by computing the average hyperproximity between all upper categories of a source path and one sense hierarchy from WordNet. Park & Kim use a different approach here, their algorithm divides the hyperproximities of each upper category by the length of the entire source category path, including the leaf node. This does not lead to a proper average, as the Park & Kim algorithm does not compute the hyperproximity between the leaf node and the sense hierarchy. Once the path-proximity between the source path and each of the possible senses of the source category term has been computed, we can determine which of the found senses fits best. This is done by selecting the sense hierarchy that has the highest average path proximity.

Candidate Path Identification. The resulting extended (split) term set of the word sense disambiguation process is used to identify candidate paths in the target taxonomy. A candidate path is a path in the target taxonomy that is marked by the algorithm as a potential target path to map the current source category to. In order to find the candidate paths, the algorithms compare the terms in the extended (split) term set with the paths in the target taxonomy.

The algorithm proposed by Park & Kim first compares the root node of the target taxonomy with the extended term set. If none of the terms in the extended term set is a substring of the currently examined category in the target taxonomy, the algorithm considers the children of the current category. Otherwise, if at least one of the terms in the extended term set is a substring of the currently examined category, that category is marked as a candidate path. In addition, the algorithm will no longer consider the children of that path as a potential candidate. The algorithm of Park & Kim assumes that if a more general category already matches the term, it is likely to be a better candidate path than a longer (more specific) path. However, this does not always hold true, as there are many composite categories in product taxonomies that split the multiple concepts in subcategories one level lower. For instance, the composite category 'Music, Movies and Games' in the Amazon.com product taxonomy has a subcategory called 'Music'. Therefore, differently than the algorithm of Park & Kim, our algorithm continues to search the entire target taxonomy for candidate paths, even when an ancestor of a path was already marked as a candidate path.

As our algorithm splits the original category name if it is a composite category, it could occur that multiple extended term sets have to be compared with the category names in the target taxonomy. This means that our algorithm has to perform the matching for each extended term set in its extended split term set. This matching process returns a true/false value for each extended term set: true if one of terms is a substring of the currently examined category term, and false if none of the terms is a substring. If at least half of the boolean values is true, we consider the path of the current target category as a candidate path.

Aggregated Path Similarity Score. Once all the candidate paths in the target taxonomy have been identified, we need to determine which one of them fits best to the source paths. In order to calculate the measure of fit, we need to calculate an aggregated similarity score for each candidate path. In the algorithm proposed by Park & Kim, the aggregated score is composed of two similarity measures, the *co-occurrence* and the *order-consistency*. Our algorithm adds an extra measure, called the *parent mapping similarity*. Furthermore, it extends the co-occurrence measure by splitting terms and using the extended (split) term set of the correct sense. The algorithm proposed by Park & Kim uses only the original term or the synonyms of all the senses for the original term for calculating these similarity measures.

The co-occurrence is a similarity that measures how well each candidate path fits to the source category path that is to be mapped. It computes the overlap between two category paths, while disregarding the order of nodes in each path. This is done by applying a lexical matching function that is based on the average of Levenshtein and Jaccard similarities to each pair of categories from the source and candidate paths. The co-occurrence is defined as:

$$\text{coOccurrence}(P_{\text{src}}, P_{\text{targ}}) = \left(\sum_{t \in P_{\text{targ}}} \frac{\text{maxSim}(t, P_{\text{src}})}{|P_{\text{targ}}|} \right) \cdot \left(\sum_{t \in P_{\text{src}}} \frac{\text{maxSim}(t, P_{\text{targ}})}{|P_{\text{src}}|} \right)$$

where P_{src} and P_{targ} are the nodes from the current source path and candidate path, respectively. The maxSim() function computes the maximum similarity between a single category name, either from the source or candidate path, and the entire path from the other taxonomy. It compares the single category name with all the nodes in the other path, using extended (split) term sets, obtained in the same way as in the candidate path selection.

The co-occurrence is useful for computing the lexical similarity between the source path and a candidate path from the target taxonomy. However, it disregards the order in which these nodes occur in the path. Therefore, we need an additional measure, which takes the order into account. Park & Kim define this measure as the order-consistency. The measure checks whether the common nodes between the paths appear in the same order. First of all, a list of matching nodes, called the *common node list*, between the two paths has to be obtained. The function common() adds a node to the list, if it can match the category term of a node, or one of the synonyms of the category term, with a node, or one of its synonyms, from the other path. In this function, all found senses from WordNet for the terms are taken into consideration. The resulting *common node list* is then used by precRel() to create binary node associations. These binary node associations denote a precedence relation between two nodes, which means that the first node occurs before the second node in the hierarchy of the source path. For every element in the *common node list*, pairs of node names from the source path are created. The consistent() function uses the precedence relations to check whether these precedence relations between two nodes also hold true for the candidate path, i.e., whether the two categories in the candidate path occur in the same order as the same two categories in the source path. If a precedence relation holds true also for the candidate path, this function returns the value 1, otherwise it returns 0. Using the aforementioned functions, the function for the order-consistency is given by:

$$\text{orderConsistency}(P_{\text{src}}, P_{\text{targ}}) = \sum_{r \in \text{precRel}(C, P_{\text{src}})} \frac{\text{consistent}(r, P_{\text{targ}})}{\binom{\text{length}(C)}{2}}$$

where P_{src} and P_{targ} are the nodes from the current source path and candidate path, respectively, and C is $\text{common}(P_{\text{src}}, P_{\text{targ}})$. The denominator in the above fraction is the number of possible combinations of two nodes, which can be obtained from the *common nodes list*. Therefore, the order-consistency is the average number of precedence relations from the source path that are consistent with the candidate path.

The co-occurrence and the order-consistency both measure the similarity between the source path and a candidate path, computing the degree of category overlap and hierarchical order similarity, respectively. However, we can also exploit our knowledge of how the parent of the source node was mapped in order to find the best candidate path. As the current source node is obviously closely related to its parent, there is a considerable chance that the best candidate path is closely related to the target category to which the parent of the source node was mapped as well. That is why we include a third measure, called the *parent mapping distance*, which is the distance (difference) in the target taxonomy between a candidate path and the path to which the parent in the source path was mapped to.

Once the various similarity scores have been calculated, we can compute the aggregated similarity score for each candidate path. The algorithm proposed by Park & Kim computes the arithmetic mean of the co-occurrence and the order-consistency. However, the arithmetic mean is not very robust to outliers. The harmonic mean is more appropriate for aggregating the similarities, as it has a bias towards very low values, mitigating the impact of large outliers. Using the overall similarity measures for each candidate path, we can determine which one of the candidates is the best. We use a threshold to determine the minimal score needed to perform a mapping.

4 Evaluation

This section compares the performance of our algorithm against the performance of the algorithm by [11] and Anchor-PROMPT [10]. We evaluate the performance of the algorithms using three real-life datasets. The largest dataset, containing over 44,000 categories, was obtained from the Open Directory Project (ODP), available at http://dmoz.org. This dataset is relatively large, which makes it interesting for evaluation purposes, as it shows how well the algorithms scale when mapping large product taxonomies. The second dataset was obtained from Amazon.com, the largest online retailer in the USA. We have selected over 2,500 different categories in total with paths that have a maximum depth of five levels. The last dataset was obtained from Overstock.com, a large online retailer from the USA. It contained just over 1,000 categories with a maximum depth of four levels and it has a comparatively broad and flat taxonomy structure with many composite categories, which makes word sense disambiguation difficult.

Table 1 shows the average results of the six mappings per algorithm. As one can notice, our algorithm performs better on both recall and the F_1-measure than Anchor-PROMPT and the algorithm of Park & Kim. The recall has increased from 16.69% for Anchor-PROMPT and 25.19% for Park & Kim to 83.66% for our algorithm. Despite the clear improvement in recall, our algorithm actually performs slightly worse on precision than the algorithm of Park & Kim. The precision dropped from 47.77% to 38.28%. The high recall of our algorithm can be attributed to the fact that we are better able to deal with composite categories. The observed improvement in recall is also due to the usage of an approximate lexical matching function instead of an exact one for order consistency. As we perform more mappings to target categories, our precision declines slightly compared to that of the algorithm of Park & Kim.

Table 1 Comparison of average results per algorithm

Algorithm	Precision	Recall	F_1-measure	Computation time
Anchor-PROMPT	28.93%	16.69%	20.75%	0.47 s
Park & Kim	47.77%	25.19%	32.52%	4.99 s
Our algorithm	38.28%	83.66%	52.31%	20.71 s

Anchor-PROMPT maps more conservatively due to the fact that it is geared towards ontology mapping in general. Making classification mistakes in product taxonomy mapping is less severe than in most ontology mapping problems, because in an e-commerce domain it is considered more important to map many categories with some imprecision rather than only mapping a few categories with high precision. This is also reflected by the fact that the optimal thresholds for both our algorithm and the algorithm of Park & Kim were found to be very low. The average optimal similarity threshold, which determines whether a mapping is performed, is 0.025 for Park & Kim and 0.183 for our algorithm. In order to obtain the optimal thresholds, we have used a subset set for each mapping data set. We can conclude from the results that algorithms that are specifically tailored to mapping product taxonomies perform better than ontology mapping algorithms (such as Anchor-PROMPT) within this domain.

5 Conclusion

This paper proposes an algorithm suitable for automated product taxonomy mapping in an e-commerce environment. Our proposed algorithm takes into account the domain-specific characteristics of product taxonomies, like the existence of composite categories and the syntactical variations in category names. The algorithm we propose is based on the algorithm of Park & Kim [11]. Similar to the Park & Kim approach, we employ word sense disambiguation techniques in order to find the correct sense of a term. Differently than the Park & Kim algorithm, we consider various lexical similarity measures for lexical matching, like the Jaccard index and the Levenshtein distance. Furthermore, our algorithm exploits the hierarchical structure of taxonomies and uses a different similarity aggregation function, in order to make the algorithm more robust against outliers.

We have shown that our algorithm performs better than other approaches when mapping product taxonomies. Its performance on mapping three real-life datasets was compared favourably with that of Anchor-PROMPT [10] and the algorithm proposed by [11]. It manages to significantly increase both the recall and the F_1-measure of the mappings. We have also argued that recall is more important than precision in the context of online shopping.

For future work, we would like to employ a more advanced word sense disambiguation technique, such as the one proposed by [5]. It would also make sense to consider using word category disambiguation, also known as part-of-speech tagging, for the words found in a category name. Often the meaning changes when the part-of-speech is different. For example, 'Machine Embroidery' and 'Embroidery Machines', referring to machine-made embroidery and a machine which makes embroidery, respectively. By differentiating between adjectives and nouns, it is possible to identify these two meanings.

References

1. Aumueller, D., Do, H.H., Massmann, S., Rahm, E.: Schema and Ontology Matching with COMA++. In: ACM SIGMOD International Conference on Management of Data 2005 (SIGMOD 2005), pp. 906–908. ACM (2005)
2. Ehrig, M., Staab, S.: QOM – Quick Ontology Mapping. In: McIlraith, S.A., Plexousakis, D., van Harmelen, F. (eds.) ISWC 2004. LNCS, vol. 3298, pp. 683–697. Springer, Heidelberg (2004)
3. Hepp, M.: GoodRelations: An Ontology for Describing Products and Services Offers on the Web. In: Gangemi, A., Euzenat, J. (eds.) EKAW 2008. LNCS (LNAI), vol. 5268, pp. 329–346. Springer, Heidelberg (2008)
4. Horrigan, J.B.: Online Shopping. Pew Internet & American Life Project Report 36 (2008)
5. Lesk, M.: Automatic Sense Disambiguation using Machine Readable Dictionaries: How to tell a Pine Cone from an Ice Cream Cone. In: 5th Annual ACM SIGDOC International Conference on Systems Documentation (SIGDOC 1986), pp. 24–26. ACM (1986)
6. Li, J.: LOM: A Lexicon-based Ontology Mapping Tool. In: 5th Workshop Performance Metrics for Intelligent Systems, PerMIS 2004 (2004)
7. Madhavan, J., Bernstein, P.A., Rahm, E.: Generic Schema Matching with Cupid. In: 27th International Conference on Very Large Data Bases (VLDB 2001), pp. 49–58. Morgan Kaufmann Publishers Inc. (2001)
8. Melnik, S., Garcia-Molina, H., Rahm, E.: Similarity Flooding: A Versatile Graph Matching Algorithm and its Application to Schema Matching. In: 18th International Conference on Data Engineering (ICDE 2002), pp. 117–128. IEEE Computer Society (2002)
9. Miller, G.A.: WordNet: A Lexical Database for English. Communications of the ACM 38(11), 39–41 (1995)
10. Noy, N.F., Musen, M.A.: The PROMPT Suite: Interactive Tools for Ontology Merging and Mapping. International Journal of Human-Computer Studies 59(6), 983–1024 (2003)
11. Park, S., Kim, W.: Ontology Mapping between Heterogeneous Product Taxonomies in an Electronic Commerce Environment. International Journal of Electronic Commerce 12(2), 69–87 (2007)
12. Shopping.com: Online Shopping Comparison Website (2011), http://www.shopping.com
13. Zhang, G.Q., Zhang, G.Q., Yang, Q.F., Cheng, S.Q., Zhou, T.: Evolution of the Internet and its Cores. New Journal of Physics 10(12), 123027 (2008)
14. Zhu, H., Madnick, S., Siegel, M.: Enabling Global Price Comparison through Semantic Integration of Web Data. International Journal of Electronic Business 6(4), 319–341 (2008)

A Case-Based Planning Mechanism for a Hardware-Embedded Reactive Agents Platform

Juan F. de Paz, Ricardo S. Alonso, and Dante I. Tapia

Computers and Automation Department, University of Salamanca. Plaza de la Merced, s/n,
37008, Spain
{fcofds,ralorin,dantetapia}@usal.es

Abstract. Wireless Sensor Networks is a key technology for gathering relevant information from different sources. In this sense, Multi-Agent Systems can facilitate the integration of heterogeneous sensor networks and expand the sensors' capabilities changing their behavior dynamically and personalizing their reactions. Both Wireless Sensor Networks and Multi-Agent Systems can be successfully applied to different management scenarios, such as logistics, supply chain or production. The Hardware-Embedded Reactive Agents (HERA) platform allows developing applications where agents are directly embedded in heterogeneous wireless sensor nodes with reduced computational resources. This paper presents the reasoning mechanism included in HERA to provide HERA Agents with Case-Based Planning features that allow solving problems considering past experiences.

Keywords. Wireless Sensor Networks, Multi-Agent Systems, Case-Based Planning, Service-Oriented Architectures.

1 Introduction

There are many situations where is useful to collect context information about users and their environment. Some of these situations include home automation, healthcare and also management scenarios, such as logistics, inventory, assembly lines and even offices. In this sense, Wireless Sensor Networks (WSNs) are widely used to obtain context information about users and their environment [1]. However, the integration of wireless devices coming from different technologies is not an easy task. Although there are plenty of technologies for implementing WSNs (e.g., ZigBee, Wi-Fi or Bluetooth), the lack of a common architecture may lead to additional costs due to the necessity of deploying non-transparent interconnection elements among different networks [2].

In this sense, the implementation of distributed architectures is presented as a solution to these problems [3]. One of the most prevalent alternatives in distributed architectures is Multi-Agent Systems (MAS), which can help to distribute resources and reduce the central unit tasks [4]. A distributed agent-based architecture provides more flexible ways to move functions to where actions are needed,

J. Casillas et al. (Eds.): Management Intelligent Systems, AISC 171, pp. 121–130.
springerlink.com

thus obtaining better responses at execution time, autonomy, services continuity, and superior levels of flexibility and scalability than centralized architectures [5].

These are some of the main motivations for developing the HERA (*Hardware-Embedded Reactive Agents*) platform [6]. In HERA, unlike other approaches, agents are directly embedded on the WSN nodes, and their services (i.e., functionalities) can be invoked from other nodes in the same WSN or other WSN connected to the former one. HERA focuses specially on devices with reduced resources to save CPU time, memory size and power consumption.

This paper presents the mechanism included in HERA to provide HERA Agents with reasoning features. This reasoning mechanism is based on Case-Based Reasoning (CBR) [7] and Case-Based Planning (CBP) models, where problems are solved by using solutions to similar past problems [8]. Solutions are stored into a case memory, which the mechanism can consult to find better solutions for new problems. HERA Agents use this mechanism to learn from past experiences and to adapt their behavior according to the context information.

The rest of the paper is organized as follows. The following section presents the problem description that essentially motivated the development of HERA and introduces the CBR and CBP models. Then, it is depicted the main characteristics and components of HERA. After that, the CBP mechanism used in HERA is described. Finally, it is presented the conclusions and the future lines of work.

2 Background and Problem Description

HERA have stemmed from the necessity to cover more efficiently several of the challenges found on systems that use WSNs in their infrastructure. The fusion of the multi-agent technology and WSNs is not easy due to the difficulty in developing, debugging and testing distributed applications for devices with limited resources. The interfaces developed for these distributed applications are either too simple or, in some cases, do not even exist, which complicates even more their maintenance. Therefore, there are researches that develop methodologies for the systematic development of Muti-Agent Systems (MAS) for WSNs [9]. ActorNet [10] is a study that describes a mobile agent platform for WSNs. Implementing agent programs over WSNs is complicated due to the limitations of the sensor nodes, their limited memory, small bandwidth and low energy autonomy. Actor-Net allows a wide range of dynamic applications, including customized queries and aggregation functions, in the sensor network platform. However, each mobile agent is only centered on a sensor node. Baker *et al.* [11] present the integration of an agent-based WSN within an existing MAS focused on condition monitoring. In this research, it is used SubSense, a multi-agent middleware platform developed to allow condition monitoring agents to be deployed onto a WSN. The architecture of the SubSense platform is based on the FIPA-defined model, but customized so that agents are embedded into sensor nodes. However, SubSense platform is not focused on working with heterogeneous WSNs and is implemented over 512KB RAM SunSPOT sensor nodes using the Java Mobile Edition (J2ME), while HERA platform can run on lightweight sensor nodes with just 8KB RAM. Furthermore, most of the works that relate Multi-Agent Systems and WSNs talk about Mobile Agents based on WSN (MAWSN). Zboril *et al.* [12] proposes WSageNt, a platform that is implemented through mobile agents running on wireless sensor nodes. WSageNt is supposed to be fault tolerant and not to be only focused on WSNs. However, it has not context-awareness features and does not contemplate the interconnection of heterogeneous WSNs.

Multi-Agent Systems can integrate Case-Based Reasoning (CBR) and Case-Based Planning (CBP) mechanisms [7] [8], which allow agents to make use of past experiences to create better plans and achieve their goals. These mechanisms provide the agents greater learning and adaptation capabilities. Case-based Reasoning (CBR) is a type of reasoning based on past experiences [7]. CBR mechanisms solve new problems by adapting solutions that have been used to solve similar problems in the past, and learn from each new experience. The primary concept when working with CBR mechanisms is the concept of case, which is described as a past experience composed of three elements: an initial state or problem description that is represented as a belief; a solution, which provides the sequence of actions carried out to solve the problem; and a final state, which is represented as a set of goals. CBR manages cases (past experiences) to solve new problems. The way cases are managed is known as the CBR cycle, and consists of four sequential phases: retrieve, reuse, revise and retain. CBP comes from CBR, but is specially designed to generate plans (sequence of actions) [8]. In CBP, the proposed solution for solving a given problem is a plan. This solution is generated by taking into account the plans applied for solving similar problems in the past. The problems and their corresponding plans are stored in a plans memory. The reasoning mechanism generates plans using past experiences and planning strategies, which is how the concept of Case-Based Planning is obtained [8]. CBP consists of four sequential stages similar to CBR stages. Problem description (initial state) and solution (situation when final state is achieved) are represented as beliefs, the final state as a goal (or set of goals), and the sequences of actions as plans. The CBP cycle is implemented through goals and plans. When the goal corresponding to one of the stages is triggered, different plans (algorithms) can be executed concurrently to achieve the goal or objective. Each plan can trigger new sub-goals and, consequently, cause the execution of new plans. CBR and CBP mechanisms have been successfully applied into Multi-Agent Systems in other research works to predict crises in business intelligence [7] or assign roles in virtual organizations of agents [13].

In this sense, HERA includes a Case-Based Planning (CBP) mechanism which allows the agents to make use of past experiences to create better plans and achieve their goals. HERA allows devices from different radio and networks technologies to coexist in the same distributed network. A totally distributed approach and the use of heterogeneous WSNs provide an architecture that expands the agents' capabilities to obtain information about the context and to automatically react over the environment.

3   The HERA Platform

HERA platform [6] facilitates the inclusion of agents into dynamic and self-adaptable heterogeneous WSNs. HERA is an evolution of the SYLPH (*Services laYers over Light PHysical devices*) platform [3] [14] and is focused specifically on devices with reduced resources to save CPU time, memory size and energy consumption. SYLPH allows the interconnection of several networks from different wireless technologies, such as ZigBee or Bluetooth. In this case, both WSNs are interconnected by a set of intermediate gateways (known as SYLPH Gateways [14]) simultaneously connected to several wireless interfaces. The underlying layers of HERA, provided by SYLPH, follow a Service-Oriented Architecture (SOA) model [2]. Unlike those approaches that integrate WSNs and MASs [9] [10] [11] [12], agents in HERA are directly embedded on the WSN nodes and their services can be invoked from other nodes in the same network or another network connected to the original one. Likewise, HERA can be executed over multiple wireless devices independently of their microcontroller or the programming language they use. This facilitates the development of context-aware capabilities into systems because developers can dynamically integrate and remove nodes on demand.

HERA implements an organization based on a stack of layers, shown in Figure 1 [6]. Each layer in a node communicates with its peer in another node through an established protocol. In addition, each layer offers specific functionalities to the immediately upper layer in the stack. The HERA layers are added over the SYLPH layers [14] and also the existent application layer of each WSN stack, allowing the platform to be reutilized over different technologies. From lowest to highest, SYLPH and HERA layers are described as follows:

- **SYLPH Message Layer (SML).** SML offers the upper layers the possibility of sending asynchronous messages between two nodes through the *SYLPH Services Protocol* (SSP). These messages specify the service invocation in a *SYLPH Services Definition Language* (SSDL) format.
- **SYLPH Services Directory Sub-layer (SSDS).** SSDS creates dynamical services tables to locate and register services in the network. A node that stores and maintains services tables is called *SYLPH Directory Node* (SDN). A node in the network can make a request to the SDN to know of a certain service.
- **SYLPH Application Layer (SAL).** SAL allows different nodes to directly communicate with each other using SSDL requests and responses that will be delivered in encapsulated SML messages following the SSP.
- **HERA Agents Layer (or just HERA).** HERA agents are specifically intended to run on devices with reduced resources. To communicate with each other, HERA agents use HERACLES, the agent communication language designed for being used under the HERA platform. There must be at least one facilitator agent in every agent platform [4]. This agent is the first created in the platform and acts as a directory for searching agents. In HERA, the equivalent of these agents is the HERA-SDN (*HERA Spanned Directory Node*).

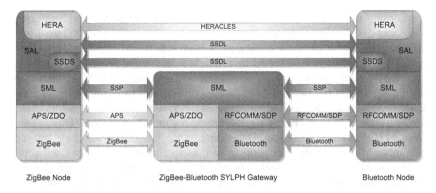

ZigBee Node   ZigBee-Bluetooth SYLPH Gateway   Bluetooth Node

Fig. 1 Layers and protocols of the HERA platform over a heterogeneous Wireless Sensor Network

Similarly, the corresponding protocols used by HERA and SYLPH are:

- **SYLPH Services Protocol (SSP).** SSP is the internetworking protocol of the SYLPH platform. SSP allows sending packets of data from one node to another regardless of the WSN to which each one belongs.
- **SYLPH Services Definition Language (SSDL).** SSDL is the IDL (Interface Definition Language) used by SYLPH. Unlike other IDLs such as WSDL (Web Services Definition Language) [15], SSDL does not use as many intermediate separating tags, and the order of its elements is fixed [14].
- **HERA Communication Language Emphasized to Simplicity (HERACLES).** HERACLES language is directly based on the SSDL language. When developing a program, programmers use the human-readable representation to define agents' functionalities. However, HERA agents transmit the more compact representation of HERACLES as frames [6].

4 HERA Case-Based Planning Mechanism

As previously mentioned, some agents in HERA integrate a Case-Based Planning (CBP) mechanism. The CBP mechanism provides the agents with greater adaptation capabilities. As it is a complex and resources demanding task, the CBP mechanism has been modeled as a service provided by a special HERA Agent, known as HERA Planning Agent, which runs on a central node (i.e., a computer or a wireless device with moderate computational resources). The main characteristics of this mechanism are described in the remainder of this section.

The CBP mechanism needs a set of HERA Agents running on a set of nodes (i.e., wireless devices). Each of the nodes is connected physically to different sensors and actuators. This way, each node in the system can transmit commands to the actuators according to the sensors measurements. Each device d_i is defined by the sensors and actuators to which it has access, as expressed in (1).

$$d_i = (S_i, A_i) \tag{1}$$

where S_i is the set of sensors and A_i is the set of actuators.

According to the values read from the sensors, each HERA Agent running in the devices makes use of the actuators in order to achieve the required goal (e.g., stop a heater when a target temperature has been achieved). Thus, the behavior of the CBP mechanism is established by a database generated from the information of the sensors and actuators. This way, when some event produces an interaction with the sensors in the devices (e.g., a user action or a variation in the environment), these devices forward the values from the sensors and actuators to a central node that runs a special HERA Agent that stores this information. This agent is known as HERA Planning Agent. Depending on the size of the whole system (i.e., the number of nodes in the network, as well as the number of sensors and actuators associated to them), this central node can be implemented as a computer with a database stored in a physical disk or as a wireless sensor node with a smaller database stored in an EEPROM or Flash memory. Each device d_j has its own cases memory. Each case of the device d_j follows the structure indicated in (2).

$$c_i^{d_j} = (V_i^{S_j}, V_i^{A_j}) \qquad (2)$$

where $V_i^{S_j}$ is the set of values from the sensors associated with the device j, and

$V_i^{A_j}$ the values associated to the actuators.

The reactive behavior of the HERA Agents is defined as a set of rules that determine the relation among the sensors and the actuators. The rules are generated by a CBP mechanism. There are two kinds of rules: *static rules* and *dynamic rules*. Static rules are pre-defined rules that have priority over dynamic rules. Static rules determine the default behavior of each node and act also as a backup of these behaviors (i.e., they are stored directly in the nodes). Dynamic rules are automatically generated from the defined cases for each of the devices. The dynamic rules are periodically updated on every run of the HERA Agents (i.e., each time they are requested to execute a task). Figure 2 shows the functioning of the CBP mechanism when a device must execute a new task.

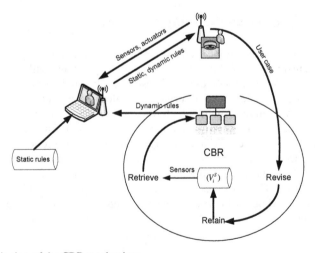

Fig. 2 Functioning of the CBP mechanism

Both static and dynamic rules are defined using the grammar shown as follows. This grammar facilitates the error detection and also the generation of native code that is actually running in the HERA Agents.

```
S::= Rules
Rules::= Rules Rule pyc |Rule pyc
Rule::= Precon then Actuator
Precon::= Precon and Sensor | Sensor
Sensor::= Ref relop Factor
Actuator::= Ref assop Factor
Factor::= nint | nflx
Ref::= id
```

In the shown grammar, the terms that start by uppercase characters correspond with non-terminal symbols (i.e., class of syntactically equivalent groupings), while the terms that start by lowercase characters correspond with terminal symbols. A non-terminal symbol is a symbol that is decomposed through a rule in the grammar, while a terminal symbol represents a symbol that is equivalent to a token. The tokens for each of the terminal nodes are defined as follows:

```
pyc=";"
then="then"
and="and"
relop="=="|"<="|">="|">"|"<"|"!="
assop="="
nint= [0-9]+
nflx= ([0-9]+)"."([0-9]*)
Id=[A-Za-z][0-9A-Za-z]*
```

On the one hand, static rules are stored in the database by the HERA Planning Agent following the grammar previously established. An example of rules could be the following:

```
movement==0 then turnlight=0;
movement==1 and light==0.25 then turnlight=1;
```

As can be seen, rules are formed by sensors, literals, comparison and logical operators, as well as the final action to be performed by the actuator.

On the other hand, dynamic rules are defined by the CBP mechanism by means of the information of the cases shown in (2). During the recovery stage the CBP mechanism recovers the information with the sensors and actuators associated to a certain device. During the reutilization stage, automatic rules are generated from this information. For the generation of automatic rules, different algorithms based on decision rules and decision trees can be used. An example of algorithm based on decision rules is M5 [16], while J48 [17] is based on decision trees.

In HERA, the J48 algorithm is used for the generation of rules [17]. The inputs of the classifier are the value of the sensors, and the output belonging to the actuator. Using this configuration the J48 is trained, thus obtaining a decision tree that represents the behaviors of the actuators. In this sense, the different actuators are chosen as leaf nodes in the decision trees, while the sensor values are placed in the intermediate nodes. There is a decision tree according to the sensors situated in the device for each actuator. The devices can select a value of the actuators according to the value of the sensors through the decision tree, and the system only has to follow the conditions in the intermediate nodes until it arrives at a leaf node.

The final schema of the decision trees would be similar to that shown in Figure 3, and rules would be generated as indicated in the schema. The generated rules follow the grammar indicated above. The decision tree generated contains all the necessary information to generate the rules according to the grammar. The system only has to select each leaf node and move up toward the root node, introducing a new condition for each movement throughout the tree.

During the revision and learning stages, the system only stores the values of the actuators and sensors when the actuators are established manually by the users. The automatic decision of the agents are not stored in the system. The system only stores the interactions of the users as it tries to adapt to their behaviors. The system does not store generated cases as it is able to predict the behaviors, and only stores a new case when a user modifies the automatic configuration.

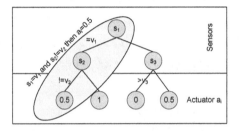

Fig. 3 Example of decision tree for the dynamic rules

For the development of the rules generator system the Weka libraries are used for the implementation of J48 algorithm. JFlex and Cup are used as lexical and syntactic analyzers. By means of Cup it is generated the native code that is then transferred and executed on chips.

5 Conclusions and Future Work

Wireless Sensor Networks provide a distributed and dynamic way to gather useful context information in a wide range of scenarios. These scenarios include business and managerial scenarios, such as inventory or assembly lines. Nevertheless, there are other situations where Wireless Sensor Networks can be successfully applied, such as home and building automation, healthcare or education.

The HERA platform allows wireless devices with reduced computational resources from different technologies to work together in a distributed way. The HERA CBP mechanism can adapt itself according to the values established by the user during its interaction with the platform. Furthermore, it facilitates the inclusion of new sensors dynamically, without need of performing offline training for each of the devices. In addition, the CBP mechanism can determine automatically the influence of the sensors to establish the final state of each of the actuators, so it is not necessary to indicate the relation among sensors and actuators.

Future work includes a experimentation stage to test the CBP mechanism and the publication of the results obtained. Furthermore, it will be designed a mechanism that will allow HERA agents to move throughout different nodes, no matter the WSN technology they use. This way, we will get, for example, HERA agents to move from a ZigBee node to a Bluetooth node through HERA, allowing the use of different programming languages and operating systems.

Acknowledgments. This project has been supported by the Spanish Ministry of Science and Technology project TIN 2009-13839-C03-03.

References

1. Marin-Perianu, M., Meratnia, N., Havinga, P., de Souza, L., Muller, J., Spiess, P., et al.: Decentralized Enterprise Systems: A Multiplatform Wireless Sensor Network Approach. IEEE Wireless Communications 14(6), 57–66 (2007)
2. Cerami, E.: Web Services Essentials: Distributed Applications with XML-RPC, SOAP, UDDI & WSDL, 1st edn. O'Reilly Media, Inc. (2002)
3. Corchado, J.M., Bajo, J., Tapia, D.I., Abraham, A.: Using Heterogeneous Wireless Sensor Networks in a Telemonitoring System for Healthcare. IEEE Transactions on Information Technology in Biomedicine 14, 234–240 (2010)
4. Wooldridge, M.: An Introduction to MultiAgent Systems, 2nd edn. Wiley (2009)
5. Sánchez, D., Isern, D., Rodríguez-Rozas, Á., Moreno, A.: Agent-based platform to support the execution of parallel tasks. Expert Systems with Applications 38, 6644–6656 (2011)
6. Tapia, D.I., Alonso, R.S., García, Ó., Corchado, J.M.: HERA: Hardware-Embedded Reactive Agents Platform. In: Pérez, J.B., Corchado, J.M., Moreno, M.N., Julián, V., Mathieu, P., Canada-Bago, J., Ortega, A., Caballero, A.F., et al. (eds.) Highlights in PAAMS. AISC, vol. 89, pp. 249–256. Springer, Heidelberg (2011)
7. Bajo, J., Borrajo, M.L., De Paz, J.F., Corchado, J.M., Pellicer, M.A.: A multi-agent system for web-based risk management in small and medium business. Expert Systems with Applications 39, 6921–6931 (2012)
8. Corchado, J.M., Bajo, J., de Paz, Y., Tapia, D.I.: Intelligent environment for monitoring Alzheimer patients, agent technology for health care. Decision Support Systems 44, 382–396 (2008)
9. Tynan, R., O'Hare, G., Ruzzelli, A.: Multi-Agent System Methodology for Wireless, Sensor Networks. Multiagent and Grid Systems 2, 491–503 (2006)
10. Kwon, Y., Sundresh, S., Mechitov, K., Agha, G.: ActorNet: an actor platform for wireless sensor networks. In: Proceedings of the Fifth International Joint Conference on Autonomous Agents and Multiagent Systems, pp. 1297–1300. ACM, Japan (2006)
11. Baker, P., Catterson, V., McArthur, S.: Integrating an Agent-Based Wireless Sensor Network within an Existing Multi-Agent Condition Monitoring System. In: 15th International Conference on Intelligent System Applications to Power Systems, ISAP 2009, pp. 1–6 (2009)

12. Zboril, F., Horacek, J., Spacil, P.: Intelligent Agent Platform and Control Language for Wireless Sensor Networks. In: Third UK Sim European Symposium on Computer Modeling and Simulation, EMS 2009, pp. 482–487 (2009)

13. Zato, C., De Paz, J.F., de Luis, A., Bajo, J., Corchado, J.M.: Model for assigning roles automatically in egovernment virtual organizations. Expert Systems with Applications (2012) (in Press), doi:10.1016/j.eswa.2012.01.185

14. Tapia, D.I., Alonso, R.S., De la Prieta, F., Zato, C., Rodriguez, S., Corchado, E., Bajo, J., Corchado, J.M.: SYLPH: An Ambient Intelligence based platform for integrating heterogeneous Wireless Sensor Networks. In: 2010 IEEE International Conference on Fuzzy Systems (FUZZ), pp. 1–8 (2010)

15. Curbera, F., Duftler, M., Khalaf, R., Nagy, W., Mukhi, N., Weerawarana, S.: Unraveling the Web services web: an introduction to SOAP, WSDL, and UDDI. IEEE Internet Comput. 6, 86–93 (2002)

16. Holmes, G., Hall, M., Frank, E.: Generating Rule Sets from Model Trees. In: Proc. of the 12th Australian Joint Conf. on Artificial Intelligence, pp. 1–12 (1999)

17. Salzberg, S.L.: C4.5: Programs for Machine Learning by J. Ross Quinlan. In: Machine Learning, vol. 16, pp. 235–240. Morgan Kaufmann Publishers, Inc. (1994)

A Linguistic Approach for Semantic Web Service Discovery

Jordy Sangers[1], Flavius Frasincar[1], Frederik Hogenboom[1],
Alexander Hogenboom[1], and Vadim Chepegin[2]

[1] Erasmus University Rotterdam, P.O. Box 1738, NL-3000 DR Rotterdam, The Netherlands
jordysangers@hotmail.com,
{frasincar,fhogenboom,hogenboom}@ese.eur.nl
[2] Tie Kinetix, P.O. Box 3053, NL-2130 KB, Hoofddorp, The Netherlands
vadim.chepegin@tieglobal.com

Abstract. We propose a Semantic Web Service Discovery framework for finding semantically annotated Web services by using natural language processing techniques. The framework searches through a set of annotated Web services for matches with a user query, which consists of keywords, so that knowledge about semantic languages is not required. For matching keywords with Semantic Web service descriptions given in Web Service Modeling Ontology (WSMO), techniques like part-of-speech tagging, lemmatization, and word sense disambiguation are used. Three different matching algorithms are defined and evaluated for their ability to do exact matching and approximate matching between the user query and Web Service descriptions.

1 Introduction

With the emergence of Web services and the Service Oriented Architecture (SOA), business process components are increasingly decoupled, while systems and business processes converge, forcing companies to change their management strategies. The usage of Web services in SOA creates a wide network of services that collaborate in order to implement complex tasks. Web services are commonly described via narrative Web pages containing information about their operations in natural languages. These Web pages contain plain text with no machine interpretable structure and hence cannot be used by machines to automatically process the descriptive information about a Web service.

Several semantic languages [1, 13, 19] have been created to aid machines in automatically processing information on Web services. These languages allow describing the functionality of services in a machine interpretable form, while original Web service descriptions contained only information about the data types and bindings as a description of a Web service functionality [5]. These Semantic Web service descriptions use ontologies to describe the behavior of a Web service by applying

J. Casillas et al. (Eds.): Management Intelligent Systems, AISC 171, pp. 131–142.
springerlink.com © Springer-Verlag Berlin Heidelberg 2012

reasoning over their semantics. The semantics described in ontologies enable systems to interpret what a Web service is doing, stimulating service discovery [3] and composition [11]. The ontologies, however, are created by humans and therefore contain natural language as ontology meta-data. This lets humans understand the concepts defined, while a system can only understand concepts and their relationships to a limited extent as specified in the ontology axioms. Natural Language Processing (NLP) techniques can therefore help in better defining the context of a Web service.

When using a single holistic ontology, machines can discover and compose Web services automatically based on the semantics defined. Using one holistic ontology is, however, hardly reachable and therefore it is impossible to reason based only on formal logic. NLP techniques can help overcoming the ambiguity problems between multiple ontologies that are being used by Semantic Web service descriptions. Service composition is often driven by people with expertise in business processes and not by technicians. Thus, end users must be able to discover these Web services based on a search query written in an easy to understand language, i.e., human language. Therefore, a discovery mechanism must be developed in such a way that Semantic Web services can be found using natural language queries.

In this paper the Semantic Web Service Discovery (SWSD) framework is proposed, which enables users to search (using keywords) for existing Web services that are described by means of a Semantic Web language for service annotation. This process consists of several steps including information extraction, word sense disambiguation, and matching the user search context and Web service context by means of a similarity measure. The result of this process is a ranked list of Web services that match the users search criteria.

This paper is structured as follows. Section 2 discusses related technologies for describing and searching Web services. Section 3 proposes the SWSD framework for Web Services Discovery based on keywords. Our implementation of the framework is described in Sect. 4. Section 5 presents an evaluation of different matching algorithms between the user input and a Web service description. Last, Sect. 6 concludes the paper and discusses future work.

2 Related Work

Current approaches for Web Service Discovery (WSD) can be divided into two types. One approach for Web service discovery is based on clustering operation parameters, while the other approach uses rich semantics for Web service discovery. These two different approaches are explained below by describing existing or proposed Web service discovery systems.

One approach for Web service discovery is by searching for similarities among different Web Service Definition Language (WSDL) service descriptions. In this way, similar operations and services can be discovered based on operation parameters, which enables searching for substitutable and composable Web services [8].

With this approach, the semantics of the Web service operations can be extracted and used for discovery purposes.

A Web service search engine that uses clustering of operation parameters is Woogle [7]. It is designed to search for similar Web service operations and composable Web service operations. Woogle computes automatically the underlying semantics of WSDL descriptions and uses these to match operations. The semantics are solely defined based on the operation parameters. However, if independent ontologies which define the Web service semantics exist, the behavior of a Web service can be known without investigating parameter names and is therefore preferable to use.

Seekda! [16] also uses clustering of operation parameters for discovery of Web services. It also extracts semantics from the WSDL files, which enables runtime exchange of similar services and composition of services. Seekda! is part of a bigger system called Service-Finder [4]. Service-Finder is a platform for service discovery where information about services is gathered from different sources such as Web pages, blogs, and Web 2.0 services. The information is automatically added to a semantic model using automatic service annotation, realizing flexible discovery of services. Service-Finder uses its own semantics for discovery and composition, therefore does not taking into account predefined semantics.

GODO [10] does not search for similar Web services, but uses a Goal-Driven approach. It consists of a repository with Web Service Modeling Ontology (WSMO) Goals and lets users state their goal by writing a sentence in plain English. A language analyzer will extract keywords from the user sentence and an existing WSMO Goal will be searched based on those keywords. The WSMO Goal with the highest match will be sent to WSMX [6], an execution environment for WSMO service discovery and composition. WSMX will then search for a WSMO Web service that is linked to the given WSMO Goal via some WSMO Mediators and return the WSMO Web service back to the user. This approach makes good use of the capabilities of the WSMO framework, but it cannot be applied for other semantic languages like OWL-S and WSMO-Lite, which do not have such goal representation elements.

The framework proposed in this paper aims to address a multitude of semantic description languages using a novel approach that combines the ontology structure with Web services similarities using natural language processing techniques for the automatic discovery of Web services. From this point of view it can be considered as a hybrid of the previously introduced approaches able to deal with a larger category of (semantic) Web services. In addition, our framework supports next to exact matching also approximate matching, improving the recall when perfect matching is not possible.

3   Semantic Web Service Discovery Framework

The SWSD framework proposes a keyword-based discovery process for searching Web services which are described using a semantic language. This search mechanism incorporates NLP techniques to establish a match between a user search query, containing English keywords, and a Semantic Web service description. It does not take into account the logic-based semantics defined in the Web service descriptions, but uses the definitions of concepts stated in imported ontologies. By making use of these definitions, the framework can establish a broader search field by also employing related concepts from the ontologies to identify the context in which the Web service is operating.

The SWSD framework assumes that there is a set of Web services described using semantic languages such as WSMO [1], WSMO-Lite [19] or OWL-S [13]. These annotations can be read by the system and words that might represent the context of the Web services will be extracted (e.g., the names of the operations or nouns and verbs stated in non-functional descriptions of concepts or conditions). These words must then be disambiguated, because words can have different senses. If the system knows the sense of the words, they can be matched with the senses disambiguated from the search query. This will result in a ranked list of Web services according to the matching degree with the user search.

The process consists of three major steps: Service Reading, Word Sense Disambiguation (WSD), and Match Making. Service Reading consists of parsing a Semantic Web service description and extracting names and non-functional descriptions of used concepts. WSD determines the senses of a set of words present in the previously extracted information. During Match Making the similarity between the different sets of senses is determined, which is subsequently used for ranking the Web services.

3.1   Semantic Web Service Reader

In order to enable a search engine to look through Web service descriptions written in different languages, several different Web service description readers are required, i.e., one for each language. A Semantic Web service reader must be able to extract various elements out of a Web service description and its used ontologies. Names and non-functional descriptions of elements such as the capabilities, conditions, and effects of the Web service help in understanding the context of the Web service, i.e., they can foster establishing the right context. The non-functional descriptions are written in natural language and thus contain a human description of the specified element. Before extracting words from a Web service description, this description has to be parsed using language-specific parsers. Subsequently, word splitting needs to be performed based on case transition, after which each word is tagged with the right Part-of-Speech (POS).

3.2   Word Sense Disambiguation

A user can represent its goal by defining two different sets of words. One set contains only nouns and the other only verbs. Because words can have multiple meanings disambiguation is needed, resulting in a set of senses, each representing a single meaning of a word. Once a set of senses from the user query and a set of senses from a Web service are established, a matching between the two can be performed.

As non-supervised WSD allows disambiguation of words without user interference, we use a variant of the SSI algorithm [15] to get the senses out of a set of words using a semantic lexicon (e.g., WordNet [14]). The algorithm disambiguates a word based on a previously disambiguated set of words and their related senses. Per sense of the word, a similarity with the senses from the context is calculated and the sense with the highest similarity is chosen. After that, the word and its chosen sense will be added to the context and another iteration is performed. This process continues until there are no ambiguous words left.

At the start of the process, a context is not yet established. In order to disambiguate meanings of the words that can have multiple senses, one first has to find the words that have only one sense (monosemous words) to initialize the context. If all the words in the set have multiple senses (polysemous words), the least ambiguous word is chosen and for each of its senses, the algorithm is simulated as if the sense was used as the starting context. Each time a new sense is added to the context, the similarity between the new sense and the context is stored. The sense which creates the highest sum of pair-wise context sense similarities (after disambiguating all words) is used for the context initialization.

Because studies have shown that the method of Jiang and Conrath [12] performs better than other semantic distance measures [2], this method is chosen to compute the semantic distance between two senses. Using this method, given two senses, a number between 0 and 1 is obtained, stating the similarity between the two senses. If this number is high, then the two senses given are close to each other and therefore are similar.

3.3   Sense Matching

After disambiguating each word gathered from the user input or a Semantic Web service description, we have obtained several different sets of senses. The framework assumes each word in the user query is equally important for the matching process and therefore the user input will contain, after the WSD, one set of senses. However, a Web service description can contain words that represent the context of the Web service better than other words. Therefore, after the WSD, several sets of senses (each having a different weight for the matching process) are computed for a Web service.

3.3.1   Level Matching

Besides matching only disambiguated senses, words that do not appear in the used semantic lexicon should also be taken into account. These words can represent important names or concepts for the discovery of Web services and hence must also be used in the matching process. For matching user input with a Semantic Web service description, the user input contains a set of words that could not be disambiguated (ws_u) and a set of senses (ss_u), and the Web service description contains multiple sets of words (mws_w) and senses (mss_w). Because the Web service description, presented in the next section, provides a number (n) of sets containing words and senses, each having a different importance for the matching process, the final similarity between the user input and the Web service input will be a weighted average of the similarities between each set of words ($mws_{wi} \in mws_w$) and senses ($mss_{wi} \in mss_w$) from the Web service description and the set of words and senses from the user input, as shown in (1). The weights (w_i) are established by means of experiments with different values and must sum up to 1 in order to make sure that the final similarity between the user query and a Web service description ranges between 0 and 1.

$$
finalSim(ss_u, mss_w, ws_u, mws_w) =
$$
$$
\sum_{i=1}^{n} w_i \times levelSim(ss_u, mss_{wi}, ws_u, mws_{wi}) \tag{1}
$$

For each set of words and senses from the Web service description, the system performs two different types of measures: one for the sense matching (shown in (3)) and one for the matching of words that could not be disambiguated (shown in (4)). These two measures will have a range between 0 (no match) to 1 (exact match) and will be combined into a single measure using a weighted average (shown in (2)). These weights are, as with the final similarity, established by means of experiments with different rates and must sum up to 1.

$$
levelSim(ss_u, ss_w, ws_u, ws_w) = w_{sense} \times senseSim(ss_u, ss_w) +
$$
$$
w_{word} \times wordSim(ws_u, ws_w) \tag{2}
$$

3.3.2 Jaccard Matching

For matching the user set of senses with a set of senses from one of the levels of a Web service description, the Jaccard matcher uses the Jaccard Index. This method is often used for computing the similarity between two sets and can so compare the different sets of senses:

$$
senseSim(ss_u, ss_w) = \frac{|ss_u \cap ss_w|}{|ss_u \cup ss_w|} . \tag{3}
$$

By dividing the number of senses which appear in both sets by the total number of senses in both sets, a similarity coefficient can be calculated. With this approach the Jaccard matcher calculates the percentage of exact matching items and can also be applied for matching the words that could not be disambiguated (not present in the semantic lexicon):

$$
wordSim(ws_u, ws_w) = \frac{|ws_u \cap ws_w|}{|ws_u \cup ws_w|} . \tag{4}
$$

3.3.3 Similarity Matching

To overcome the fact that for calculating similarity values only exact matching items are used, the similarity matcher uses a similarity based approach for matching different sets of senses or non-disambiguated words. Using this approach, words that are almost identical are not considered to be a mismatch, but an almost match. The same applies for sense matching. If two senses are closely related, their similarity value will approach 1. This approach allows more flexible matching between different items than previously considered.

For calculating a similarity between two sets of senses, the same similarity function as in WSD is applied. Equation (5) describes how the similarity between the user set of senses (ss_u) and a Web service set of senses (ss_w) is computed. The average of the similarity between each sense (s_u) from the user set of senses, and the Web service set of senses (s_w) is computed. The average of the similarity between each sense (s_w) from the Web service set of senses, and the user set of senses (s_u) is added to that to provide a symmetric match.

$$senseSim(ss_u, ss_w) = \sum_{s_u \in ss_u} \frac{senseScore(s_u, ss_w)}{|ss_u| + |ss_w|} + \sum_{s_w \in ss_w} \frac{senseScore(s_w, ss_u)}{|ss_u| + |ss_w|} \tag{5}$$

The similarity between a sense (s_a) and a set of senses (ss_b) is determined by the maximum similarity between the sense and one of the senses (s_b) from the other set. Equation (6) shows this computation.

$$senseScore(s_a, ss_b) = \text{argmax}_{s_b \in ss_b} senseNorm(s_a, s_b) \tag{6}$$

Because the similarity distance method from Jiang and Conrath can give any value between 0 and infinity as a result and a range between 0 and 1 is preferred for quantifying the degree of match, a logarithmic function must be used to transform the values of the similarity. Using Equation (7), exact similar senses will result in 1 as resulting similarity and a total mismatch between senses will result in 0:

$$senseNorm(s_a, s_b) = 1 - e^{-sim(s_a, s_b)}. \tag{7}$$

For matching the sets of non-disambiguated words, the Levenshtein Distance metric is used. This metric calculates the total number of edit operations that needs to be done in order to transform one word to another. The similarity between two sets of words is done in the same way as when comparing two sets of senses. The only difference is that instead of the similarity function from WSD, now the Levenshtein Distance is applied.

The similarity function for calculating the similarity between the user set of words (ws_u) and a Web service set of words (ws_w) is described in (8). Equation (9) shows how the similarity between a word and a set of words is computed.

Finally, (10) describes how the Levenshtein Distance is used for comparing two words, where *maxLength* is the number of tokens of the longest word that is being compared. If this formula returns a negative value, which means that too many updates had to be done to change one word into another word, a value of 0 will be used to indicate a total mismatch.

$$wordSim(ws_u, ws_w) = \sum_{w_u \in ws_u} \frac{wordScore(w_u, ws_w)}{|ws_u| + |ws_w|} +$$

$$\sum_{w_w \in ws_w} \frac{wordScore(w_w, ws_u)}{|ws_u| + |ws_w|} \tag{8}$$

$$wordScore(w_a, ws_b) = \text{argmax}_{w_b \in ws_b} wordNorm(w_a, w_b) \tag{9}$$

$$wordNorm(w_a, w_b) = 1 - 2 \times \frac{levenshtein(w_a, w_b)}{maxLength(w_a, w_b)} \tag{10}$$

4 Semantic Web Service Discovery Engine

Our implementation of the SWSD approach, the Semantic Web Service Discovery Engine, allows users to search for semantically annotated Web services on an existing repository by defining a set of keywords. It is able to handle Web services that are annotated using the WSMO [1] framework. Based on the modularity of the implementation, the engine can be extended with readers that can parse other Semantic Web service languages.

A WSMO Web service reader and a WSMO Ontology reader have been implemented in Java using the WSMO4J [9] API. After reading the Web service files, the found concepts are used for scanning the ontologies for their descriptions. Based on a full identifier, the reader can search for a concept. If a concept is found, the non-functional definition, attributes, and related concepts can be used for WSD. The engine uses seven different levels of information about the Web service, each of which has a different associated importance for the matching process. Importance is expressed using weights (established after examining multiple Semantic Web service descriptions and experimenting manually with non-normalized weights ranging from 1 to 10), which, after normalization, sum up to 1. The different levels and their associated weights are:

- Non-functional description and name of the Web service, 7/27, (direct relation);
- Non-functional descriptions and names of concepts used by Web service, 5/27, (direct relation);
- Non-functional descriptions of properties of capabilities of the Web service, 4/27, (direct relation);
- Non-functional descriptions and names of superconcepts of the concepts used by the Web service, 4/27;
- Non-functional descriptions and names of subconcepts of the concepts used by the Web service, 3/27;

- Non-functional descriptions and names of concepts related via attributes with concepts used by the Web service, 3/27 (indirect relation);
- Non-functional descriptions and names of attributes of concepts used by the Web service, 1/27 (indirect relation).

The names and non-functional descriptions of the entities returned from the readers undergo an additional NLP step: nouns and verbs are extracted from the non-functional descriptions using the Stanford POS tagger [17] and words are split if they consist of case-transitions. Subsequently, our WSD implementation makes use of the WordNet [14] API and also the JWordNetSim [18] API for similarity calculation between two WordNet senses.

5 Evaluation

This section covers the evaluation of different matching algorithms that can be used for Semantic Web service discovery. The algorithms described in Sect. 3 are implemented in the SWSD engine and are evaluated using a set of predefined queries and sets of preferred Web services related to one of the queries. The evaluated matching algorithms are: simple, Jaccard-based, and similarity matching. The simple algorithm uses the Jaccard similarity only for lexical representations of words. It does not make use of NLP and thus it is used as a baseline to check whether the NLP steps add value to the discovery of Semantic Web services. The Jaccard-based and similarity matching algorithms work as described in Sect. 3.

In order to evaluate the performance of the three matching algorithms, we have conducted 61 separate experiments, in which we used 14 Web services semantically annotated using WSMO. The tests can be divided into two types: 33 tests have been done to measure the matching performance of the algorithms using queries that have been designed to search for Web services that are present in the repository. The other 28 tests have been done to measure the performance using queries that have been designed to search for Web services that are not present in the repository. In the latter case, a number of similar Web services from the repository have been used to test how well the algorithms can discover similar Web services.

Testing with 61 queries and three matching algorithms results in 183 PR graphs. As a thorough analysis of all graphs is cumbersome, PR graphs consisting of average precision values for the recall points are created. This enables the comparison of all the different algorithms at once. However, the testing is done with lists of preferred Web services that can vary in the number of listed services. For testing, lists with two to five preferred Web services have been used. Because these variations in number of Web services cause different recall values, average precision values could only be calculated for queries that have the same amount of preferred Web services.

The performances of the different matching algorithms are visualized by eight different PR graphs. The four PR graphs that are displayed in Fig. 1 show the average results for the exact matching tests. For each of the four different numbers of

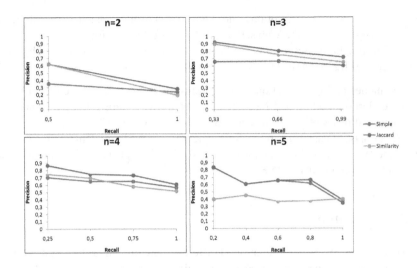

Fig. 1 PR graphs for discovery of exact matching services

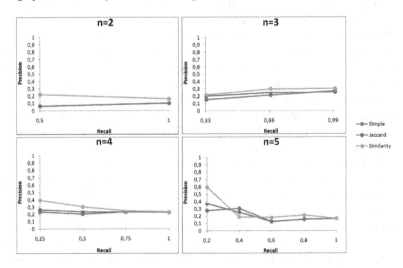

Fig. 2 PR graphs for discovery of similar services

preferred returned Web services (*n*) a PR graph is created. The four PR graphs that are shown in Fig. 2 show the average results for the approximate matching tests.

From the different PR graphs that are shown in Fig. 1, we can make two observations. First, the Jaccard-based algorithm has in most cases a higher precision for the first half of the graph than the simple and the similarity algorithm. Second, all algorithms have about the same precision to provide a full recall. This means that to provide all the preferred Web services to the user, they need about the same amount of Web services to be displayed. Hence, the user has to browse through a number of services – the same for each algorithm – to find the last preferred service. However, as the Jaccard-based algorithm displays a higher precision for a low recall, the Jaccard-based algorithm provides at least some of the preferred Web services in an earlier stage to the user then the other algorithms. It can therefore be seen as the best algorithm to discover exact matching Web services.

From the different PR graphs that are shown in Fig. 2, we can make the observation that in case of non-exact matching, the similarity algorithm performs overall better for discovery of similar Web services than the Jaccard-based and simple matching algorithms. In most of the cases the precision lines of the similarity algorithm are above the line of the Jaccard-based algorithm.

6 Conclusions and Future Work

In order to facilitate managers to compose Web services for business processes, we proposed the SWSD framework, which is a keyword-based discovery process for searching Web services that are described using semantically enriched annotations. It makes an intensive use of natural language processing techniques and a WordNet-based similarity measure for matching keywords written in natural language with semantic Web services described using semantic languages, hereby supporting managers to discover relevant Web services using natural language queries.

Our implementation of the SWSD framework can search for WSMO Web services based on user search words for similar matches. A matching score is computed based on the similarity between the words in the user query and a Web service description. After experimenting with different matching functions, we found that Jaccard matching is performing best for discovering exact matching Web services, while matching using a similarity-based approach gives the best results for finding similar Web services.

As the SWSD engine is currently limited to WSMO descriptions, as future work, the SWSD engine could be extended in such a way that it has the ability to read more annotation formats, e.g., WSMO-Lite. Another limitation of the proposed framework is the lack of detailing in Web service contexts, which can be tackled in future work by retrieving additional information from WSDL files. Finally, the weights that are used by the matching algorithms, which are currently established by means of experiments, could be determined by making use of artificial intelligence techniques such as neural networks or Bayesian networks, to optimize the discovery process.

References

1. de Bruijn, J., Bussler, C., Domingue, J., Fensel, D., Hepp, M., Keller, U., Kifer, M., Konig-Ries, B., Kopecky, J., Lara, R., Lausen, H., Oren, E., Polleres, A., Roman, D., Scicluna, J., Stollberg, M.: Web Service Modeling Ontology (WSMO). W3C Member Submission (June 3, 2005), http://www.w3.org/Submission/WSMO/

2. Budanitsky, A., Hirst, G.: Semantic distance in WordNet: An experimental, application-oriented evaluation of five measures. In: WordNet and Other Lexical Resources: Applications, Extensions and Customizations, NAACL 2001 Workshop (2001)

3. Bussler, C., Cimpian, E., Fensel, D., Gomez, J.M., Haller, A., Haselwanter, T., Kerrigan, M., Mocan, A., Moran, M., Oren, E., Sapkota, B., Toma, I., Viskova, J., Vitvar, T., Zaremba, M., Zaremba, M.: Web Service Execution Environment (WSMX). W3C Member Submission (June 3, 2005), http://www.w3.org/Submission/WSMX/

4. Cefriel, Seekda!, Ontoprise, University of Sheffield: Service-Finder (2010), http://www.service-finder.eu/

5. Christensen, E., Curbera, F., Meredith, G., Weerawarana, S.: Web Services Description Language (WSDL). W3C Note (March 15, 2001), http://www.w3.org/TR/wsdl

6. DERI Galway: Web Service Execution Environment (2010), http://www.wsmx.org/

7. Dong, X., Halevy, A., Madhavan, J., Nemes, E., Zhang, J.: Similarity Search for Web Services. In: 30th International Conference on Very Large Data Bases (VLDB 2004), vol. 30, pp. 372–383 (2004)

8. Ernst, M.D., Lencevicius, R., Perkins, J.H.: Detection of Web Service Substitutability and Composability. In: International Workshop on Web Services — Modeling and Testing (WS-MaTe 2006), pp. 123–135 (2006)

9. EU IST, FIT-IT: WSMO4J API (2010), http://wsmo4j.sourceforge.net/

10. Gomez, J.M., Rico, M., Garcia-Sanchez, F., Bejar, R.M., Bussler, C.: GODO: Goal driven orchestration for Semantic Web Services. In: 1st Workshop on Web Services Modeling Ontology Implementations (WIW 1004), CEUR Workshop Proceedings, vol. 113 (2004)

11. Hikimpour, F., Sell, D., Cabral, L., Domingue, J., Motta, E.: Semantic Web Service Composition in IRS-III: The Structured Approach. In: 7th IEEE International Conference on E-Commerce Technology (CEC 2005), pp. 484–487. IEEE Computer Society (2005)

12. Jiang, J., Conrath, D.: Semantic Similarity Based on Corpus Statistics and Lexical Taxonomy. In: International Conference Research on Computational Linguistics (ROCLING X), pp. 19–33 (1997)

13. Martin, D., Burstein, M., Hobbs, J., Lassila, O., McDermott, D., McIlraith, S., Narayanan, S., Paolucci, M., Parsia, B., Payne, T., Sirin, E., Srinivasan, N., Sycara, K.: OWL-S. W3C Member Submission (November 22, 2004), http://www.w3.org/Submission/OWL-S/

14. Miller, G.A., Beckwith, R., Fellbaum, C., Gross, D., Miller, K.: Wordnet: An on-line lexical database. International Journal of Lexicography 3(4), 235–244 (1990)

15. Navigli, R., Velardi, P.: Structural Semantic Interconnections: a Knowledge-Based Approach to Word Sense Disambiguation. In: IEEE Transactions on Pattern Analysis and Machine Intelligence, vol. 27, pp. 1075–1086. IEEE Computer Society (2005)

16. Semantic Technology Institute: Seekda (2009), http://seekda.com/

17. The Stanford Natural Language Processing Group: Stanford Log-linear Part-Of-Speech Tagger (2009), http://nlp.stanford.edu/software/tagger.shtml

18. The University of Sheffield: Pure Java WordNet Similarity Library (2010), http://nlp.shef.ac.uk/result/software.html

19. Vitvar, T., Kopecky, J., Fensel, D.: WSMO-Lite: Lightweight Semantic Descriptions for Services on the Web. In: 5th IEEE European Conference on Web Services (ECOWS 2007), pp. 77–86. IEEE Computer Society (2007)

Applying Multi-objective Optimization for Variable Selection to Analyze User Trust in Electronic Banking

F. Liébana-Cabanillas[1], R. Nogueras[2], F. Muñoz-Leiva[3], I. Rojas[1], and A. Guillén[2]

[1] Former Director of Distribution Channels in Caja Rural Granada, Dpt. of Marketing and Market Research, University of Granada, Spain
[2] Dpt. of Computer Technology and Architecture, University of Granada, Spain
[3] Dpt. of Marketing and Market Research, University of Granada, Spain

Abstract. The potential fraud problems, international economic crisis and the crisis of confidence in markets have affected financial institutions, which have tried to maintain customer trust in many different ways. To maintain the trust level in financial institutions, the implementation of electronic banking for customers has been considered a successful strategy. However, the parameters that define user trust have not been analysed in detail due to the lack of experience and the recent use of e-banking. This paper aims to determine which variables are relevant to user trust by applying machine learning techniques as multi-objective genetic algorithms for the preparation of business strategies to improve confidence and profitability. The algorithms have been tuned following the indications given by experts and their results have been validated by them, setting a level of reliability. There is also a comparison among different fitness functions used in the evolution process that are able to rank the subset of variables encoded by the individuals.

1 Introduction: The Economic Crisis and Trust in the Financial Sector

The behaviour of the financial system against the economic crisis has been different among the countries within the European Union. While many international institutions focused their interest on credit and risk transfer, neglecting customer service, the banking sector continued to have an extensive network of offices through which to distribute financial products and to foster close client relationships. This very competitive environment forced banks to strictly control costs, which has made the financial system one of the world's most efficient [1]. Despite these advantages, the Spanish financial system was also in a precarious position particularly due to its exposure in real estate. In the latter part of the 90's and in the early part of the last decade there was an excess supply of real estate and therefore a large demand for financing. This situation forced financial institutions to go to wholesale markets since domestic markets did not have the resources to cover as much investment as was

J. Casillas et al. (Eds.): Management Intelligent Systems, AISC 171, pp. 143–152.
springerlink.com

being generated. Due to this and the pressing international crisis, the government and the Central Bank had to intervene different economies, among them, the Spanish.

The Spanish financial sector has already started to change as a result of this situation thanks to the Bank Restructuring Fund (FROB[1]), restructuring banks and strengthening the resources of credit institutions. In this complicated situation, the Spanish financial system has had to make technological improvements to reduce costs and optimize investments. Of all the available tools used to achieve these objectives, electronic banking has been the most widely implemented.

From the 90s to the present, electronic banking has become the distribution channel with the greatest potential for financial institutions [2]. Currently, the majority of companies offer their customers access to most of their services through this channel. Therefore, electronic banking has become a crucial service by which to gain customer satisfaction and loyalty and establish closer customer relationships, thereby meeting user expectations [3] and [4].

Thus, the primary alternative channel to the traditional bank branch is electronic banking as it has many advantages for customers including convenience, global access, availability, cost and time-savings, information transparency, choice and comparison, customization, and financial innovation [5] and [6]. However, this service also has some drawbacks, mainly related to trust and security. But trust, together with satisfaction, is considered one of the key elements in building long-term relationships, a fundamental business strategy in the current economic situation [7] and [8].

By the time of offering a client the e-banking services, it would be useful to know in advance if that client could be more attracted or not to the use of it. To do so, data mining and machine learning techniques can be applied to data bases in order to identify which attribute determine the use trust and if there is a model that could predict it. In this paper, these two phases are studied and the results have been validated using an expert committee. Therefore, the rest of the paper is structured as follows: Section 2 will describe the fieldwork and how the data was collected, Section 3 will introduced the algorithms and models that were applied as well as the novelties added to classical algorithms. Then, Section 4 will present the results and in Section 5, Conclusions will be drawn.

2 Perceived Trust in Electronic Banking and Fieldwork

The behaviour of users of electronic banking (Orange Foundation, 2010: 118) is characterized by prudence. To this end, users periodically review on-line bill movements, do security checks for electronic transactions and connections, avoid access from public computers, do not provide personal information by e-mail or phone, and log off bank web sites before closing browsers. However, some groups of customers are still reticent about such services. Regarding electronic commerce in general,

---

[1] The Bank Restructuring Fund was established by Royal Decree-Law 9 / 2009 of June 26, 2009.

consumers show more concern about the use of banking services strictly speaking, when the amount of money potentially exposed to fraud is significantly larger, than with other types of services or organizations.

Every year consumers lose an increasing amount of money through internet fraud. According to internet Fraud Watch (http://www.fraud.org) directed by the National Consumers League, consumers lost approximately 18.82 million dollars through fraud in 2010, significantly higher than the $5.79 million lost in 2004. On the average, the losses per person varied from $293 in 1999 to $2,165.15 in 2007. The Rivest, Shamir and Adleman (RSA, 2010) laboratories identified 281,000 phishing attacks in January 2010 aimed at financial institutions of any size.

In virtual environments trust is affected by the security and privacy problems [9] and [10]. However, it seems that these problems are being overcome because according to the Survey on Equipment and Information and Communication Technology Use in Households (National Institute of Statistics - NIE, 2011) in Spain, 72.4% of e-commerce transactions are made on a regular basis although users still shy away from using electronic banking for the reasons noted above.

2.1 Information Collection

The survey was conducted between September and October 2009. Participation in the survey was voluntary and was presented to the user once the authenticated party signed on to the website of a national saving bank in southern Spain.

The survey sample size was 1,081 completed questionnaires by individual visitors, but the final number of questionnaires used for this research was reduced to 946.

Those questionnaires completed by users with juridical personality were eliminated in order to only analyse the behaviour of individuals, or natural persons (see tables 1 and 2).

Table 1 Technical data. * For the estimation of a ratio, where $P = Q = 0.5$ and a confidence level of 95%, under the principles of simple random sampling

Population: Internet bank users.
Sample Frame: Users online Banking.
Type of Sampling: Simple random sampling.
Sample Size: 946 valid cases.
Sample Error*: 3,19%
Date of Field work: September and October 2009

The literature shows that the level of consumer trust toward a website depends on a number of factors, including the perceived reputation of the website [11]; site characteristics (web design, information availability, ease when navigating the site, privacy and security especially in those places where you can perform financial

Table 2 Respondent characteristics

Items	Data	Frecuency	Percent	Cumulative	(%) Cumulative
Gender	Male	634	67,02%	634	67,02%
	Female	312	32,98%	946	100,00%
Age	16-25	56	5,92%	56	5,92%
	26-35	324	34,25%	380	40,17%
	36-45	277	29,28%	657	69,45%
	46-65	259	27,38%	916	96,83%
	>65	30	3,17%	946	100%

transactions) [13]; and consumer characteristics [6]. In our research we analysed a total of 33 variables grouped into three clusters, socio-demographic, economic-financial and beliefs (trust).

In order to determine the relevance of each variable (see table 3), two criteria were selected: Delta Test and Mutual Information. The subsection below describes both of them.

Table 3 Variables Analyzed

Var. number	Type	Variable
1		Office
2		Geographic Region
23		Age
24	Socio-demographic	Gender
25		Mobile Telephone
26		E-mail
27		Zip code
28		Province
3		Profitability per Customer in 2010 per entity
4		Profitability per Customer in 2009 per entity
5		Average Liability Balance within Client Account 2010
6		Average Liability Balance within Client Account 2009
7		Average Liability Balance outside of Client Account in 2010
8		Average Liability Balance outside of Client Account in 2009
9		Average Balance of Client Assets in 2010
10		Average Balance of Client Assets in 2009
11		Number of products purchased in 2010
12		Number of products purchased in 2009
13		Linked products per client in 2010
14		Linked products per client in 2009
15	Economic- Financial	Business Volume per Client in 2010
16		Business Volume per Client in 2009
17		Customer profitability in 2010
18		Customer profitability in 2009
19		Direct Deposit for Paychecks
20		Direct Deposit for Pensions
21		Debit Card
22		Credit Card
28		Months of Experience with Ruralvía
29		Number of Operation on Ruralvia in 2010
30		Number of Operation on Ruralvia in 2009
31		Total Euro Amount of Operations on Ruralvia in 2010
32		Total Euro Amount of Operation on Ruralvia in 2009
	Behavioural	Trust

3 Models and Algorithms

This section formulates the problem tackled formally and will present the techniques applied to solve it. A new selection operator is discussed which improves the optimization results and the computational costs.

In order to identify which elements are the most important variables regarding the customer's trust and in order to design accurate models, a pre-processing of the data should be made by variable selection.

The problem of variable selection should be stated as: Given a set of N input/output pairs $(\mathbf{x}_i^j, y_i)$ where $i = 1...N, j = 1...d, \mathbf{x}_i^j \in R^d$ and $y_i \in R$ it is desired to obtain a subset of variables where the cardinality of it and the validation error are minimums chosen from a Pareto.

In order to solve this problem, Genetic Algorithms (GAs) where adapted in such a way that they keep a Pareto of non-dominated solutions defining a new class of GAs: Multi-objective GAs (MOGAs). Among the MOGAs, the NSGA-II (Non-dominated Sorting Genetic Algorithm) [14] [15] which is an updated version of the classical NSGA, i.e., multi-objective optimization algorithm from the field of Evolutionary Computation. The main goal of the NSGA-II algorithm is to find the best individuals of a population of candidate solutions according to the Pareto front by performing a sorting procedure that considers the different objectives to be optimised. NSGA-II has been succesfully applied to a wide variety of problems providing an excellent performance.

3.1 Multi-objective Optimization

Multi-objective optimization is the process of optimizing two or more objectives subject to certain constraints. A multi-objective optimization problem (MOP) has a set of n decision variables (x), a set of k objective functions $(y = f(x))$ and a set of m inequality constraints $(g(x))$ and a p equality constraints $(h(x))$, and objective functions and constraints depends on the n decision variables. More formally:

Optimize $y = f(x) = \{f_1(x), f_2(x), ..., f_k(x)\}$ subject to $g(x) = \{g_1(x), g_2(x), ..., g_m(x)\} \geq 0$ and $h(x) = \{h_1(x), h_2(x), ..., h_p(x)\} = 0$

where $x = \{x_1, x_2, ..., x_n\} \in X$ and $y = \{y_1, y_2, ..., y_k\} \in Y$ and the decision vector is x, the decision space is X, the objective vector is y and the objective space is Y. We assume that a solution to this problem can be described in terms of a decision vector $(x_1, x_2, ..., x_n)$ in the decision space X. A function $f : X \rightarrow Y$ evaluates the quality of a specific solution by assigning it an objective vector $(y_1, y_2, ..., y_k)$ in the objective space Y. Therefore the problem consists in finding x with the best value for $f(x)$. The set of all decision vectors which satisfies the $m + p$ constraints is named Feasible Solution Set and denoted as X_f.

An decision vector x_1 is said to dominate another decision vectors x_2 $(x_1 < x_2)$ if no component of x_1 is greater (smaller) than the corresponding component of x_2 and at least one component is smaller (greater). This concept is known as Pareto dominance: $\forall i \in \{1, 2, ..., k\}, f_i(x_1) \leq f_i(x_2) \wedge \exists j \in \{1, 2, ..., k\} \mid f_j(x_1) < f_j(x_2)$

The set of all optimal solutions in the decision space X is in general denoted as the Pareto set $X^* \subseteq X$ and it is defined as: $P^* = \{x \in X_f \mid \neg \exists x\prime \in X_f \wedge x\prime \succ x\}$ and its image in objective space as Pareto front $Y^* = f(X^*) \subseteq Y_1$.

3.2 Multi-objective Selection Genetic Algorithm: MSGA

The NSGA-II has a good behaviour although it is quite expensive when measuring the computation time. As data bases become larger, this aspect should be kept in mind when choosing an algorithm. Another element that could be improved of this algorithm when applied to Variable Selection (VS) is to reduce the size of the Pareto, allowing to exploit more convenient solutions that include too many variables.

In order to avoid the cost of the non-dominated sort but keeping the MO aspect, a new selection operator within the GA has been defined. Selecting the binary tournament selection, one of the parents is selected considering the quality of the subset of variables and the other one is chosen considering the number of variables.

As in VS the solutions with a high number of variables are not desired (even if they provide the best optimization criterion), another operator has been introduced into the algorithm. On each generation, the individuals that have more than α variables are discarded. The α value should be selected manually considering the expert's opinion.

The results provided by a classical GA with these two elements ends up in better results and smaller computation times.

3.3 Delta Test

This method [19] is able to perform an estimation of the noise between input/output pairs, therefore, it is a good indicator of how precise a model can approximate a data set without overfitting these data. The application to the variable selection problem is quite straight forward: the solution is to find the subset of variables that provides the smallest value of the Delta Test (DT) [20].

The DT for a set of input vectors $X = \{\mathbf{x}_k\}$ and their output $Y = \{y_k\}$ with $k = 1...n$ is defined as:

$$\delta_{n,k} = \frac{1}{2n} \sum_{i=1}^{N} (y_i - y_{nn[i,k]})^2 \tag{1}$$

where $nn[i,k]$ is the index of the k-th nearest neighbour to x_i usually according to the Euclidean distance. Since $\delta_{n,1} \approx \sigma_e^2$, where σ_e^2 is the variance of the noise in the output, $\delta_{n,1}$ can be used as an estimation of the minimum mean squared error that can be obtained by a model without overfitting.

The main drawback of this methodology is its lack of robustness when the number of input samples is not large because the convergence to σ_e^2 is achieved when increasing n.

3.4 Mutual Information

The concept of Mutual Information (MI), also known as cross-entropy has been used already to solve the problem of selecting or identifying the most relevant variables from a set of input-output pairs, showing a good performance. Let $X = \{\mathbf{x}_k\}$ and $Y = \{y_k\}$ for $k = 1...n$, then the MI between X and Y can be understood as the amount of information that the subset of variables X provide over the output variable Y and its formulation is $I(X,Y) = H(Y) - H(Y|X)$ where $H(Y)$ is the entropy of Y and $H(Y|X)$ is the conditional entropy that measures the uncertainty of Y given a known X. Thus, we can obtain a numerical value that measures the relevance of X.

For the case where the variables are continuous, following Shannon formulation, the entropy can be defined as:

$$H(Y) = -\int \mu_Y(y) \log \mu_Y(y) dy, \tag{2}$$

where $\mu_Y(y)$ is the marginal density function. This function can be defined as the joint between the probability density functions of X and Y ($\mu_{X,Y}$), this is: $\mu_Y(y) = \int \mu_{X,Y}(x,y) dx$.

Therefore, once it is known the value of $H(Y)$, to obtain the MI value it is necessary to compute $H(Y|X)$, that, for the continuous case and reformulating it using the properties of the entropy, is defined as:

$$I(X,Y) = \int \mu_{X,Y}(x,y) \log \frac{\mu_{X,Y}(x,y)}{\mu_X(x)\mu_Y(y)} dxdy. \tag{3}$$

Then, to obtain the MI value it is only needed to estimate the joint probability density function (PDF) between X and Y. This value can be obtained using methods based on the k-NN [21] or in Parzen Window [22].

4 Experimental Results

Several experiments where carried out in order to find the best subset of variables that characterise client's trust. The classical NSGA-II was applied using the three criteria described in the previous section as well as the MSGA considering two values of α. The parameter setting for the GAs, that were using binary encoding, was:

- Population size: 50,100 and 150.
- Crossover: two-points ; Probability: 0.85
- Mutation: single gene level operator ; Probability: 0.1
- Selection: binary tournament for NSGA-II and the MO one for MSGA
- Stop criterion: no modification of the Pareto front for several iterations

4.1 Algorithms Comparison

Table 4 shows the results obtained after executing the three algorithms commented in the previous section using the three different criteria. The values represent the mean value of the Delta Test and the Mutual Information computed using k-NN and Parzen window provided by the best individual (using only one criterion) in the population after several runs.

Table 4 Results obtained for the three GAs implemented. The value represents the result obtained for each criterion and in the next column, the subset of variables that provide that result. In bold are the best results (DT lower is better, MI higher is better).

NSGA-II		
DT	3.1e-2 (5e-3)	5,10,12,14,15,19,21,22,23,25,29,30
MI-Parzen	1.06e-2 (1e-3)	11,20,27
MI-kNN	1.06e-2 (1e-3)	5,11,20,27

MSGA		
DT	3.27e-2 (1e-3)	2,4,5,11,14,19,21,22,23,25,27,28,31
MI-Parzen	**1.77e-2** (1e-3)	27
MI-kNN	**1.46e-2** (6e-3)	2,5,6,7,14,19,20,24,25,30

MSGA- $\alpha = 10$		
DT	**2.92e-2** (6e-3)	5,10,12,14,15,19,21,22,23,25,29,30
MI-Parzen	1.12e-2 (1e-3)	4,27
MI-kNN	1.35e-2 (1e-3)	2,4,6,7,8,9,17,19,20,22,26

4.2   Expert's Validation

In order to validate the quality of the results obtained, a committee of five experts, with background in different financial entities, was consulted. The validation process consisted in four stages: personal interview, method and data evaluation, results evaluation, and feedback. The experts background was at least 10 years of experience and they have been working in comercial tasks in the last three. The ages were in the interval [34,46].

The interview had the aim of explaining the experts how the algorithms worked and the criteria they use to determine if a subset of variables is good or not. Afterwards, the results obtained in the experiments were given as input to the experts that had to evaluate in a Likert scale (1-7).

The results of the expert's opinions is shown in table 5. As this table shows, the best algorithm is the MSGA providing satisfactory results both using DT and MI using k-NN. This last criterion was the most valuable for the experts.

Table 5 Expert's punctuation

Algorithm	Method	Expert 1	Expert 1	Expert 1	Expert 1	Expert 1	Mean
	DT	3	4	4	5	4	4
NSGA-II	MI-Parzen	3	4	4	4	5	4
	MI-kNN	5	5	6	5	5	5.2
	DT	5	4	4	5	5	4.6
MSGA	MI-Parzen	1	1	1	1	1	1
	MI-kNN	6	4	4	5	4	4.6
	DT	5	5	6	5	6	5.4
MSGA $\alpha = 10$	MI-Parzen	3	2	3	2	2	2.4
	MI-kNN	4	5	6	6	5	5.2

5 Conclusions and Implications for Management

As the studies show, e-banking seems to make a difference for the costumers to select a bank or to make them keep their savings in it. This paper aims to identify which characteristics define the customers' trust in order to improve the most common operations and to provide more information about e-baking to certain costumers.

Several multi-objective algorithms were evaluated, including a new modifciation that allows the algorithm to obtain accurate results and a reasonable number of variables. Furthermore, three criteria commonly used to perform variable selection were compared and all the results were evaluated and validated by experts in the field.

The results obtained by the algorithms and the opinion of the experts coincide in that the best multi-objective algorithm is the proposed MSGA and the best criteria to perform variable selection is mutual information computed using the k-NN algorithm.

References

1. Álvarez, J.M.: La banca española ante la actual crisis financiera. Estabilidad Financiera 15, 23–38 (2008)
2. Karjaluoto, H., Mattila, M., Pento, T.: Factors underlying attitude formation toward online banking in Finland. International Journal of Bank Marketing 20(6), 261–272 (2002)
3. Hsu, S.H.: Developing an index for online customer satisfaction: Adaptation of American Customer Satisfaction Index. Expert Systems with Applications 34, 3033–3042 (2008)
4. Berrocal, M.: Fidelización y Venta Cruzada. Informe Caja Castilla La Mancha (2009)
5. Delgado, J., Nieto, M.J.: Incorporación de la tecnología de la información a la actividad bancaria en España: La banca por Internet. Estabilidad financiera, Banco de España 3, 85–105 (2002)
6. Muñoz-Leiva, F.: La adopción de una innovación basada en la Web. Tesis Doctoral. Departamento de Comercialización e Investigación de Mercados, Universidad de Granada (2008)
7. Lam, S.Y., Shankar, V., Murthy, M.K.: Customer Value, Satisfaction, Loyalty, and Switching Costs: An Illustration from a Business-to-Business Service Context. Journal of the Academy of Marketing Science 32(3), 293–311 (2004)
8. García, N., Sanzo, M.J., Trespalacios, J.A.: Can a good organizational climate compensate for a lack of top management commitment to new product development? Journal of Business Research 61, 118–131 (2008)
9. Ha, H.Y.: Factors Influencing Consumer Perceptions of Brand Trust Online. Journal of Product and Brand Management 13(5), 329–342 (2004)
10. Laroche, M., Yang, Z., Mcdougall, G.H.G., Bergeron, J.: Internet Versus Bricks-and-Mortar Retailers: An Investigation Into Tangibility and Its Consequences. Journal of Retailing 81(4), 251–267 (2005)
11. Muñoz-Leiva, F., Luque-Martínez, T., Sanchez-Fernandez, J.: How to improve trust toward electronic banking. Online Information Review 34(6), 907–934 (2010)
12. Flavián, C., Guinalíu, M.: Un análisis de la influencia de la confianza y del riesgo percibido sobre la lealtad a un sitio web: el caso de la distribución de servicios gratuitos. Revista Europea de Dirección y Economía de la Empresa 16(1), 159–178 (2007)

13. Flavián, C., Guinalíu, M., Gurrea, R.:: Análisis empírico de la influencia ejercida por la usabilidad percibida, la satisfacción y la confianza del consumidor sobre la lealtad a un sitio web. In: XVI Encuentros de Profesores Universitarios de Marketing, pp. 209–226. Esic (2004)
14. Deb, K., Pratap, A., Agarwal, S., Meyarivan, T.: A Fast and Elitist Multiobjective Genetic Algorithm: NSGA-II. IEEE Transactions on Evolutionary Computation 6(2) (2002)
15. Srinivas, N., Deb, K.: Multi-objective Optimization using Nondominated sorting in Genetic Algorithms. Evolutionary Computation 2(3), 221–248 (1994)
16. Eirola, E., Liitiainen, E., Lendasse, A., Corona, F., Verleysen, M.: Using the Delta Test for Variable Selection. In: ESANN 2008 Proceedings, European Symposium on Artificial Neural Networks - Advances in Computational Intelligence and Learning, Bruges, Belgium (2008)
17. Gallant, S.I.: Perceptron-based learning algorithms. IEEE Transactions on Neural Networks 1(2), 179–191
18. Guillen, A., Sovilj, D., Lendasse, A., Mateo, F., Rojas, I.: Minimising the Delta Test for Variable Selection in Regression Problems. International Journal High Performance Systems Architecture 1(4) (2008)
19. Pi, H., Peterson, C.: Finding the Embedding Dimension and Variable Dependencies in Time Series. Neural Computation 6(3), 509–520 (1994)
20. Lendasse, A., Corona, F., Hao, J., Reyhani, N., Verleysen, M.: Determination of the Mahalanobis matrix using nonparametric noise estimations. In: ESANN, pp. 227–232 (2006)
21. Kraskov, A., Stögbauer, H., Grassberger, P.: Estimating mutual information. Physics Review (June 2004)
22. Kwak, N., Choi, C.H.: Input Feature Selection by Mutual Information Based on Parzen Window. IEEE Trans. Pattern Analysis and Machine Intelligence 24(12), 1667–1671 (2002)

A Context-Aware Mobile Recommender System Based on Location and Trajectory

Manuel J. Barranco, José M. Noguera, Jorge Castro, and Luis Martínez

Department of Computer Sciences, University of Jaén.
Campus Las Lagunillas s/n, Jaén, 23071, Spain
{barranco,jnoguera,jcastro,martin}@ujaen.es

Abstract. Recommender systems have typically been used in tourism applications to filter out irrelevant information and to provide personalized recommendations to the users. With the advent of mobile devices and ubiquitous computing, RSs have begun to incorporate Location Based Services (LBS) into mobile tourism guides to provide users with interesting points of interest (POIs) according to their contextual information, mainly physical location. In this paper, we propose a context-aware system for mobile devices that incorporates some implicit contextual information that is scarcely used in the literature: the user's speed and his trajectory. This system has been specifically crafted to assist travelling users by providing them with smart and personalized POIs along their route taking into account their current location and driving speed.

1 Introduction

Recommender systems (RSs) are software tools that help people to find relevant items in large databases according to their interests, needs or tastes [3, 20, 25]. These systems filter information, removing irrelevant items and providing the best ones according to the user preferences. Traditional RSs build a user profile by analyzing the user's past actions. Most of these systems suppose that the user's necessities and tastes do not vary over time. However, this is not absolutely true as the user's necessities or preferences may change depending on different factors: user's mood, season, physical position, etc.

Context-aware recommender systems (CARSs) are a new trend in recommender systems [2, 4] that take into account these aspects. They aim to provide personalized recommendations according to both the users' profiles and their current contextual conditions. In other words, if you are in London, a CARS may recommend you "*Fifteen*" to have dinner according to your preferences, but if you are in Paris, obviously "*La Tour d'Argent*" would be a better choice than "*Fifteen*", unless you have a private plane to fly from Paris to London before dinner, even though the latter is closer to your preferences.

J. Casillas et al. (Eds.): Management Intelligent Systems, AISC 171, pp. 153–162.
springerlink.com   © Springer-Verlag Berlin Heidelberg 2012

In previous works we proposed REJA [21, 27], a web-based RS for restaurants in the province of Jaén, Spain. More recently, we also proposed in [22] a mobile CARS based on REJA that took advantage of the additional information that mobile devices provide, mainly their ubiquitous nature and the implicit knowledge of the user's physical location obtained by its GPS (Global Positioning System). However, additional contextual information such as direction, trajectory, speed, etc, was not considered, in spite of the fact that they are very interesting context features for tourism purposes.

In this contribution we propose a new context-aware filtering technique that enhances our previous work by adding two new contextual information elements into the recommendation process: the travel speed and trajectory of the user. This new proposal has been specifically designed to support on-the-move users using mobile devices and travelling aboard automobiles in interurban scenarios.

Our system recommends the most interesting POIs that will soon be found along their routes according to their preferences, location, speed and trajectory, and neglects those POIs that have been left behind by users according to their current trajectory and speed. This system would increase the input of personalized marketing and the loyalty of users to the enterprises (e.g., restaurants) related to the system.

The rest of the paper is organized as follows. Section 2 presents current research in mobile CARSs. In Section 3, our novel contextual filtering technique is presented. Section 4 describes a prototype that implements our technique and discusses its behavior. Section 5 concludes the paper.

2 Related Work

RSs have been widely studied in the literature, and there are several types that differ according to the techniques they use: demographic RSs [17], content-based RSs [19, 23], collaborative filtering RS [11], knowledge based RSs [7] and utility based RSs [13]. However, the RSs most commonly used in real world solutions are hybrid systems that combine two or more of the aforementioned techniques in order to overcome their own limitations and drawbacks, see for example [8, 21].

The recent increasing interest in mobile technologies and *Location Based Services* (LBSs) with the dramatic improvement of mobile devices (smartphones, tablets...) and cellular communications has led to new challenges and opportunities arising in the field of RSs. Since GPS-enabled mobile devices are sensitive to the physical conditions of the user (location, orientation, etc.), a new trend in RSs has arisen aiming to exploit this contextual information. CARSs exploit such information, suggesting items according to context-aware knowledge about the user: geographical location, mood, day of the week, season, etc.

Recent surveys by Adomavicius et al [4] and Ricci [26] provide an overview of these techniques. Several studies have proved that contextual information improves both the quality of the recommendations and user satisfaction when the system is employed [4, 5, 12].

One of the most typical applications of CARSs is to mobile tourism guides [26]. Traditional mobile tourism guides [1, 9, 24, 29] took advantage of contextual information, mainly the physical position of the user, but were not able to recommend POIs according to user preferences. Recently, different proposals have been suggested taking into account both contextual information and user preferences, following either content-based [18, 28, 30] or collaborative filtering [6, 14, 15, 16, 22] approaches. Context-aware information can be incorporated into the recommendation process following different approaches. Adomavicius et al. [4] described the following approaches: (i) *Contextual pre-filtering*, firstly, information is filtered using contextual information to select relevant items. These items are provided to a traditional RS as an input, which selects the best ones according to the user's preferences. (ii) *Contextual post-filtering*, a traditional RS processes all the available items and generates a list of interesting items according to the user's preferences. Contextual information is then used to adjust the list. (iii) *Contextual modeling*, in this approach, contextual information is used directly in the recommendation process.

3   Location and Trajectory Aware Recommendation Approach

Our proposal consists of using users' contextual information to provide more accurate recommendations regarding their current situation. Thus, we propose a basic architecture that combines two subsystems: a) a context-aware filter; and b) a traditional recommender technique. Both elements are independent, the proposed context-aware filtering can be used to add context-awareness to any traditional (i.e. non-context aware) RS.

This section describes our proposed context-aware filtering approach. This filter determines the user's physical position, speed and trajectory and uses them to guide the recommendation process.

3.1   Contextual Filtering

In previous works [22] we proposed a contextual prefiltering based on the user's position. This filtering considered an *area of interest* (AOI) defined by the current user's location P and a radius R_{outer}, see Figure 1a. R_{outer} was a user-defined parameter that expressed how far she was willing to travel to visit a POI. Outside this area, the items were not considered and discarded by the filter. Consequently, they were not recommended by the RS. Here, we define a new contextual filtering that enhances our previous solution by considering two new contextual variables: speed and trajectory.

Consider the following scenario: a user is moving along a motorway and wants to stop at a restaurant, see Figure 1a. If we employ an AOI centered on her current location to discard distant items, recommended restaurants may be located in the opposite direction to that in which she is heading. We overcome this issue by calculating an *adjusted AOI* that is centered on a new point P', located on the road

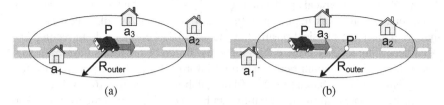

Fig. 1 Displacing the AOI according to the speed and the user-defined parameter R_{outer}

ahead of the user, according to the current travel direction, see Figure 1b. The faster the user moves, the more this new adjusted AOI will be displaced. Therefore, the system is able to anticipate interesting POIs that the user will pass by in the near future.

This contextual filtering has been designed for cascade hybridizing with any other traditional RS either as a contextual pre-filtering or as a contextual post-filtering [4].

3.2 Calculation of the Adjusted Area of Interest

In what follows, we describe in further detail how to calculate the adjusted AOI previously described. Let us suppose a user located at a geographical position P. In order to calculate the adjusted AOI, we must translate the point P to obtain a new point P', so that P' is located ahead of the user (according to her direction). Also, as the distance from P to P' depends on the user's speed, the performance of the following 2D geometric transformation [10] on P is required:

$$P' = P + T \tag{1}$$

Where $T(t_x, t_y)$ is the translation amount needed to move $P(x, y)$ to $P'(x', y')$. In order to determine T, we proceed as follows. First we calculate the current direction of the user. Several methods can be used for this purpose. For example, we could use the GPS receiver to sample a set of way-points obtained at regular intervals of time. A cubic parametric curve could then be calculated using these points as restrictions. The parametric tangent vector of this curve at the user's current location gives us the velocity of the user, which can be used to determinate her direction. See [10] for a thorough study on curves representation.

The previous technique, although powerful, is computationally expensive for a mobile device. Therefore, we propose a simpler technique to approximate the user's trajectory that only requires two points: the current and the previous locations, P and P_{prev}, provided by the GPS by sampling the user's location at regular intervals of time. We use them to determine a direction vector $\overrightarrow{v} = P - P_{prev}$. After normalizing $\overrightarrow{v}$, we obtain the unitary vector $\overrightarrow{u}$ that estimates the user's trajectory.

With this vector, we can determine T as:

$$T = \overrightarrow{u} \, d \, R_{outer} \tag{2}$$

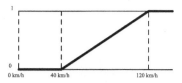

Fig. 2 d depends linearly on the user's speed

Where $\vec{u}$ and R_{outer} have already been defined, and $d \in [0,1]$ is a fuzzy parameter that depends on the user's speed. We define d as follows:

$$d = \begin{cases} 0 & if & speed < a \\ \frac{x-a}{b-a} & if\ speed \geq a\ and\ speed < b \\ 1 & if & speed \geq b \end{cases} \tag{3}$$

Where $speed$ is the average user's speed at the point P. Considering the speed limits in most countries, we propose $a = 40km/h$, $b = 120km/h$, see Figure 2. If the user's speed is lower than 40 km/h, we suppose she is in an urban environment, and therefore, the AOI is centered on her current location. As her speed increases, the AOI is progressively translated forward. Since 120 km/h is a common speed limit, we do not consider translating the circle further.

By applying equation (1), the AOI around the user is translated according to the user's current trajectory and speed a maximum distance of R_{outer}, as shown in Figure 3.

4   System Overview and Discussion

We have built an operational prototype that implements the ideas described in the previous Section. Our goal is to study the behavior of our proposal under real world contextual conditions. Our prototype is aimed at providing recommendations of restaurants in the province of Jaén, Spain.

The implemented system follows a classical client-server paradigm that comprises two elements: the recommender server and the mobile clients. The logic of the RS runs on a dedicated remote server. In our implementation, the filtering method described in Section 3 has been used as a contextual pre-filtering in cascade with a traditional hybrid RS, known as REJA, described in [21, 27]. Figure 4 portrays the system. The client, on the other hand, consists of a mobile application that is installed on a mobile device (e.g. a smartphone). This application is in charge of interacting with the user, processing contextual conditions and requesting recommendations from the server when needed.

Prior to using the system, users are required to install the mobile application on their devices. When the application is launched, the user has to provide his

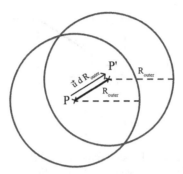

Fig. 3 The AOI is translated from P to P', according to the user's speed and trajectory

log-in data and the desired value of R_{outer}. Then, the built-in GPS receiver (currently included in most mobile phones) is used to determine the user's location and speed.

The computation of the user's travel direction and speed described in Section 3 is carried out entirely by the mobile device. This means that the server is unaware of the current location of the user, P, and only requires P' to compute the recommendations. This reduces the workload on the server and enhances scalability. Consequently, to request a recommendation list from the server, the client computes P' according to the user's location P, her speed and direction, and transmits it to the server. The user's unique ID and the value of R_{outer} are also transmitted. In response, the server generates an XML file containing the recommended items, their coordinates, recommendation values and other descriptive data. This file is downloaded via HTTP. Once the initial recommendation list has been downloaded, the mobile device continues to track the user's speed and trajectory, and issues new requests whenever the user moves a distance greater than a predefined threshold. In this way, recommendations are updated in real time as the user travels. Recommended items are portrayed on a map-based interface built on top of *Google Maps*.

Figure 5 illustrates the behavior of our system running on *iOS* (*iPhone*) under different contextual circumstances. These images are screenshots of the user interface of our mobile client application. The user location P is depicted on the map by a

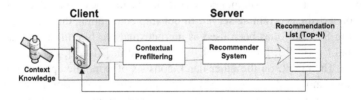

Fig. 4 In our prototype, contextual information is incorporated into the RS by means of a contextual pre-filter

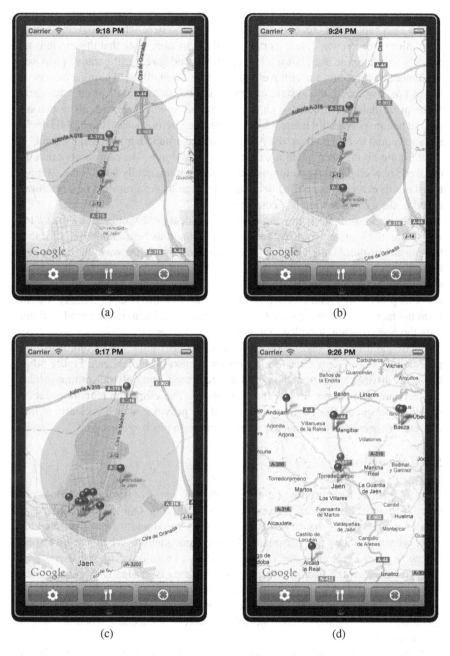

(a)

(b)

(c)

(d)

Fig. 5 Screenshots of the implemented prototype. The green pin marks the user's location, whereas the red pins are recommended restaurants. The blue circle represents the area within R_{outer} (2km). Traveling south along a motorway at a) 40 km/h, b) 80 km/h, and c) 120 km/h. d) Non context-aware version. The user location remains constant in all images.

green pin, whereas recommended restaurants are represented by red pins. The blue shaded area represents the AOI considered by the system when providing recommendations, see Section 3, and its radius is $R_{outer} = 2km$. Note that this circle is not actually shown to the user, but it has been introduced here for illustrative purposes.

Figure 5a represents a user travelling south by car at 40 km/h along a motorway. Given this low speed, the system assumes that the user is either in an urban environment or that she is likely to change directions soon. Therefore, all restaurants around her are potentially interesting. Under these circumstances, the mobile device makes P equal to P', and thus, the AOI is centered around her physical location.

In Figure 5b the user speed is assumed to be 80 km/h. We can observe that under these new contextual conditions, the system automatically adjusts the AOI and translates its center P' according to the travel direction. As a result, a new restaurant ahead of the user's location is recommended.

In Figure 5c, the user is moving south very quickly (120km/h) along the motorway. Here, the system avoids recommending restaurants behind the user's location by translating the AOI according to her direction in such a way that the user is located just on the border of the area, i.e. $|PP'| = R_{outer}$. We can see that several restaurants located in the city are recommended, but on the contrary, the restaurant that was previously recommended in Figures 5a and b has disappeared. This stems from the fact that now the system has more items available to recommend, and thus only the best ones are provided to the user.

Finally, and for the sake of completeness, we also show the recommendations provided by our system after turning off its context-awareness, see Figure 5d. In this case, the user's location and speed are neglected, and only her profile is taken into account when computing recommendations. Therefore, the recommendations provided are spread all over the geography of the province, which is clearly of little interest to a user travelling in a car and looking for a place to eat soon.

5   Conclusions

In this paper, we have proposed a novel method to improve recommendations provided by classical RSs by using a context-aware filtering. Our technique defines an area of interest (AOI) that depends on the user's speed and travel direction. Only POIs within this area are considered for recommendation. When the user speed is low (e.g., inside a city), the center of the area coincides with the user's position. However, as the user speed increases, the area is gradually displaced ahead in the estimated direction of the user. In this way, those POIs left behind by the user are excluded from the recommendation process. By contrast, more distant POIs that are in her estimated future trajectory are taken into account when making the recommendations.

A prototype has also been developed to show the possibilities of using CARSs on mobile devices. This type of system might facilitate the marketing processes of companies and increase their impact due to its flexibility, accuracy and personalization. Specifically, our proposal would contribute to the success and customer loyalty of tourism platforms.

With regard to future works, we plan to study smarter techniques to estimate the future direction of the user, e.g., an interpolation of the user location in the future based on cubic parametric curves. Different shapes and sizes of the AOI can also be considered. We also plan to use GIS (Geographic Information System) technologies to apply path techniques to vector maps of the road network in order to use estimations of the time needed to reach the POIs instead of the Euclidean distance.

Acknowledgements. This work has been partially supported by the *Ministerio de Ciencia e Innovación* of Spain, the *Junta de Andalucía* and FEDER funds through the research projects TIN2009-08286 and AGR-6581.

References

1. Abowd, G.D., Atkeson, C.G., Hong, J., Long, S., Kooper, R., Pinkerton, M.: Cyberguide: a mobile context-aware tour guide. Wirel. Netw. 3, 421–433 (1997)
2. Adomavicius, G., Sankaranarayanan, R., Sen, S., Tuzhilin, A.: Incorporating contextual information in recommender systems using a multidimensional approach. ACM Transactions on Information Systems 23(1), 103–145 (2005)
3. Adomavicius, G., Tuzhilin, A.: Toward the next generation of recommender systems: a survey of the state-of-the-art and possible extensions. IEEE Transactions on Knowledge and Data Engineering 17(6), 734–749 (2005)
4. Adomavicius, G., Tuzhilin, A.: Context-aware recommender systems. In: Ricci, F., Rokach, L., Shapira, B., Kantor, P.B. (eds.) Recommender Systems Handbook, pp. 217–253. Springer, US (2011)
5. Baltrunas, L., Ludwig, B., Peer, S., Ricci, F.: Context relevance assessment and exploitation in mobile recommender systems. Personal and Ubiquitous Computing, 1–20 (2011)
6. Biuk-Aghai, R., Fong, S., Si, Y.-W.: Design of a recommender system for mobile tourism multimedia selection. In: 2nd International Conference on Internet Multimedia Services Architecture and Applications, IMSAA 2008, pp. 1–6 (2008)
7. Burke, R.: Knowledge-based recommender systems. Encyclopedia of Library and Information Systems 69(32) (2000)
8. Burke, R.: Hybrid recommender systems: Survey and experiments. User Modeling and User-Adapted Interaction 12(4), 331–370 (2002)
9. Cheverst, K., Davies, N., Mitchell, K., Friday, A., Efstratiou, C.: Developing a context-aware electronic tourist guide: some issues and experiences. In: Proc. of the SIGCHI Conference on Human Factors in Computing Systems, CHI 2000, New York, USA, pp. 17–24 (2000)
10. Foley, J.D., van Dam, A., Feiner, S.K., Hughes, J.F.: Computer graphics: principles and practice, 2nd edn. Addison-Wesley Longman Publishing, Boston (1990)
11. Goldberg, D., Nichols, D., Oki, B.M., Terry, D.: Using collaborative filtering to weave an information tapestry. Communications of the ACM 35(12), 61–70 (1992)
12. Gorgoglione, M., Panniello, U., Tuzhilin, A.: The effect of context-aware recommendations on customer purchasing behavior and trust. In: Proc. of the Fifth ACM Conference on RS, RecSys 2011, pp. 85–92. ACM, New York (2011)
13. Guttman, R.H.: Merchant differentiation through integrative negotiation in agent-mediated electronic comerce. Master's thesis, School of Architecture and Planning, Program in Media Arts and Sciences, Massachusetts Institute of Technology (1998)

14. Horozov, T., Narasimhan, N., Vasudevan, V.: Using location for personalized poi recommendations in mobile environments. In: Proceedings of the International Symposium on Applications on Internet, pp. 124–129. IEEE CS, USA (2006)
15. Huang, H., Gartner, G.: Using context-aware collaborative filtering for poi recommendations in mobile guides. In: Advances in Location-Based Services. Lecture Notes in Geoinformation and Cartography, pp. 131–147. Springer, Heidelberg (2012)
16. Kenteris, M., Gavalas, D., Mpitziopoulos, A.: A mobile tourism recommender system. In: IEEE Symposium on Computers and Communications (ISCC), pp. 840–845 (2010)
17. Krulwich, B.: Lifestyle finder: intelligent user profiling using large-scale demographic data. AI Magazine 18(2), 37–45 (1997)
18. Kuo, M.-H., Chen, L.-C., Liang, C.-W.: Building and evaluating a location-based service recommendation system with a preference adjustment mechanism. Expert Systems with Applications 36(2, Part 2), 3543–3554 (2009)
19. Martínez, L., Pérez, L., Barranco, M.: A multi-granular linguistic content-based recommendation model. International Journal of Intelligent Systems (2007) (in press)
20. Martínez, L., Pérez, L., Barranco, M., Mata, F.: A multi-granular linguistic based-content recommender system model. In: 10th Int. Conf. on Fuzzy Theory and Technology (2005)
21. Martínez, L., Rodríguez, R.M., Espinilla, M.: Reja: A georeferenced hybrid recommender system for restaurants. In: IEEE/WIC/ACM International Joint Conferences on Web Intelligence and Intelligent Agent Technologies, WI-IAT 2009, vol. 3, pp. 187–190 (2009)
22. Noguera, J.M., Barranco, M.J., Segura, R.J., Martínez, L.: A mobile 3d-gis hybrid recommender system for tourism. Technical report, University of Jaén, Spain, TR-1-2012 (2012)
23. Pazzani, M., Muramatsu, J., Billsus, D.: Syskill webert: Identifying interesting web sites. In: Proceedings of the Thirteenth National Conference on Artificial Intelligence, AAAI 1996, vol. 1, pp. 54–61. AAAI Press (1996)
24. Poslad, S., Laamanen, H., Malaka, R., Nick, A., Buckle, P., Zipl, A.: Crumpet: creation of user-friendly mobile services personalised for tourism. In: 2nd Int. Conf. on 3G Mobile Communication Technologies, pp. 28–32 (2001)
25. Resnick, P., Varian, H.: Recommender systems. Association for Computing Machinery. Communications of the ACM 40(3), 56–58 (1997)
26. Ricci, F.: Mobile recommender systems. International Journal of Information Technology and Tourism 12(3), 205–231 (2011)
27. Rodríguez, R., Espinilla, M., Sánchez, P., Martínez, L.: Using linguistic incomplete preference relations to cold start recommendations. Internet Research 20, 296–315 (2010)
28. Saiph Savage, N., Baranski, M., Elva Chavez, N., Hllerer, T.: I'm feeling loco: A location based context aware recommendation system. In: Advances in Location-Based Services. Lec. Notes in Geoinformation & Cartography, pp. 37–54. Springer, Heidelberg (2012)
29. van Setten, M., Pokraev, S., Koolwaaij, J.: Context-aware recommendations in the mobile tourist application COMPASS. In: De Bra, P.M.E., Nejdl, W. (eds.) AH 2004. LNCS, vol. 3137, pp. 235–244. Springer, Heidelberg (2004)
30. Yang, W.-S., Cheng, H.-C., Dia, J.-B.: A location-aware recommender system for mobile shopping environments. Expert Systems with Applications 34(1), 437–445 (2008)

Software Applications and Prototypes

MarkiS: A Marketing Intelligent System Software Application for Causal Modeling

Francisco J. Marín[1], Jorge Casillas[2], and Francisco J. Martínez-López[3]

[1] Dept. Computer Science and Artificial Intelligence, Univesity of Granada, 18071
Granada, Spain
xmarin@correo.ugr.es
[2] Dept. Computer Science and Artificial Intelligence and Research Center on Information
and Communication Technology (CITIC-UGR), University of Granada, 18071 Granada,
Spain
casillas@decsai.ugr.es
[3] Dept. Management, University of Granada, 18071 Granada, Spain
fjmlopez@ugr.es

Abstract. MarkiS is a software platform that uses intelligent systems for knowledge extraction from large marketing databases. MarkiS allows the marketing expert to model, learn and analyze marketing models using two different genetic algorithms for learning fuzzy systems with multi-item variables. Using these intelligent systems the expert can obtain more valuable information about the model and improve his/her decisions. MarkiS functioning is divided into five steps: the creation of the model, the presentation of the dataset, the edition of the variables and items of the model, the learning of the fuzzy rule-based systems that explain the model, and the analysis of these fuzzy systems.

Keywords: causal modeling, fuzzy systems, genetic algorithms, marketing, software.

1 Introduction

Understanding customers is a core element in businesses' success. For this reason, it is necessary to take into account their opinions before designing marketing strategies and actions. A practical way to obtain these opinions is through questionnaires compounded by a collection of items used to gather variables of interest for practitioners and researchers. These questionnaires usually represent the articulation of a set of variables that need to be measured to solve certain decisional and/or research problems. Also, it is habitual, especially in academic surveys, which said variables are the result of a subjacent theoretical model that relations all of them.

When marketing practitioners have to process and analyze large amounts of information in a short time, the major problem that companies have to deal with is the lack of appropriate levels of knowledge to take the right decisions. Conventional analysis systems, based on statistical methods, are valid to explain the models but

J. Casillas et al. (Eds.): Management Intelligent Systems, AISC 171, pp. 165–174.
springerlink.com

are not accurate enough, taking into account the current competitive requirements. For this reason, marketing practitioners are increasingly demanding new accurate and illustrative techniques that facilitate their decision processes. With this aim, new artificial intelligence techniques for knowledge extraction (KDD) in large marketing databases have been developed.

The use of knowledge extraction algorithms allows (automatically) finding accurate information about the relational structure among the database's variables, taking a theoretical model of base as an a priori information. The outputs provided by said algorithms improve the decision of the experts, as they are able to provide detailed information on the relations about variables along all their value range. Since the first expert systems [1, 2], algorithms based on different branches of the artificial intelligence have been tested on this field: neural networks [3, 4], case-based reasoning [5], fuzzy systems [6, 7], etc.

Casillas and Martínez-López [6, 7] developed two machine learning methods based on two genetic algorithms for learning fuzzy systems with multi-item variables, obtaining interesting results. The former tries to get complete fuzzy systems with a good precision degree while maintaining a good legibility, while the latter tries to find individual interesting patterns with a good description level. These algorithms have several options to modify their behavior and need a previous step to preprocess the dataset and the initial fuzzy systems. Also, to visualize the output results, the user needs some external programs. This may be a problem for a user without knowledge about the algorithms. For this reason, to promote their use in the professional (also in the academic) field, it has been necessary to create a software that includes these algorithms and facilitates their use to the experts: MarkiS.

MarkiS is a cross-platform application that gathers all the steps in a friendly way, from the creation of the model to the learning and analysis of the fuzzy systems that explain it, in order to facilitate the adaptation to the learning algorithms presented in [6, 7]. This software allows the experts easily using these algorithms, without worrying about the complexities of the preprocessing step, since it is internally handled by the program. However, MarkiS maintains the configurability options offered by the algorithms, so a user can apply it in all its potential, also with the support of a complete user guide.

In Section 2, the functioning of MarkiS is detailed, while Section 3 presents some concluding remarks and further work.

2 MarkiS Description

MarkiS covers step by step the entire process of generation, learning and analysis of the marketing models, structured in five steps including: the creation of the model, the presentation of the data set, edition of the items and variables of the model, the learning of the fuzzy models and their subsequent analysis.

- **Model creation**: The first step is performed importing the questionnaire data in csv format (comma-separated values) that can be created with spreadsheet programs like Excel or LibreOffice Calc, creating the variable set and assigning the different items to the variable set. Once the variable set has been created, the user can add the variables to the model and set relations between them in the model tab (Section 2.1).

- **Dataset viewer**: After that, the expert can analyze the data set with several ordering and filtering options in the dataset viewer tab (Section 2.2).
- **Variable editing**: The expert can check and eventually modify some properties of the variables and the items before starting the learning process. The variable editing tab presents useful information and several options for the variables of the model and the items assigned to the variables (Section 2.3).
- **Learning**: The next step is the learning algorithm that will derive the fuzzy rule-based systems (FRBSs) that explain the relations between variables. MarkiS includes the algorithms and their options in the learning tab (Section 2.4).
- **Fuzzy System Analysis**: After the learning is performed, the analysis tab (Section 2.5) presents information about the FRBSs (both data base and rule base) and several quality measures. The expert can also modify the fuzzy sets of the data base and adding or removing rules from the fuzzy rule base in order to analyze the effects on the learned FRBSs. Plots of the predicted surfaces can be also analyzed for a better understanding of the fuzzy model.

2.1 Model Creation

The first step is the creation of the model. This step starts in MarkiS importing a csv file with the values of the itemset, being each column a single item of the questionnaire. The csv file can also include a header row with short names for the items. After that, the expert has to create the set of variables of the model and assign each item to its variable. However, MarkiS offers an option to generate the variable set and assign automatically the items to their respective variable, including this information in the header row of the csv file with the format "variable name#item name." Although is not necessary to use the entire set of items in the causal model, MarkiS will warn the expert if some items are unassigned.

After the set of variables has been created, the user can add these variables to the causal model and operate with them. The construction of the causal model is made in a graphical way, allowing the user adding or removing variables from the causal model, editing the variable properties (Section 2.3), grouping several first order variables into a second order variable (or ungrouping a second order variable into a set of first order variables), and creating or deleting links between variables. An example of a complete model is presented in Figure 1 (which represents the model proposed by Novak, Hoffman, and Yung for web consumer behavior analysis [8]):

Relations among variables in the causal model are analyzed with FRBSs. This FRBS list is updated whenever a new link is added or removed. After the expert has finished the building of the causal model, each fuzzy system of the FRBS list can be automatically obtained in the learning step.

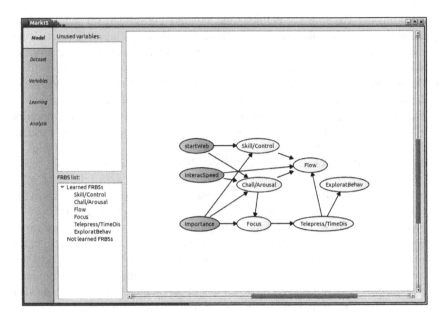

Fig. 1 Model screenshot

2.2 Dataset Viewer

The dataset viewer tab (see Figure 2) collects the values of the itemset. Items are presented as columns of a table, and are grouped by its parent variable. The table offers some ordering and filtering options for facilitating an a priori analysis of the data:

- **Ordering options**: Rows can be ordered in ascending or descending order for a single item or for a set of items. The set of items is sorted by order of selection: the rows are sorted using the first item selected and, for the equal values, the ordering is performed with the next selected item.
- **Filtering options**: Like in Excel, the dataset viewer offers some filters to present only interesting values to the expert. For the selected item the user can filter a range of values so these values will not appear in the table; or filtering all values except the values in the range.

2.3 Variable Editing

The variable editing tab presents information about the variables that form the causal model and their set of items. A screen capture is shown in Figure 3.

Variable properties are presented in the *variable details* group and collects information like the variable name, group and type. The variable group property identifies the variable as a simple, first order variable or a second order group of variables.

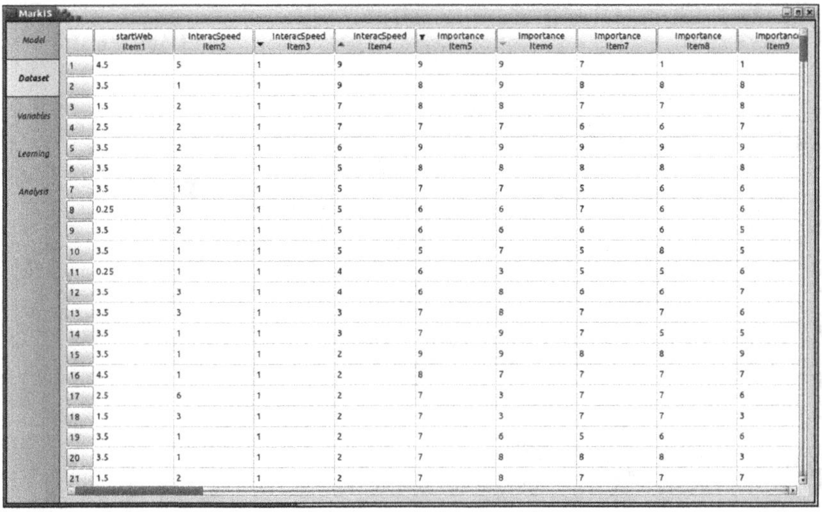

	startWeb Item1	InteracSpeed Item2	InteracSpeed Item3	InteracSpeed Item4	Importance Item5	Importance Item6	Importance Item7	Importance Item8	Importance Item9
1	4.5	5	1	9	9	9	7	1	1
2	3.5	1	1	9	8	9	8	8	8
3	1.5	2	1	7	8	8	7	7	8
4	2.5	2	1	7	7	7	6	6	7
5	3.5	2	1	6	9	9	9	9	9
6	3.5	2	1	5	8	8	8	8	8
7	3.5	1	1	5	7	7	5	6	6
8	0.25	3	1	5	6	6	7	6	6
9	3.5	2	1	5	6	6	6	6	5
10	3.5	1	1	5	5	7	5	8	5
11	0.25	1	1	4	6	3	5	5	6
12	3.5	3	1	4	6	8	6	6	7
13	3.5	3	1	3	7	8	7	7	6
14	3.5	1	1	3	7	9	7	5	5
15	3.5	1	1	2	9	9	8	8	9
16	4.5	1	1	2	8	7	7	7	7
17	2.5	6	1	2	7	3	7	7	6
18	1.5	3	1	2	7	3	7	7	3
19	3.5	1	1	2	7	6	5	6	6
20	3.5	1	1	2	7	8	8	8	3
21	1.5	2	1	2	7	8	7	7	7

Fig. 2 Dataset editing screenshot

Likewise, the type of the variable can be endogenous or exogenous whether it is or is not determined, respectively, by other/s variable/s of the model. The items of the variable are also listed on this group and can be activated or deactivated by the expert. Only the active items will be used in the learning and analysis processes; said items will be used to calculate the Cronbach's alpha of the multi-item variable, also presented in the variable details group. The last parameters of the variable details group are fuzzy parameters for the learning step: the number of fuzzy sets and the universe of discourse of the fuzzy variable. The universe of discourse is the [min, max] range where the fuzzy variable is applied, while the number of fuzzy sets is the number of labels that form the linguistic variable.

For each item of the variable, the expert can find in the *item details* group statistical information as minimum, maximum, mean and standard deviation values and the histogram of the item values. Also, the user can reverse the values of the items. See that Likert-type (and differential semantic) scales have a subjacent bipolar assessment approach (e.g. totally disagree-totally agree) so, to avoid a response inertia when respondent is filling the questionnaire, it is usual introducing some reversed statements/items. The invert option is included to set, when necessary, all the items' scales of a variable in the same direction.

2.4 Learning

Once the causal model has been created, the fuzzy systems that explain the relations between variables have to be learned. As we said in Section 1, MarkiS includes two genetic algorithms for fuzzy learning with multi-item variables.

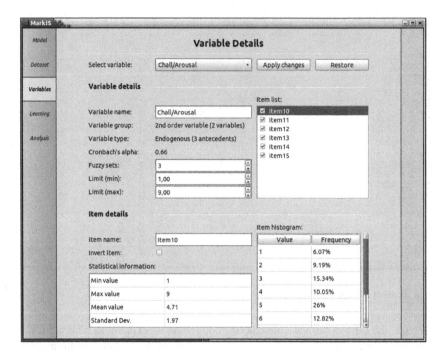

Fig. 3 Variable editing screenshot

The former is a Pittsburgh-style genetic algorithm for predictive induction [6]. This algorithm encodes the entire set of rules into each individual of the population, in order to obtain precise and compact sets of rules. Thus, the learned rules can predict the model reliably while the description of the model is sufficiently legible.

The latter is a Michigan-style genetic algorithm for descriptive induction [7]. This algorithm tries to obtain descriptive rules with a high quality individually, instead of a precise set of rules. In this case, the model description is the main objective instead of the prediction of the variables, explaining the complexities of the model using simple, understandable rules with great precision. Michigan-style genetic algorithms encode one rule into each individual of the population, generating the set of rules using the entire population.

Both algorithms have the same learning parameters. The set of parameters can be divided into two groups: *general parameters* and *genetic parameters*. In the first group the user can set options like the type of algorithm (Pittsburgh or Michigan), a seed value for the random algorithm, the division of the dataset in training examples and test examples, and the number of iterations of the genetic algorithm. Internal parameters of the genetic algorithm are collected in the second group. These parameters define the behavior of the genetic process. Among these options the user can find the crossover and mutation probabilities, the number of individuals of the population and two options for operating with fuzzy sets: fusion (fusion of two sets

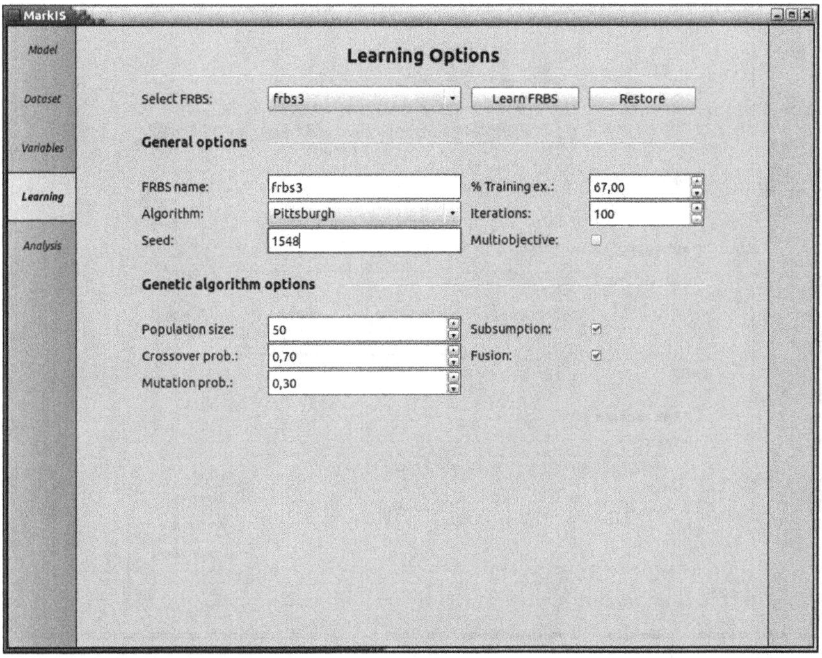

Fig. 4 Learning options

into one) and subsumption (addition of one fuzzy set in the fuzzy variable). When these options are activated, the initial number of fuzzy sets can vary.

MarkiS includes all these options in the learning tab (See Figure 4). Parameters can be set for a single fuzzy system or for the entire set of FRBSs of the model.

2.5 Fuzzy Rule-Based System Analysis

In the last step, the analysis tab (Figure 5) presents to the expert information about the performance and behavior of the learned FRBSs. The number of antecedents and rules and the mean square error obtained over the dataset is collected in the quality summary of the fuzzy system. Better fuzzy systems present lower values in mean square error and number of rules, specially when the Pittsburgh learning algorithm is used. Also, information about the quality of each rule of the fuzzy rule set is presented in the summary table of the rules. This information is very useful in general, but it is specially useful for the FRBSs learned with the Michigan algorithm. The quality of the rules is measured through three values: support, confidence and #Cases. Support (in the range $[0,1]$) measures the representation degree of the rule in the dataset, while confidence (also in the range $[0,1]$) measures the accuracy of the fuzzy rules. High values for support mean that the rule is more general and represents a higher portion of the sample; high values for confidence mean that the

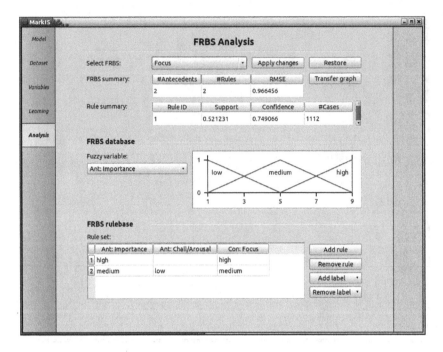

Fig. 5 Analysis screenshot

rule is reliable. The last value (#Cases) is the number of examples matched by the antecedent of the rule.

Information about the FRBS components is also collected, divided into two groups: *data base information* and *rule base information*. The former presents graphical information about the fuzzy sets that form each variable of the FRBS (antecedents and consequents): the number of fuzzy sets and their shapes. The latter collects the rules of the rule base, presenting the labels of the antecedents and consequents that form each rule (if there is more than one label for a variable, each label is separated with the disjunction "O").

A plot of the transference function of the fuzzy systems is also presented in this tab. It represents the predicted value of the consequent according to the values of the antecedents. Mark*i*S shows plots for fuzzy systems with one and two antecedents (2D and 3D plots, Figure 6), three antecedents (an array of density plots) and four antecedents (a matrix of density plots).

The expert can edit both the data base and the rule base of the FRBSs in order to tune them, adding the expert knowledge about the model. The data base can be edited adding new sets in a variable, removing sets or editing the shape of a existing set by changing it from a triangular shape to a trapezoidal shape or viceversa, or adjusting any of the vertices of the shape. On the other hand, the rule base can be edited adding new rules or removing an existing one. The labels of a rule can also be edited, adding a label or removing one. New values for mean square error, support,

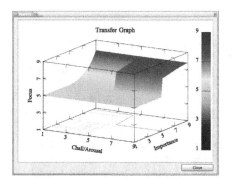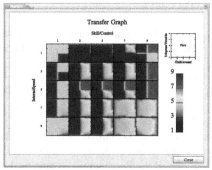

Fig. 6 Transfer graphs for FRBSs with two and four antecedents

confidence and #Cases will be calculated for the edited FRBS. The transference function will also vary with the changes made in the FRBS.

3 Conclusion and Further Work

A software for analyzing causal models in marketing has been presented. This program allows an expert with little knowledge about machine learning to work with several soft computing algorithms for fuzzy learning with multi-item variables. These methods are rarely used in the professional field, dominated by conventional statistical methods, since they are quite recent and need some previous knowledge about machine learning. However, their use in the professional field can lead to a better knowledge on the customers' opinions and consequently better market decisions taken by the marketing expert/manager.

MarkiS reduces the complexities of these algorithms and the needs of previous knowledge, since the expert starts with a dataset with the answers to the questionnaire, creating the whole model and the relations between variables easily. MarkiS handles internally the complexities of the preprocessing, while offering the expert all the configurability of the algorithms. Finally, the user can analyze the quality of the learned fuzzy systems and edit them to add his/her expert knowledge about the marketing problem uner analysis (articulated with a model) in a simple way.

Finally, it should be highlighted that, though MarkiS has been developed to be applied in a marketing context, it is equally valid in modeling contexts that use theoretical models to structure the database, and also with a similar measurement philosophy to the one used here as a base for the experimentation. This means that MarkiS could be also used by management and business modelers, in general. Likewise, it could be applied in non-business areas where these kind of theoretical models are also used, as psychology or sociology, for instance.

As further work, new algorithms recently developed will be included in the software as well as new features, among them, translation to other languages (MarkiS is currently available only in English).

References

1. Sisodia, R.S.: Marketing information and decision support systems for services. The Journal of Services Marketing 6(1), 51–64 (1992)
2. Wierenga, B., Van Bruggen, G.T.: Marketing management support systems: Principles, tools and implementation. Kluwer Academic Publishers (2000)
3. Fish, K.E., Johnson, J.D., Dorsey, R.E., Blodgett, J.G.: Using an artificial neural network trained with a genetic algorithm to model brand share. Journal of Business research 57(1), 79–85 (2004)
4. Boone, D.S., Roehm, M.: Retail segmentation using artificial neural networks. International Journal of Research in Marketing 19, 287–301 (2002)
5. Burke, R.R.: Reasoning with empirical marketing knowledge. International Journal of Research in Marketing 8, 75–90 (1991)
6. Casillas, J., Martínez-López, F.J.: Mining uncertain data with multiobjective genetic fuzzy systems to be applied in consumer behaviour modelling. Expert Systems with Applications 36, 1645–1659 (2009)
7. Martínez-López, F.J., Casillas, J.: Marketing Intelligent Systems for consumer behaviour modelling by a descriptive induction approach based on Genetic Fuzzy Systems. Industrial Marketing Management 38, 714–731 (2009)
8. Novak, T., Hoffman, D., Yung, Y.: Measuring the customer experience in online environments: A structural modelling approach. Marketing Science 19(1), 22–42 (2000)

Prosaico: An Intelligent System for the Management of a Sports Facility

E. Mosqueira-Rey[1], D. Prado-Gesto[2], A. Fernández-Leal[2], and V. Moret-Bonillo[2]

[1] University of A Coruña, Faculty of Computer Science, Campus de Elviña s/n, 15071, A Coruña, Spain
eduardo@udc.es
[2] University of A Coruña
{diego.prado,afleal,civmoret}@udc.es

Abstract. Prosaico is an intelligent system used for personalised sport advising in sports facilities. Personalised sport advising consists of advising the users of a sports facility when choosing the most appropriate type of exercise given their condition and their circumstances. To introduce personalised sport advising in publicly-funded facilities we have to face several problems, for instance, low instructor/user ratio, consistency in the recommendations between instructors and/or the attendance of users with health issues. Our solution is to improve and automate the management of information in a way which enhances the quality of the advising service. In order to do that it is necessary to develop an intelligent system for the management of the sports facility, focusing on the automation of the generation of work plans. This intelligent system is defined as a constraint-satisfaction problem.

1 Introduction

Prosaico is an intelligent system used for personalised sport advising in sports facilities. Personalised sport advising consists of advising users of the sports facility when choosing the most appropriate type of exercise given their physical condition and their circumstances.

We can summarise personalised sport advising in the following phases:

- **Information reception:** whereby the user is interviewed by an instructor on topics such as availability and preferences. This phase also includes an estimation of the user's physical capabilities and needs and the selection of an objective between several objectives, previously characterised by the technical director or administrator of the sports facility.
- **Processing of the information:** In this phase the instructor designs a work plan consisting of a set of exercise programmes which are repeated weekly depending on user availability.
- **Transmission:** Transmission to the user of the aforementioned work plan and the monitoring of its completion, along with the associated recommendations.

J. Casillas et al. (Eds.): Management Intelligent Systems, AISC 171, pp. 175–184.

1.1   Context of the Problem

First of all it is important to say that we are dealing with personalised sport advising in publicly-funded sports facilities, which have different requirements to those of private owned facilities. In our case the work described in this paper is a joint research project between the LIDIA laboratory of the University of A Coruña (Spain) and the company Sidecu Gestión S.A. (a part of the Supera group). Sidecu works in the management of sports facilities, mainly in the public sector, and at present manages around 20 facilities in Spain.

To introduce personalised sport advising in this kind of publicly-funded facilities we have to face several problems:

- **Low instructor/user ratio:** Affordable prices are a prerequisite, so the number of users is very high and it is not financially possible to have a high number of sport instructors per user.
- **Consistency in the recommendations:** It is conceivable that the same user could systematically receive different recommendations depending on the instructor, which could damage user trust. Moreover, because the criteria followed by the instructors are essentially intuitive, experience-based and are not recorded anywhere, it is impossible to compare their validity.
- **Health issues:** The increasing number of users with functional and/or structural bodily alterations should be taken into account. Naturally, these alterations in the user's physical conditions significantly enhance the above mentioned problems.

All these problems have an impact on the quality of customer service because less time is dedicated to interviewing users, making decisions, and designing and explaining work plans. We can identify two approaches to solving these problems: increasing the number of instructors or decreasing the number of users. The first option is economically unfeasible because publicly-funded facilities require affordable prices. The second option is more feasible through the use of personal trainers, i.e. a paid service on top of the standard fees of the sports facility. Even though the latter option is effective, it has the drawback of causing additional costs to the users and creating income-based *classes* of users, which is something a public facility should aim to avoid.

1.2   Our Proposed Solution

Our solution is to improve and automate the management of information in a way that enhances the quality of the advising service. Thus, additional costs are avoided and it becomes possible to correctly attend to the user's needs with regards to their health and physical well-being. In order to do that, it is necessary to develop an intelligent system for the management of the sports facility, focusing on the automation of the generation of work plans.

This automation of work plans not only implies better time attending to users, it also implies that the generation of work plans is now controlled by the administration of the sports facility and is less dependent on the instructor who is attending to the user. Thus a user can be attended to by different instructors but follow the same planning for reaching the desired objective (that normally needs to link several work plans as the user progresses). The functioning of the system is described in the next section.

2 Functioning of Prosaico

The functioning of the system is presented in the activity diagram of Figure 1. The first step is performed by the administrator of the sports facility indicating the type of exercises that can be done in the facility (fitness room, aquatic area, etc.). The administrator must also identify and describe the objectives that will be used as a desired goal for generating work plans for the users.

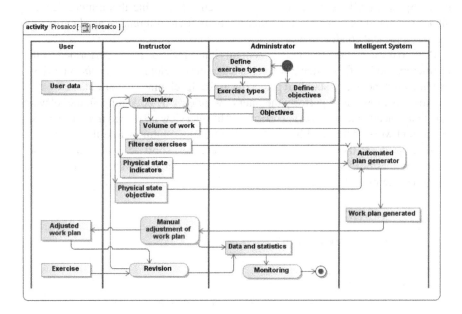

Fig. 1 Functioning of Prosaico in the process of personalised sport advising

The next step is developed by the instructor of the sports facility who interviews the user about his/her physical condition, preferences, desired objectives, etc. From this interview we can obtain the volume of work that the user is going to perform weekly at the facility, a list of exercises filtered by user preferences, several indicators about his/her current physical state and a description of the user's objective.

All the latter information is the input for the intelligent system [3] that, as output, obtains a personalised work plan for the current user. The instructor can then make minor adjustments to the work plan and presents it to the user. After a few weeks a revision interview is done in order to check the fulfilment of the work plan by the user. After this revision a new interview is needed to record the the user's progression and to plan and to plan the next phase towards the achievement of the initial objectives.

The administrator of the facility can, at any moment, consult data and obtain statistics in order to monitor the functioning of the system and the instructors. In the next subsections some of the most important parts of this process are described in more detail.

2.1 Interview

Performing an interview is a very time-consuming task. The first time the user enters the sports facility a questionnaire such as the Physical Activity Readiness Questionnaire [6] must be done to detect health problems. Following this a more detailed interview should be carried out in order to obtain user preferences, availability, objectives, physical condition, etc.

Prior to the use of the Prosaico system this interview was freely performed by the instructors and briefly written down. This led to the developing of *ad-hoc* interviews and hindered the use of standard protocols in the developing of work plans.

Prosaico allows us to perform the interview quickly by collecting all the relevant data that is needed to perform a work plan. The interview is divided in four sections: (1) current work, developed by the user (Figure 2) (2) future work, that the user is planning to do at the sports facility, (3) preferences, sport activities that the user prefers or does not like (Figure 3) and, (4) objectives the user wants to achieve with his or her exercise.

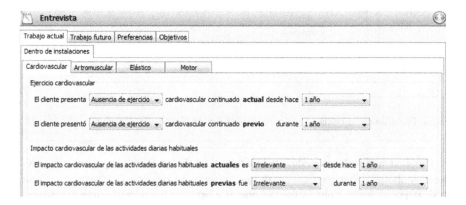

Fig. 2 Current physical exercise performed by the user in the sports facility (in Spanish)

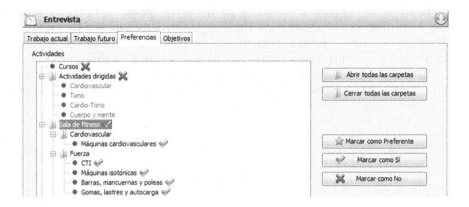

Fig. 3 Preferences of the user (in Spanish)

2.2 Physical State

To develop a work plan we must know the starting condition and the final condition we are trying to reach. The starting condition is represented by the user's current physical state. This state is characterised by ten physical indicators that can be estimated with the information that the users give in the interview about their past and current activities. These indicators are classified in four categories:

- **Cardiovascular:** Estimates the functional level of the cardiovascular system of the user (heart, veins and arteries), which represents the organism capacity of satisfying the increments in energy demands during physical effort.
- **Arthromuscular:** Estimates the functional level of the non-muscular components of the user's locomotive system, which indicate his or her ability to tolerate and assimilate the stimuli caused by mechanical loads. It is composed of three measures for the three areas of the body (upper, lower and torso).
- **Vascular:** Estimates the functional level of the user's musculature, which represents the muscles capacity for making good use of the resources provided through the cardiovascular system during medium/high physical effort. It is also composed of three indexes.
- **Elastic:** Estimates the user's control over his or her body and it is also divided in three indexes for the three parts of the body.

All these indexes are calculated using a method that was derived from experience of dealing with users in sports facilities for several years. More details of these calculations can be obtained in [4].

2.3 Objectives

Objectives represent the final condition we are trying to reach with the exercise done at the sports facility. They are defined by the administrator and are crucial for the definition of work plans.

First of all, an objective defines the target values that the user is trying to reach with the performance of a work plan. These target values are indicated for each of the ten physical condition indexes defined previously. Along with the target values the minimal values that the user must have for each physical condition index are also defined and that enables him or her to perform the planned exercises in a correct way. If one of the user indexes is below this minimal value the first step is to plan exercises to reach this minimum (Figure 4).

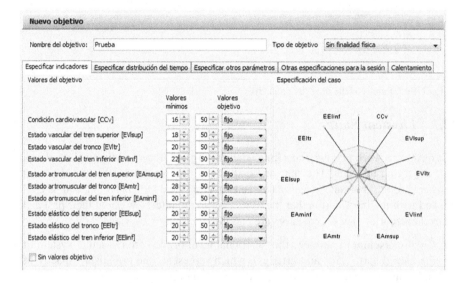

Fig. 4 Minimum and target values for each physical condition index in the definition of a given objective (in Spanish)

With these three values representing physical condition (minimum values, target values and current state) we will be able to calculate the amount of physical exercise that users should perform to improve the indexes in order to reach the desired goal.

But the objective not only defines the goal to reach, it also defines the time distribution of the workout session. This time distribution indicates how the session must be divided into cardiovascular, strength or stretching work. This distribution is the same for everyone who wishes to complete this objective but it is only approximate, since the final division will depend on the schedule that the user indicates in the previous interview with the instructor (Figure 5).

Finally, the main contribution of specifying the objectives in this way is the creation of work blocks. The administrator is responsible for describing the characteristics of the work blocks based on his/her experience. Depending on the user's physical profile (current values of the indicators), the administrator indicates the duration of the blocks, the intensity of the work, the number of repetitions, the incompatible groups of muscles (those muscles that should not be exercised in the

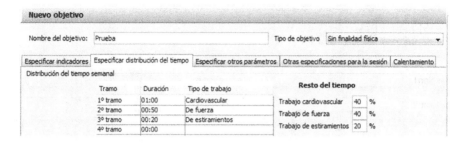

Fig. 5 Time distribution of workout sessions for cardiovascular, strength and stretching work in the definition of an objective (in Spanish)

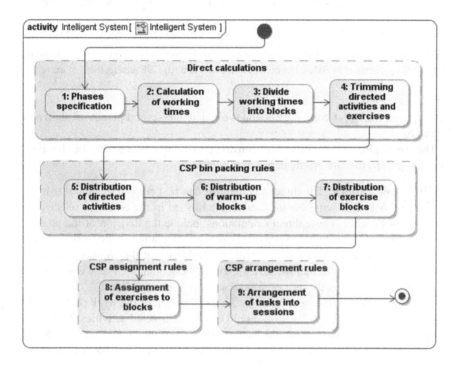

Fig. 6 Phases for the automated construction of a work plan

same session), etc. After that, the blocks will be created serving as inputs of the intelligent system.

2.4   Constraint-Satisfaction Problem

The automatic construction of a personalised work plan for the user is structured as a Constraint-Satisfaction Problem (CSP) [1]. In order to simplify the process of

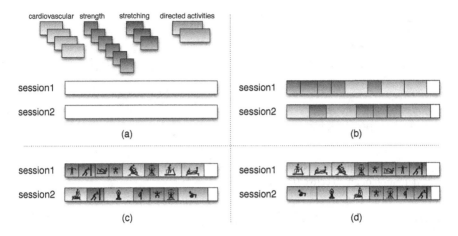

Fig. 7 The four phases of the automated construction of a work plan: (a) Direct calculations, (b) CSP bin packing, (c) CSP assignment of exercises, and (d) CSP arrangement of exercises

developing a work plan this construction is divided into several phases, as shown in Figure 6.

These nine phases in the process of construction of the work plan can be classified into four groups that are briefly depicted in Figure 7.

- **Direct calculations:** The direct calculations can be performed without using a CSP and only following the indications established in the objective selected as target for the user. These direct calculations include the division of the process of reaching the objective into phases (1) and the calculation of the working times needed to reach the first phase specified (2). After that the working times are divided in work blocks that we have to put in a work plan (3). These work blocks are filtered following the user preferences and availability (4).
- **CSP bin packing:** The next phase in the construction of the work plan is a bin packing problem [5] in which we should distribute directed activities (5), warm-up activities (6) and exercise blocks (7). This part of the system was developed using the Choco library [2]. The idea is to redistribute the work blocks following a series of constraints such as "the amount of cardiovascular work should be evenly distributed among the different days".
- **CSP assignment of exercises:** Once the exercise blocks are distributed in the different days, specific exercises should be assigned to these blocks taking into account constraints such as the type of exercise, the duration of the block, the preferences of the client, etc.
- **CSP arrangement of exercises:** The last part of the construction of a work plan consists of the arrangement of the different exercises taking into account arrangement rules such as "aquatic exercises should be placed at the end of the session".

The output of the intelligent system will be a work plan consisting in a list of activities to be performed by the client. These activities are characterised by a set of parameters, such as: session, execution time, level of intensity, type of exercise, etc.

3   Conclusions and Future Work

The Prosaico system described here aims to develop a system for a better and more individualised attention to users who go to sports facilities to try to recover, maintain or improve their health, physical condition or athletic performance. The central part of the system is an intelligent system that automatically generates work plans tailored to the user's characteristics, objectives and preferences. The use of the system involves the following improvements in the daily activities of a sports facility:

- Gathering more information from interviews in a reduced amount of time. This would help to establish more precise recommendations, which is necessary in order to correctly attend to the needs of the different types of users.
- Shortening the amount of time that the users need to wait to receive their work plan and giving more time to the instructors to attend to users' needs.
- Making theoretical predictions over time on the evolution of the physical conditions of users and being able to compare them with their actual evolution.
- Unifying and improving prescription criteria by establishing debate procedures within a facility (among the instructors), and between different facilities (among technical directors and administrators). This is done by analysing the acceptance of the automatic results obtained by the system.
- Contributing to the continuing training of the instructors and making the service more autonomous (it is currently too dependent on who the instructor is).
- Making the development and monitoring of work plans less dependent on instructors and more controlled by the administrator of the sports facility. The users can then be attended to by different instructors, something that it is important in an area with high employee turnover.
- Centralising all the data will make it possible to analyse it in order to find trends and establish user profiles. These profiles will be useful both for prescription purposes (differences in predicted results according to age, physical condition, motor capabilities, weekly work load, etc.) and for internal management (service use depending on age, gender, physical condition, etc).

As a limitation of the intelligent system we can mention that the quality of the results is highly dependent on the quality of the input. That means that, although the system was designed to make the creation of work plans less dependent on instructors, these instructors are still a vital part of the process, performing the interview, interpreting user's needs and objectives and translating them to the system and finally fine-tuning the generated, automated work plan.

As future work it is planned to extend the Prosaico system to include information about diseases and health disorders in the user's physical state. The system will mainly focus on those diseases that have a wide prevalence in the population and for which the prescription of physical exercise represents an advance in the health of the person who is affected by them (for example, cardiovascular problems, bone fractures and osteoporosis, backache, etc.). This new information will be used to develop a work plan that takes into account these conditions and allows the instructors to attend to these users adequately while complying with the limitations imposed by both the fitness and health state.

Therefore, the new project will cover the space between the intervention of physical activity professionals (dedicated to prevent the occurrence of these diseases in apparently healthy people) and health professionals (who treat individuals who already have the above problems). This will mean a real advance in the treatment of non-serious dysfunctions related to these diseases.

Acknowledgements. This research has been jointly funded by the Xunta of Galicia (Spain) and Sidecu Gestion S.A. under project 09SIN027E.

References

1. Dechter, R.: Constraint processing. Morgan Kaufmann (2003), ISBN 1-55860-890-7
2. Jussien, N., Rochart, G., Lorca, X.: The CHOCO constraint programming solver. In: CPAIOR 2008 Workshop on Open-Source Software for Integer and Contraint Programming (OSSICP 2008), Paris, France (June 2008)
3. Mosqueira-Rey, E., Fernández-Leal, A., Rodríguez-Poch, E., Moret-Bonillo, V., Alonso-Ríos, D., Soto-Martínez, E.: CommonKADS Modelling of an Intelligent System for Personalised Sports Advising. In: The 13th IASTED International Conference on Artificial Intelligence and Soft Computing (ASC 2009), Palma de Mallorca, Spain, September 7-9, pp. 174–181 (2009)
4. Mosqueira-Rey, E., Moret-Bonillo, V., Rodrguez-Poch, E., Alonso-Ríos, D., Soto-Martínez, E.: Physical Condition Evaluation within the Scope of an Intelligent System for Sports Advising. In: The Ninth IASTED International Conference on Artificial Intelligence and Applications (AIA 2009), Innsbruck, Austria, February 17-18, pp. 54–59 (2009)
5. Shaw, P.: A constraint for bin packing. In: Wallace, M. (ed.) CP 2004. LNCS, vol. 3258, pp. 648–662. Springer, Heidelberg (2004)
6. Thomas, S., Reading, J., Shephard, R.J.: Revision of the Physical Activity Readiness Questionnaire (PAR-Q) Canadian. Journal of Sport Sciences 17(4), 338–345 (1992)

Automatic Extraction of the Real Organizational Hierarchy Using JADE

Mihai Horia Zaharia[1], Alexandru Hodorogea[2], and Gabriela Maria Atanasiu[3]

"Gheorghe Asachi" Technical University
mike@cs.tuiasi.ro, alexhodorogea@yahoo.com,
gabriela.atanasiu@gmail.com

Abstract. Nowadays the globalization process is a reality. As a result, the creation of new transnational corporations has increased. Any corporation of this magnitude requires a very complex staff hierarchy to properly function. As a result, typical problems in bureaucracy management have developed that usually occurred in highly developed economies. These problems have been researched at a theoretical level for a very long time. Certain results have been obtained by using enterprise resource planning systems that ensure transparency in any point of the process where the document processing flows. There are moments when a tight control must be exercised as in reducing personnel schema. This must be done without decreasing the organizational efficiency. Anyway, it is very hard to control a structure when one also uses a part of it to generate the executive reports necessary. In this paper a solution that may help to increase the control over bureaucracy is presented. The solution is used to generate software based on agents in order to find and extract the real organization structure by doing automatic analysis of all document workflows. Probably the most important advantage of the solution is that it is not under the control of the audited structure and consequently the results will not be modified because the lower staff will always try to protect themselves.

Keywords: distributed computing, artificial intelligence, business process re-engineering.

1 Introduction

The IT involvement in the business process has existed from the early beginnings of the domain. The most used approaches were database related. But nowadays there are a lot of methodologies or tools used in making business process assessment and control from the artificial intelligence area as, for example, the use of the expert systems. The workflow management systems (WMF) have also been used for the last decades. It represents software tools used as decision support instruments into an organization Workflow. They can be implemented as tools for simulations of a proposed business level (Zaharia 1999 and Dumas 2005).

J. Casillas et al. (Eds.): Management Intelligent Systems, AISC 171, pp. 185–196.
springerlink.com

One of the components involved in this system is the associated document flow that must also be handled. This job enters in the area of Enterprise Content Management (ECM). This represents a formalized means of organizing and storing documents, as well as other information related content related to the organizations processes. The term is larger and refers to strategies, methods, and tools used throughout the lifecycle of the content. There are a lot of large software platforms in the market used to handle the document flow (Pelz-Sharpe 2012).

Therefore, we have business ideas followed by business rules. Then, a business flow is proposed and analyzed until it fits in the originating ideas. Finally the most unpleasant part is handled by the ECM systems. So, until now it seems that all is under control because the IT helps the management staff to handle large amounts of data. The ideal situation would be when the staff used in implementing the business control will perfectly fit over the initial model. Unfortunately, in real life the situation differs when the dimension of the organization is in excess of 1000 people. Because one may control the data flow but one cannot control the people involved in the system very efficiently. In fact, the control can be absolute but the result of this approach can be seen in the eastern European countries problems related to the total inability of having efficient governance (Vodrážka 2009), even some time after when the original conditions have dissipated. So, if there is a reasonable degree of freedom, the bureaucracy will quickly begin to change the original model by the use of replication and sometimes new proposed positions in the structure are declared as being essential to proper work.

As we have seen until now, there are no approaches used in doing a reverse engineering over an organizational schema because it is presumed as being well controlled, therefore the approach is futile. This paper proposes a method and a tool for automatically extracting the organizational structure and roles extraction in the case of very large organizations.

Because bureaucracy is used to control the rest of the structures, it is very hard to be controlled. The organizational culture of bureaucracy is strongly dependent on the current social and economic environment (Cameron and Quinn 2011). So, in an information society, the bureaucracy continues to survive almost unmodified but having a more comfortable life due to the use of software instruments.

The complexity of a modern organization (corporate or governmental level) makes it almost impossible to do an efficient global assessment. As a result, the analyses are usually focused on a small set of problems (Hood 2011).

Anyhow, it seems that during the last decades the extensive use of quality standards (like ISO 9001) has brought about some changes in designing the assessment methods. This process is more visible when concrete aspects (e.g. security auditing) are used where the engineers are partially or totally involved.

In the area of humanist approaches, such organizational culture analyses or fine tuning of the organizations to the desired goals - the classical approaches - are still used because they are still efficient (Lynch and Cruise 2005).

The notion of control usually involves both positive and negative actions and constraints. If in terms of positive actions (e.g. financial rewards or promotion) they are always in excess, the complexity of the structure makes it almost impossible to apply all the negative ones (e.g. demotion) continually, in order to maintain the balance and create a fine tuning.

Nowadays, most methods in changing the organizational culture imply the use of indirect approaches. We need to rethink the ways in doing that by developing complementary direct approaches in order to improve their efficiency. Even when legislative and congress control is used, the bureaucracy proves that it can anticipate the actions and sometimes can even modify those reactions partially (McGrath 2011). To achieve a correct control we need instruments that may help us handle the large amount of data about the organization itself, not only about how it works. The latter one is controlled by the use of professional ERP solutions and the former one is assumed to be known. If we take into account Weber's attributes of bureaucracy (for example the predisposition of staff to grow above necessary levels) (Weber 1911) it is clear that it will always find ways to fight with negative constraints. The problem of controlling the organizational redundancy can be also be tricky (Streeter 1992). As a result, we need to have some tools to analyze the real structure of the organization, not the presumed one.

2 The Method

In what follows we propose a method for automatic extraction of the real organizational structure in terms of hierarchical and functional positions and real job description and also its real place in the hierarchy.

At first sight this approach may seem futile. Yet, we consider it to be necessary because the full control of a large organization is almost impossible. One reason is related to the individual rights granted by the constitutions but, this is not a dominant motive. The real reason is related to the organization need to be very flexible and dynamic in order to survive. To do that, information technologies (IT) are used. Because of the globalization process, the organization will have a continuously increasing speed of development. In order to achieve that flexibility, larger levels of autonomy are granted to lower staff levels of the organization and only a partial control is assured on the middle levels.

The proposed method is independent of the organization type. It can be applied to retake control (if needed) over oversized bureaucracy. The motives for doing that may vary from one organization to another. Nowadays, the most obvious reason will be related to the global crisis that drives everyone into a serious cost control problem. Of course, the method cannot be applied without a careful analysis. It is not recommended to be used by itself. Best results appear when the stakeholder identification and also the load per clerk quantum analysis will be made. Based on these assumptions the method can provide a good solution to decrease the staff dimension without decreasing its organizational and economic efficiency too much.

The main idea is very simple. Most modern organizations handle the document workflow using ERP (Enterprise Resource Planning) such as MySAP ERP produced by SAP AG (see at: www.sap.com). Of course, there are other ERP solution providers like Oracle, BAAM, RAMCO People soft and JD Eduards, but nowadays the leader of the market remains the SAP solution which is also the largest business software in Europe.

The solution is based on the assumptions that the targeted organization has a well defined pattern of the documents and that metadata are used. The last assumption is not mandatory because the existence of a pattern will avoid most of the data extraction problems related to natural language, so a metadata will be quickly and correctly generated in most cases (Lee et al 1999).

2.1 Metadata Generation

Because the analyzed documents have a distinct pattern and also use a well defined set of terms, the extraction of the needed ones is almost trivial because not all the content of the document must be analyzed. The company must provide the set of used patterns and also a large dictionary with typical terms and much other information related to the personnel (such as unique ID, full name, rank, attributions and location). After the needed metadata is extracted from the document (in case that the document does not have one already) or just verified for all required information, then we can go on to the next step, which is the organizational hierarchy extraction.

2.2 Organization Hierarchy Extraction

The metadata will provide at least the following types of information:

- the origin of the document (the issuing);
- the destination of the document (who will receive it in order to approve, check and pass, or just solve);
- the aim of the document (it is a request for goods, a demand for approval or a result of a request of some previous document).

To do the extraction of the real organizational hierarchy, first we will construct the staff relation/interaction graph using the previously mentioned information.

Each node of the graph will contain at least a name and a list of operations executed by the employee. The arcs are directly obtained by analyzing the pair (point of origin, destination(s)). Let us analyze the creation of the operations list.

From the organization dictionary, we already have a distinct zone with the possible operations (the set where to search) and one or more keywords that can be found in the metadata. The metadata of the documents will be analyzed piece by piece and the graph will be created.

After finishing the process, we will have a graph that will contain at least the following information about each employee:

- the relations of each employee with others in the system and the frequency of using each channel with the specific use;
- each list of used keywords about the current solved type of jobs;
- time stamp for all the work.

In the next phase, the graph analysis may begin. We have two directions. One is to extract the real organizational tree. In the dictionary we will have equivalence between a job description and a list of keywords or attributes. As a result, using some form of logic, we will simply identify the dominant executed operations using a histogram (or fuzzy logic) and replace the executed operations list with the name of the job.

The other direction is to analyze the load of the clerk and also the real redundancies that exist at various organizational levels. The program will not make any assessment, it will just provide the analysis results. From an IT point of view it is easy to make supplementary analysis and generate suggestions to transform the application into a fully decision assisted instrument. Yet, this is a very sensible problem from a human point of view. The higher staff may consider it as a menace even if it is not so. Thus, full decision power will remain with the staff. It is simple to evaluate the load of the employee because we can easily select only the monthly processed documents and see their type (so the complexity of the job) and the number for each type.

3 First Implementation of the Concept

The decisional factors usually need an overview of the problem in order to properly manage the complex situations that may appear. But from time to time they also need as much detail as possible. As a result any decision support tool must provide full information in the background leaving to the staff the possibility of selecting the desired level of detail when consulted.

When we begin to design an application we must select the used computing model. In this case the dimension of the problem is large. Also the data is distributed globally in most cases. As a result, the distributed approach will be an implicit selection. The aim is to obtain the full organizational real graph. The graph algorithms are usually executed either sequentially, or on parallel machines. Of course, there are a lot of distributed algorithms for graph processing. Unfortunately, they will involve higher amounts of communication between nodes. As a result the efficiency will be low when the communication channels do not have high performance in terms of speed and quality of service (QoS). But the input data and control are inherently distributed. One solution may be to use the corporate distributed database or ERP system to hide all of the problems and to develop a centralized application. As we have already mentioned, this is not feasible. So, a distributed application is needed. Yet, we have the constraint of graph processing. A hybrid approach is recommended. So, the application will be distributed when it gathers data. The local processing will be done and then the results will be fed into a central point where the final image of the graph and further processing will be done. A modular structure for the application is shown in Fig. 1.

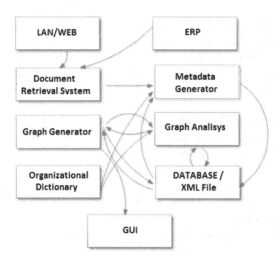

Fig. 1 Application structure

Here one can see that there is a module used in document retrieval. There are various ways of handling internal documents in corporations. The application will need to have abilities to make a crawling both in the LAN as in WAN and also to have the possibility to directly interrogate the ERP used by the organization. Regarding the database module we can mention that the use of a database management system must be carefully analyzed and will strongly depend on the number of supplementary mining techniques that may or may be not applied over the raw data. Another solution may be to use an XML file to store the data.

To elect the software technology used in implementation of the solution we have two major options: to design from scratch or to use some existing dedicated frameworks. Of course, even in the case of designing from scratch, the new technologies used in distributed computing also involve some frameworks like J2EE or .NET.

Yet, the problem specification leads us to the idea that in the life cycle we will need to add complex data mining techniques. As a result the best idea will be to use a mobile agent framework (that can be upgraded easily to mobile intelligent agents). We chose JADE (Java Agent DEvelopment framework) because it is an environment independent framework used for the development of applications. It is based on the peer-to-peer intelligent autonomous agent architecture. JADE enables the creation of multi-agent systems, including agents running on wireless networks and mobile devices (Bellifemine et al 2007). JADE is a real framework (not used only just for simulation), so real applications can be developed by using it. It offers many tools with the intention of simplifying multi-agent systems (MAS) and distributed system development allowing for most of the development process to move on to the agent level, rather than the framework level. Using agent frameworks usually raises some security related problems. In this particular case, only the final information will be submitted outside the local node to the center. This information is smaller and can be easily secured using a thin crypto-graphic layer if the corporation will consider it necessary.

3.1 Agents

Automating the task at hand is achieved by extending the functionalities of the agent distributed computing framework and dedicate the generic *ProcessingAgent* entity to two distinct roles:

- Analysis agent: it constructs the final graph representation and extracts relevant information from its structure.
- Information retrieval agent: it explores a given search space, retrieves document metadata information (existing or based on content from the documents), creates a set of indexes in order to represent the information and passes that to the *AnalysisAgent* for processing.

The analysis agent begins by first receiving the exploration domain, in the form of a list of hyper-text protocol (HTTP) resources which it, then, splits into individual sub-lists of resources that are passed as arguments to the retrieval agents upon creation. This ensures benefits both in terms of speed – parallel processing – and mobility.

After receiving the exploration domain, the analysis agent issues requests to create information retrieval agents and distributes target resources among them.

Once the retrieval agents have finished their jobs and have reported back to the analysis agent, the resulting indexes are combined into a global index which is used to create the graph used for analysis.

At this point, the analysis agent can either open a graphical user interface (GUI) for the end-user to view the graph, its associated information and run specific algorithms on it or do some sort of post-processing activity, such as save the graph for later usage or pass the information to another program or agent. The state machine for *AnalysisAgent* is presented in Fig. 2.

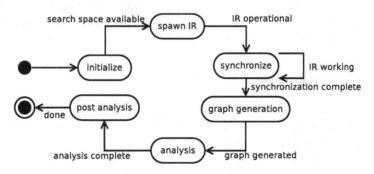

Fig. 2 *AnalysisAgent* State Machine Diagram

Once the retrieval agent has been created and the scope of its exploration has been established, the framework will distribute it to one of the nodes from the system where it will wait for its turn to begin processing.

Once the processing signal has been fired, the agent will start its execution unit which contains a sequential web crawler. The web crawler has a basic architecture (Manning et al 2008) consisting of:

- a fetching component: it retrieves a resource of a given universal resource locator (URL);
- a domain name service (DNS) client: it finds the network address of a given domain name;
- a parsing component: it extracts meaningful information from the fetched data such as links to other resources, content, various meta-content elements;
- an exploration queue: also known as the "url frontier", this component holds in a first in – first out (FIFO) manner the resources that the crawler will explore;
- an exploration log: it helps to eliminate exploring the same resource multiple times.

After parsing each resource, the information is passed on to an indexing component which stores it in an easy-to-access format. The *InformationRetrievalAgent* state machine diagram is presented in Fig. 3.

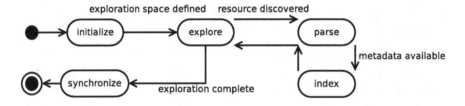

Fig. 3 *InformationRetrievalAgent* State Machine Diagram

The fetching module also implements a variant of the robot exclusion protocol (see at www.robotstxt.org), in case some portions of the domain should be skipped from exploration for various reasons.

3.2 Agent Collaboration

Since the task does not require special optimization techniques that would otherwise impose complex collaboration schemes in terms of agent communication, a simple master-slave scheme was selected.

In this particular case, the analysis agent takes up the role of the master of a group of information retrieval agents and, thus, controls their spawning rate and processing load.

After each retrieval, the agent finishes its job and it initiates a synchronization step inside which it sends the resulting data to its master analysis agent which adds it to its internal data structure. After all of the retrieval agents have finalized their synchronization steps, they finish execution and the analysis agent begins analyzing and graph construction.

Actual synchronization is embedded inside of a synchronization behavior, one for the master and one for the slave. Other synchronization mechanisms that might need implementation in the future can be embedded in the same fashion.

3.3 Graph Construction

The construction of the final version of the graph begins with analyzing the documents from the exploration domain. The preferred means of analysis at this level is through already existent metadata XML files which, in this case, are assumed to have a resource name of *target_resource_name.xml*.

If such a file does not exist, an attempt to generate one will be made, utilizing the dictionary provided by the organization prior to the analysis and leveraging the standard format of the analyzed documents.

For the time being and for the purposes of this material, the metadata files are assumed to be always there.

Utilizing this kind of information, the agent constructs an index relating the documents which are described by the same *task_id* with the same *receiver_id* and the same *author_id*. This index is later utilized in constructing the organizational graph.

In working with graphs, the application utilizes the "Java Universal Network/Graph Framework" (JUNG) which simplifies the process of working with graphs of all kinds by providing easy-to-use methods for graph creation, visualization, analysis and so forth (see at: jung.sourceforge.net).

In the organizational graph, a vertex will represent a person from within the organization and will appear, when visualized, as a circle decorated with his or her name and an edge will appear as a line with an arrow indicating the direction of the relationship, decorated by the names of all the tasks that move from the starting vertex of the edge to the target one. When right clicking the mouse on any component of the graph, additional information can be seen relating to that component as it is presented in Fig. 4.

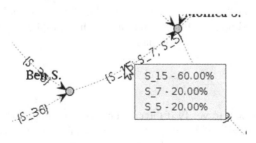

Fig. 4 Statistics and vertex information

Right-clicking on a vertex will reveal details about the person's employee iden-tification (ID) as it appears in the organizational dictionary, his/her location and department. Right-clicking on an edge will reveal statistical information about the tasks relative to the analyzed documents. Clicking on one of the statistics will re-veal more detailed information such as the actual number of documents in which the selected tasks appears.

The mechanism behind displaying such information is pluggable and readily available in the JUNG framework.

If needed, based upon the generated graph, the agent can extract and display its minimum spanning tree assuming that the root of the tree has been determined prior to this, for example, based upon the organization's dictionary.

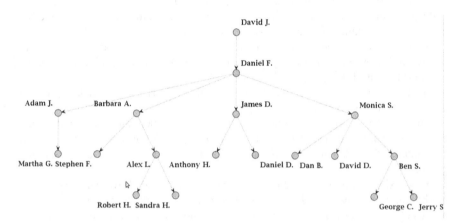

Fig. 5 Organization Hierarchy

The tree is also a directed graph, however, unlike what happens in the real life in the organizational graph, the meaning of the direction of the edges is that of "execution of influence" rather than "executes tasks for". One example of final hierarchy extraction is presented in Fig. 5

The graph can also be exported to a file for later analysis. This is done partly by utilizing JUNG's own mechanisms and partly by specifying which portions of the vertex and edge data structures will be exported and how, manually. The resulting file will be in the GraphML format (see at: graphml.graphdrawing.org) which is XML based and thus, per se, easy to handle.

4 Conclusions

The main problem was to verify if the concept could be implemented as it was in-itially designed. The use of JADE gives us a reasonable development time. In this solution a dedicated graph library has been used. This will be replaced in the fu-ture by a custom-designed library due to the need of performance boosts in case of handling large amounts of documents. Using an agent framework gives us the possibility to utilize various artificial intelligence related approaches to continually improve the performance of the systems (e.g. load balancing techniques) needed, due to the problem dimension.

From business people points of view, the instrument was designed to enter into the category of computer assisted decision support class of applications. The business world has already used this type of instruments especially the ones based on data mining or information retrieval. The presented ones were designed to help (in terms of speeding the process and also of increasing the precision of analysis) higher level staff of an organization in the process of regaining control if needed over the bureaucracy in case that some problems were detected. The IT involvement is required because of the dimension of data flows that must be processed and also because it requires a part of bureaucracy to create necessary executive reports. As a result, there is no chance to have a precise tune in this situation. The use of software will remove the need to use parts of organization bureaucracy in the process, so its efficiency will be significantly increased. This instrument can also be used to better choose the people who must be dispensed from an organization without perturbing its efficiency too much when budget cuts are inevitable. Another advantage of the method and tool is total confidentiality of the process.

Acknowledgements. This work was partially funded from the project E-Fast GA nr. 212109/2008.

References

Bellifemine, F.L., Caire, G., Greenwood, D.: Developing Multi-Agent Systems with JADE. John Wiley & Sons, Inc. (2007)

Cameron, K.S., Quinn, R.E.: Diagnosing and Changing Organizational Culture: Based on the Competing Values Framework, 3rd edn. John Wiley & Sons, Inc., Jossey-Bass (2011)

Dumas, M., Aalst, W., Hofstede, A.: Process-Aware Information Systems: Bridging People and Software through Process Technology. Wiley & Sons (2005)

Hood, C.: The Blame Game: Spin, Bureaucracy, and Self-Preservation in Government. Princeton University Press, Oxford (2011)

Lee, T., Chams, M., Nado, R., Madnick, S., Siegel, M.: Information integration with attribution support for corporate profiles. ACM, USA (1999) 1-58113-146-1/99/0010

Lynch, T.D., Cruise, P.L.: Handbook of Organization Theory and Management: The Philosophical Approach, 2nd edn. CRC, Taylor & Francis LLC, Chico, USA (2005)

Manning, C.D., Raghavan, P., Schtze, H.: Introduction to Information Retrieval. Cambridge University Press, New York (2008)

McGrath, R.J.: Legislatures, Courts, and Statutory Control of the Bureaucracy across the U.S. States. In: The 12th Annual State Politics and Policy Conference in Houston, Texas, (February 16 - 18, 2011),
http://2012sppconference.blogs.rice.edu/files/2012/02/McGrath-2012-SPPC.pdf (accessed February 10, 2012)

Pelz-Sharpe, A.: Document Management (ECM) Market Analysis (2012),
http://www.realstorygroup.com/Research/Advisory/Download/116 (accessed in 2012)

Streeter, L.C.: Redundancy in Organizational Systems. Social Service Review 66(1), 97–111 (1992) ISSN 0037-7961

Vodrážka, M.: Social Capital in the Czech Republic and Public Policy Implications. Central European Journal of Public Policy 3(1), 66–88 (2009), http://www.cejpp.eu/upload/journals/1/issues/cejpp_09_03_01.pdf (accessed in 2012)

Weber, M.: Bureaucracy. In: Gerth, H.H., Mills, C.W. (eds.) From Max Weber, pp. 196–244. Routledge, London (1911)

Zaharia, M.H.: Theory and Practice. In: Distributed Industrial Workflows Simulations, research report on December IAT Gelsenkirchen Germany (1999)

Integration of a Proximity Detection Prototype into a VO Developed with PANGEA

Carolina Zato, Alejandro Sánchez, Gabriel Villarrubia, Javier Bajo, and Sara Rodríguez

Departamento Informática y Automática, Universidad de Salamanca, Salamanca, Spain
{carol_zato,asanchezyu,gvg,jbajope,srg}@usal.es

Abstract. This article presents a proximity detection prototype that uses ZigBee technology developed using the agent's platform PANGEA (Platform for Automatic coNstruction of orGanizations of intElligent Agents). PANGEA is an agent platform to develop open multiagent systems, specifically those including organizational aspects such as virtual agent organizations. The platform allows the complete management of organizations and offers tools to the end user. Due to the specific characteristics of this prototype, PANGEA is the perfect candidate to develop the prototype that will be included in the future in an integral system primarily oriented to facilitate the integration of people with disabilities into the workplace.

Keywords: Zigbee, proximity detection, virtual organizations, agent platform, multiagent system.

1 Introduction

Within the field of technologies specifically developed to facilitate the lives of people with disabilities, there have been many recent advances that have notably improved their ability to perform daily and work-related tasks, regardless of the type and severity of the disability. This article presents a proximity detection prototype, specifically developed for a work environment, which can facilitate tasks such as activating and personalizing the work environment; these apparently simple tasks are in reality extremely complicated for some people with disabilities.

This prototype has been developed using PANGEA. There are many different platforms available for creating multiagent systems that facilitate the work with agents [10][11][12][13]; however our aim is to have a tool that allows to create an increasingly open and dynamic multiagent system (MAS). This involves adding new capabilities such as adaption, reorganization, learning, coordination, etc. Virtual Organizations (VOs) of agents [8] [9] emerged in response to this idea; they include a set of agents with roles and norms that determine their behavior, and represent a place where these new capabilities will assume a critical role. Possible organizational topologies and aspects such as communication and

J. Casillas et al. (Eds.): Management Intelligent Systems, AISC 171, pp. 197–204.
springerlink.com

coordination mechanisms determine in large part the flexibility, openness and dynamic nature that a multiagent system can offer. All these features are taken into account with the new platform PANGEA, for this reason the prototype has been integrated into a MAS developed with the mentioned platform.

The rest of the paper is structured as follows: The next section introduces the PANGEA platform used in the development of this prototype. Section 3 presents the most important characteristics of the prototype. Section 4 explains the case study and finally, in section 5 some conclusions are presented.

2 PANGEA Overview

PANGEA is a service oriented platform that allows the MAS to create with it to take the maximum advantage of the distribution of resources. To this end, all services are implemented as Web Services. This makes it possible for the platform to include both a service provider agent and a consumer agent, thus emulating a client-server architecture. The provider agent knows how to contact the web service; once the client agent's request has been received, the provider agent extracts the required parameters and establishes the contact. Once received, the results are sent to the client agent.

This leads us to one of the most important features that characterize the platform; PANGEA allows to create and integrate any kind of agent, does not matter the language, the functionality or other characteristics. The own agents of the platform are implemented with Java, nevertheless the agents of the detection prototype are implemented in .NET and nesC.

Using PANGEA, the platform will automatically launch the following agents:

- OrganizationManager: the agent is responsible for the actual management of organizations and suborganizations. It is responsible for verifying the entry and exit of agents, and for assigning roles. To carry out these tasks, it works with the OrganizationAgent, which is a specialized version of this agent.
- InformationAgent: the agent is responsible for accessing the database containing all pertinent system information.
- ServiceAgent: the agent is responsible for recording and controlling the operation of services offered by the agents.
- NormAgent: the agent that ensures compliance with all the refined norms in the organization.
- CommunicationAgent: the agent is responsible for controlling communication among agents, and for recording the interaction between agents and organizations.
- Sniffer: manages the message history and filters information by controlling communication initiated by queries.

Initially, the platform creates a general VO but thanks to the organizational concepts related to this model, it is possible to create suborganizations. This is the reason why the OrganizationAgents are included, its function is to help the

OrganizationManager to support all the suborganizations. Each suborganization or work unit is automatically provided with an OrganizationAgent by the platform during the creation of the suborganization. This OrganizationAgent is similar to the OrganizationManager, but is only responsible for controlling the suborganizationn, and can communicate with the OrganizationManager if needed. If another suborganization is created hierarchically within the previous suborganization, it will include a separate OrganizationAgent that communicates with the OrganizationAgent from the parent organization. These agents are distributed hierarchically in order to free the OrganizationManager of tasks. This allows each OrganizationAgent to be responsible for a suborganization although, to a certain extent, the OrganizationManager can always access information from all of the organizations. Each agent belongs to one suborganization and can only communicate with the OrganizationAgent from its own organization; this makes it possible to include large suborganizational structures without overloading the AgentManager. All of the OrganizationAgents from the same level can communicate with each other, unless a specific rule is created to prevent this.

Following this modeling way, the detection proximity prototype is included as a suborganization called DetectionProximityOrganization with his own OrganizacionAgent. In the next section, all the architecture will be explained.

3   Integration of the Detection Proximity Prototype

ZigBee sensors are used to deploy the detection prototype. ZigBee is a low cost, low power consumption, two-way wireless communication standard that was developed by the ZigBee Alliance [5]. It is based on the IEEE 802.15.4 protocol [2], and operates on the ISM (Industrial, Scientific and Medical) band at 868/915MHz and a 2.4GHz spectrum. Due to this frequency of operation among devices, it is possible to transfer materials used in residential or office buildings while only minimally affecting system performance [1]. Although this system can operate at the same frequency as Wi-Fi devices, the possibility that it will be affected by their presence is practically null, even in very noise environments (electromagnetic interference). ZigBee is designed to be embedded in consumer electronics, home and building automation, industrial controls, PC peripherals, medical sensor applications, toys and games, and is intended for home, building and industrial automation purposes, addressing the needs of monitoring, control and sensory network applications [5]. ZigBee allows star, tree or mesh topologies. Devices can be configured to act as network coordinator (control all devices), router/repeater (send/receive/resend data to/from coordinator or end devices), and end device (send/receive data to/from coordinator) [6]. One of the main advantages of this system is that, as opposed to GPS type systems, it is capable of functioning both inside and out with the same infrastructure, which can quickly and easily adapt to practically any applied environment.

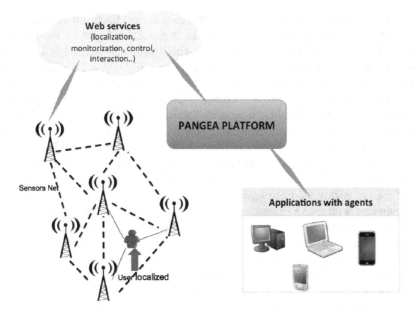

Fig. 1 Diagram of the Zigbee network

3.1 Architecture

The agents involved in the specialized suborganization developed inside PANGEA are:

- ZigbeeManagerAgent: it manages communication and events and is deployed in the server machine.
- ProfileManagerAgent: it is responsible for managing user profiles and is also deployed in the server machine.
- ClientComputerAgent: these are user agents located in the client computer and are responsible for detecting the user's presence with ZigBee technology, and for sending the user's identification to the ZigbeeManagerAgent. These agents are responsible for requesting the profile role adapted for the user to the ProfileManagerAgent.
- DatabaseAgent: the detection proximity system uses a database, which stores data related to the users, sensors, computer equipment and status, and user profiles. It can also communicate with the InformationAgent of PANGEA.
- ZigBeeCoordinatorAgent: it is an agent included in a ZigBee device responsible for coordinating the other ZigBee devices in the office. It is connected to the server by a serial port, and receives signals from each of the ZigBee tags in the system.

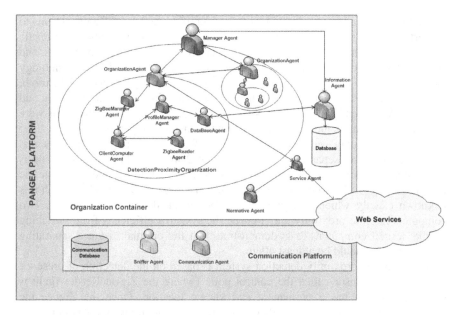

Fig. 2 System architecture

- ZigBeeReaderAgent: these agents are included in several ZigBee devices that are used to detect the presence of a user. Each ZigBeeReaderAgent is located in a piece of office equipment (computer).

Every user in the proposed system carries a Zigbee tag, which is detected by a ZigBeeReaderAgent located in each system terminal and continuously in communication with the ClientComputerAgent. Thus, when a user tag is sufficiently close to a specific terminal (within a range defined according to the strength of the signal), the ZigBeeReaderAgent can detect the user tag and immediately send a message to the ClientComputerAgent, which is coordinated by the ZigBeeCoordinatorAgent.

The system uses a LAN infrastructure that uses the wake-on-LAN protocol for the remote switching on and off of equipment. Wake-on-LAN/WAN is a technology that allows a computer to be turned on remotely by a software call. It can be implemented in both local area networks (LAN) and wide area networks (WAN) [4]. It has many uses, including turning on a Web/FTP server, remotely accessing files stored on a machine, telecommuting, and in this case, turning on a computer even when the user's computer is turned off [7].

4   Case Study

This paper presents a proximity detection system that it is used by people with disabilities to facilitate their integration in the workplace. The main goal of the system is to detect the proximity of a person to a computer using ZigBee technology. This allows an individual to be identified, and for different actions to be performed on the computer, thus facilitating workplace integration: automatic switch on/off of the computer, identifying user profile, launching applications, and adapting the job to the specific needs of the user. Thanks to the Zigbee technology the prototype is notably superior to existing technologies using Bluetooth, infrareds or radiofrequencies, and is highly efficient with regards to detection and distance. Additionally, different types of situations in a work environment were taken into account, including nearby computers, shared computers, etc.

In our Case Study we have a distribution of computers and laptops in a real office environment, separated by a distance of 2 meters. The activation zone is approximately 90cm, a distance considered close enough to be able to initiate the activation process. It should be noted that there is a "Sensitive Area" in which it is unknown exactly which computer should be switched on; this is because two computers in close proximity may impede the system's efficiency from switching on the desired computer. Tests demonstrate that the optimal distance separating two computers should be at least 40cm.

The proposed proximity detection system is based on the detection of presence by a localized sensor called the control point (where the ZigBeeReaderAgent is deployed), which has a permanent and known location. Once the Zigbee tag carried by the person has been detected and identified, its location is delimited within the proximity of the sensor that identified it. Consequently, the location is based on criteria of presence and proximity, according to the precision of the system and the number of control points displayed.

The parameter used to carry out the detection of proximity is the RSSI (Received Signal Strength Indication) [14], a parameter that indicates the strength of the received signal. This force is normally indicated in mW or using logarithmic units (dBm). 0 dBm is equivalent to 1mW. Positive values indicate a signal strength greater than 1mW, while negative values indicate a signal strength less than 1mW.

Under normal conditions, the distance between transmitter and receiver is inversely proportional to the RSSI value measured in the receiver; in other words, the greater the distance, the lower the signal strength received. This is the most commonly used parameter among RTLS [15].

RSSI levels provide an appropriate parameter for allowing our system to function properly. However, variations in both the signal transmission and the environment require us to define an efficient algorithm that will allow us to carry out our proposal. This algorithm is based on the use of a steps or measurement levels (5 levels were used), so that when the user enters the range or proximity indicated by a RSSI level of -50, the levels are activated. While the values received are less than the given range, each measurement of the system activates a level. However, if the values received fall outside the range, the level is deactivated. When the maximum number of levels has been activated, the system interprets this to mean that the user is within the proximity distance of detection and wants to use the computer equipment. Consequently, the mechanisms are activated to remotely switch on both the computer and the profile specific to the user's disability.

The system is composed of 5 levels. The tags default to level 0. When a user moves close to a marker, the level increases by one unit. The perceptible zone in the range of proximity gives an approximate RSSI value of -50. If the user moves away from the proximity area, the RSSI value is less than -50, resulting in a reduction in the level. When a greater level if reached, it is possible to conclude that the user has remained close to the marker, and the computer will be turned on.

On the other hand, reaching an initial level of 0 means that the user has moved a significant distance away from the workspace, and the computer is turned off.

5   Conclusions

This prototype is specifically oriented to facilitate the integration of people with disabilities into the workplace. The detection and identification of a user makes it possible to detect any special needs, and for the computer to be automatically adapted for its use. This allows the system to define and manage the different profiles of people with disabilities, facilitating their job assimilation by automatically switching on or off the computer upon detecting the user's presence, or initiating a procedure that automatically adapts the computer to the personal needs of the user.

Thanks of the PANGEA platform, the prototype could be designed and deployed easily since the platform itself provides agents and tools for the control and management of any kind of open MAS or VO. Moreover, the platform allows to deploy different agents, indeed those included into the mobile devices and performs the communication with the agents embedded into the Zigbee sensors.

The prototype is part of complete and global project in which different tools for helping disabled people will be included. Using the platform PANGEA, that models all the services as Web Services and promotes scalability, the addition in the future of all those services that conform the global project will be easier.

Acknowledgements. This project has been supported by the Spanish CDTI. Proyecto de Cooperación Interempresas. IDI-20110344.

References

1. Huang, Y., Pang, A.: Comprehensive Study of Low-power Operation in IEEE 802.15.4. In: Proceeding of the 10th ACM Symposium on Modeling, Analysis and Simulation of Wireless and Mobile Systems, Chaina, Crete Island, Greece (2007)
2. Singh, C.K., et al.: Performance evaluation of an IEEE 802.15.4 Sensor Network with a Star Topology (2008)
3. Universidad Pontificia de Salamanca [En línea] (2011),
   http://www.youtube.com/watch?v=9iYX-xney6E
4. Lieberman, P.: Wake on LAN Technology, White paper (2011),
   http://www.liebsoft.com/pdfs/Wake_On_LAN.pdf
5. ZigBee Standards Organization: ZigBee Specification Document 053474r13. ZigBee Alliance (2006)

6. Tapia, D.I., De Paz, Y., Bajo, J.: Ambient Intelligence Based Architecture for Automated Dynamic Environments. In: Borrajo, D., Castillo, L., Corchado, J.M. (eds.) CAEPIA 2007, vol. 2, pp. 151–180 (2011)
7. Nedevschi, S., Chandrashekar, J., Liu, J., Nordman, B., Ratnasamy, S., Taft, N.: Skilled in the art of being idle: reducing energy waste in networked systems. In: Proceedings of the 6th USENIX Symposium on Networked Systems Design and Implementation, Boston, Massachusetts, April 22-24, pp. 381–394 (2009)
8. Ferber, J., Gutknecht, O., Michel, F.: From Agents to Organizations: An Organizational View of Multi-agent Systems. In: Giorgini, P., Müller, J.P., Odell, J.J. (eds.) AOSE 2003. LNCS, vol. 2935, pp. 214–230. Springer, Heidelberg (2004)
9. Foster, I., Kesselman, C., Tuecke, S.: The anatomy of the grid: Enabling scalable virtual organizations. Int. J. High Perform. Comput. Appl. 15(3), 200–222 (2001)
10. Agent Oriented Software Pty. Ltd. JACK™ Intelligent Agents Teams Manual. Agent Oriented Software Pty. Ltd. (2005)
11. Hübner, J.F.J.: Moise+ Programming organisational agents with Moise+ & Jason. In: Technical Fora Group at EUMAS 2007
12. Giret, A., Julián, V., Rebollo, M., Argente, E., Carrascosa, C., Botti, V.: An open architecture for service-oriented virtual organizations. In: Braubach, L., Briot, J.-P., Thangarajah, J. (eds.) ProMAS 2009. LNCS, vol. 5919, pp. 118–132. Springer, Heidelberg (2010)
13. Galland, S.: JANUS: Another Yet General-Purpose Multiagent Platform. In: Seventh AOSE Technical Forum, Paris (2010)
14. Zhao, J., Zhang, Y., Ye, M.: Research on the Received Signal Strength Indication Location Algorithm for RFID System. In: International Symposium on Communications and Information Technologies, ISCIT 2006, pp. 881–885 (2006)
15. Park, D.-J., Choi, Y.-B., Nam, K.-C.: RFID-Based RTLS for Improvement of Operation System in Container Terminals. In: Asia-Pacific Conference on Communications, APCC 2006, pp. 1–5 (2006)

Marketing and Consumer Behavior

Approximating the Pareto-front of Continuous Bi-objective Problems: Application to a Competitive Facility Location Problem

J.L. Redondo[1], J. Fernández[3], J.D. Álvarez[2], A.G. Arrondoa[3,*], and P.M. Ortigosa[4]

[1] Dpt. of Computer Architecture and Technology, University of Granada
jlredondo@atc.ugr.es
[2] Dpt. of Computer Sciences and Languages, University of Almería, ceiA3
jhervas@ual.es
[3] Dpt. of Statistics and Operations Research, University of Murcia
{josefdez,agarrondo}@um.es
[4] Dpt. of Computer Architecture and Electronics, University of Almería, ceiA3
ortigosa@ual.es

Abstract. A new general multi-objective optimization heuristic algorithm, suitable for being applied to continuous multi-objective optimization problems is proposed. It deals with the problem at hand in a fast and efficient way. It combines ideas from different multi-objective and single-objective optimization evolutionary algorithms, although it also incorporates new devices which help to reduce the computational requirements, and also to improve the quality of the provided solutions. To show its applicability, a bi-objective competitive facility location and design problem is considered. This problem has been previously tackled through exact general methods, but they require high computational effort. A comprehensive computational study shows that the heuristic method is competitive, being able to reduce, in average, the computing time of the exact method by approximately 98%, and offering good quality in the final solutions.

1 Dealing with Inner and Outer Competition: A Planar Bi-objective Location Problem

Competitive location deals with the problem of locating facilities to provide a service (or goods) to the customers (or consumers) of a given geographical area where other competing facilities offering the same service are already present or will enter the market in the near future. Many competitive location models are available in literature. However, the literature on multi-objective competitive location models is rather scarce. This is in part due to the fact that single-objective competitive location problems are difficult to solve, and considering more than one objective makes the problem nearly intractable.

---

* Corresponding author.

J. Casillas et al. (Eds.): Management Intelligent Systems, AISC 171, pp. 207–216.
springerlink.com   © Springer-Verlag Berlin Heidelberg 2012

In this paper, we propose a new multiobjective evolutionary algorithm whose aim is to obtain a good approximation of the set of all solutions for any nonlinear multiobjective optimization problem, and as quick as possible. So, the method, called FEMOEA, is not only suitable for the particular competitive facility location problem we deal with in this paper, but also for any other continuous multiobjective optimization problem. In this sense, we feel that FEMOEA can be a helpful tool for any decision-maker facing a complex problem with several conflicting objectives.

As a particular example, we revisit here the bi-objective problem described in [3]. A franchise wants to increase its presence in a given geographical region by opening one new facility. Both the franchisor (the owner of the franchise) and the franchisee (the actual owner of the new facility to be opened) have the same objective: maximizing their own profit. However, the maximization of the profit obtained by the franchisor is in conflict with the maximization of the profit obtained by the franchisee.

In the model, the *demand* is supposed to be *inelastic*, i.e., fixed regardless of the condition of the market. So we deal with *essential* goods (e.g. food). A related model considering elastic demand (for inessential goods) can be found in [10]. Furthermore, we assume that the demand is concentrated at n demand points, whose locations p_i and *buying power* w_i are known. The location f_j and quality of the existing facilities are also known. We consider that demand points split their buying power among all the facilities proportionally to the *attraction* they feel for them. The attraction (or utility) function of a customer towards a given facility depends on the distance between the customer and the facility, as well as on other characteristics of the facility which determine its *quality*. The location and the quality of the new facility are the variables of the problem.

The following notation will be used throughout this paper:

Indices
i index of demand points, $i = 1, \ldots, n$.
j index of existing facilities, $j = 1, \ldots, m$.

Variables
x location of the new facility, $x = (x_1, x_2)$.
α quality of the new facility ($\alpha > 0$).

Data

p_i	location of the i-th demand point.
w_i	demand (or buying power) at p_i.
f_j	location of the j-th existing facility.
d_{ij}	distance between p_i and f_j.
α_{ij}	quality of f_j as perceived by p_i.
$g_i(\cdot)$	a non-negative non-decreasing function.
$\alpha_{ij}/g_i(d_{ij})$	attraction that p_i feels for f_j.
γ_i	weight for the quality of the new facility as perceived by p_i.
k	number of existing facilities that are part of one's own chain (the first k of the m facilities are assumed in this category, $0 < k < m$).

Miscellaneous

d_{ix} distance between p_i and the new facility x.

$\gamma_i \alpha / g_i(d_{ix})$ attraction that p_i feels for x.

From the previous assumptions, the total market share attracted by the franchisor is

$$M(x, \alpha) = \sum_{i=1}^{n} w_i \frac{\dfrac{\gamma_i \alpha}{g_i(d_{ix})} + \sum_{j=1}^{k} \dfrac{\alpha_{ij}}{g_i(d_{ij})}}{\dfrac{\gamma_i \alpha}{g_i(d_{ix})} + \sum_{j=1}^{m} \dfrac{\alpha_{ij}}{g_i(d_{ij})}}.$$

We assume that the operating costs for the franchisor pertaining to the new facility, as well as the revenues from the royalties, are fixed. In this way, the profit obtained by the franchisor is an increasing function of the market share that it captures. Thus, maximizing the profit obtained by the franchisor is equivalent to maximizing the market share that it captures. This will be the first objective of the problem.

The second objective of the problem is the maximization of the profit obtained by the franchisee, to be understood as the difference between the revenues obtained from the market share captured by the new facility minus its operational costs. The market share captured by the new facility (franchisee) is given by

$$m(x, \alpha) = \sum_{i=1}^{n} w_i \frac{\dfrac{\gamma_i \alpha}{g_i(d_{ix})}}{\dfrac{\gamma_i \alpha}{g_i(d_{ix})} + \sum_{j=1}^{m} \dfrac{\alpha_{ij}}{g_i(d_{ij})}}$$

and the profit is given by the following expression,

$$\pi(x, \alpha) = F(m(x, \alpha)) - G(x, \alpha),$$

where $F(\cdot)$ is a strictly increasing function which determines the expected sales (i.e., income generated) for a given market share $m(x\alpha)$, and $G(x, \alpha)$ is a function which gives the operating cost of a facility located at x with quality α.

The problem considered is

$$\begin{cases} \max M(x, \alpha) \\ \max \pi(x, \alpha) \\ \text{s.t.} \quad d_{ix} \geq d_i^{\min} \ \forall i \\ \qquad \alpha \in [\alpha_{\min}, \alpha_{\max}] \\ \qquad x \in R \subset \mathbb{R}^2 \end{cases} \tag{1}$$

where the parameters $d_i^{\min} > 0$ and $\alpha_{\min} > 0$ are given thresholds, which guarantee that the new facility is not located over a demand point and that it has a minimum level of quality, respectively. The parameter $\alpha_{\max}$ is the maximum value that the quality of a facility may take in practice. By R we denote the region of the plane where the new facility can be located.

If we set $y = (x, \alpha)$, $f_1(y) = -M(x, \alpha)$, $f_2(y) = -\pi(x, \alpha)$ and we denote S its feasible set, then (1) can be rewritten as

$$\begin{aligned} \min \ & \{f_1(y), f_2(y)\} \\ \text{s.t.} \ & y \in S \subseteq \mathbb{R}^n \end{aligned} \quad (2)$$

When dealing with multi-objective problems we need to clarify what 'solving' a problem means. Some widely known definitions to explain the concept of solution of (2) follow.

Definition 1. If y_1 and y_2 are two feasible points and $f_l(y_1) \leq f_l(y_2), l = 1, 2$, with at least one of the inequalities being strict, then we say that y_1 *dominates* y_2 (or that $f(y_1)$ *dominates* $f(y_2)$),

Definition 2. An objective vector $z^* = f(y^*) \in Z$ is said to be *non-dominated* iff there does not exist another vector which dominates it. The set Z_N of all non-dominated vectors is called the *non-dominated set* or *Pareto-front*.

Ideally, solving (2) means obtaining the whole Pareto-front. To the extent of our knowledge, only two exact general methods, namely, two interval branch-and-bound methods (see [2, 3]) have been proposed in literature with that purpose for the general nonlinear biobjective problem (2), but they are time consuming and have large memory requirements. The use of (meta)heuristics may allow to obtain good approximations of the Pareto-front. In particular, genetic and evolutionary algorithms have proved to be good tools to cope with (2), see for instance [1, 12]. However, they are designed to obtain a good approximation of the Pareto-front given a budget in the number of function evaluations, regardless the CPU time needed for that. In this paper, we present a new Fast and Efficient Multi-Objective Evolutionary Algorithm (FEMOEA) whose aim is to obtain a good approximation of the Pareto-front as quick as possible. In the computational studies we compare it with the best of the interval branch-and-bound methods, iB&B (see [3]).

2 A New Method for Approximating the Pareto-Front

FEMOEA is an evolutionary algorithm initially devised to solve any multi-objective optimization problem. Its main objective is to provide a set of well-distributed and non-dominated solutions as fast as possible.

The most important concept in FEMOEA is that of subpopulation. A subpopulation is defined by a center and a radius. The center is a solution and the radius is a positive number, which determines the subregion of the search space covered by that subpopulation. The radius of a subpopulation is given by a decreasing exponential function, varying from R_1 to R_L (input parameters), which are the given largest and smallest radii, respectively. For a detailed description on how to compute the radius see [9].

Apart from the center and the radius, a subpopulation has two attributes which are related to the objective space: the non-domination rank (d_{rank}) and the crowding

---

Algorithm 1. Algorithm FEMOEA

---

1 Init_subpopulation_lists
2 While stopping rule is not satisfied
3   Create_new_subpopulations(*evals*)
4   If ((length(*population_list*) > M)   Select_subpopulations(*population_list*)
5   If ((length(*external_list*) > M)   Select_subpopulations(*external_list*)
6   Improve_subpopulations(*population_list*)
7   If ((length(*external_list*) > M)   Select_subpopulations(*external_list*)
8   Improve_subpopulations(*external_list*)

---

distance (c_{dist}) (see [1] for an in-detail description of these values). The former indicates the number of subpopulations which dominates that particular subpopulation, whereas the latter is an estimation of the density of solutions surrounding a particular solution in a population.

During the process, two lists of subpopulations are kept by FEMOEA, each with a maximum size M, (another input parameter). M refers to the desired number of solutions in the final approximation of the Pareto-front. The first list, named *population_list*, is composed of M diverse subpopulations with different attributes, i.e. various radii, non-domination ranks and crowding distances. FEMOEA is in fact a method for managing this list (i.e. creating, deleting and improving subpopulations). The second list, called *external_list*, can be understood as a deposit to keep non-dominated solutions. This external archive is also used in other algorithms described in literature [6, 8].

Initially, a set of diverse subpopulations is created in the initialization phase. After this procedure, the FEMOEA main loop starts, which basically consists of three procedures: creating, improving and selecting subpopulations. This loop is executed until a stopping condition is fulfilled. For the problem at hand, the algorithm stops if either a considerable improvement has not been obtained among consecutive approximations (placed in *external_list*) of the Pareto-front or a maximum number of levels (cycles or generations) L (an input parameter) is achieved.

A global description of FEMOEA is given in Algorithm 1. In the following, the different key stages in the algorithm are described:

- *Init_subpopulation_lists*: In this procedure, as many subpopulations as the parameter M indicates are created. The center of the subpopulation is randomly computed, while the radius will be the one associated at level 1, R_1. The *population_list* is initialized from this set of subpopulations, while the *external_list* will consist only of the corresponding non-dominated subpopulations.
- *Create_new_subpopulations(evals)*: For every subpopulation in the *population_list*, random trial points on the area defined by its radius are created, and for every pair of trial points, the midpoint of the *segment* connecting the pair is computed. Furthermore, for each new candidate solution, the closest point in the *external_list* is calculated. In this context, the closest point is that from which the minimum Euclidean distance (in the objective space) can be obtained. Then, a new point

is computed in the segment joining the candidate solution with its closest point. If the mid-way point dominates the candidate solution, then it will be included on the *population_list* as new subpopulation. On the contrary, if the candidate solution is the one which dominates, it will be the one inserted on the population. Additionally, if the two points are *non-comparable* (not one dominates the other), then they both will be inserted as new subpopulations. A radius is assigned to each new subpopulation, whose value is the one associated with the current level. Furthermore, both the non-domination ranks and the crowding distances associated to each new subpopulation are computed. The *population_list* is then sorted according to the *crowded comparison operator* [1]. Afterwards, we check whether any of the non-dominated subpopulations in the *population_list* deserves to be included in the *external_list*. Only those which are not dominated by any element from the *external_list* will be inserted. In case a subpopulation in the *external_list* is dominated by one of the new inserted elements, it is removed.

The parameter *evals* indicates the number of function evaluations allowed to create a new offspring per subpopulation. For the problem at hand, we have set $evals = 20$.

- *Select_subpopulations(list)*: If *list* reaches its maximum allowable capacity, a decision has to be made to determine which individuals should be removed. The selection strategy used in this work is based on the *crowded comparison operator* [1], which selects the most *preferable* subpopulations as follows: Between two subpopulations with different non-domination ranks, the one with the lowest rank is preferred. In case of ties, the subpopulation with the highest crowded distance (i.e., the one located in a region with fewer number of points) is preferred.
- *Improve_subpopulations(list)*: For the problem at hand, a gradient-based improving method has been designed, which tries to find a direction along which both objective functions can be improved. It has been applied to both subpopulation lists (*population_list* and *external_list*) at different stages of the algorithm.

The improving method is applied to every subpopulation of the *list*. If as a consequence of the optimization procedure, a new point dominates the center of the original subpopulation, the center is replaced, the subpopulation keeps the same radius value. In case non-comparable solutions are obtained, the procedure varies depending on whether the *population_list* or the *external_list* are considered:

- For *population_list*: Each non-comparable solution is compared pairwise to the elements of the *external_list*. If this new solution is dominated by an individual from such a list, it will be discarded; otherwise, the solution will be stored in the *external_list*. Furthermore, the *external_list* is updated taking the non-dominated solutions of the *population_list* into account.
- For *external_list*: The non-comparable solutions are discarded.

As final step in both cases, if any of the elements from the *external_list* is dominated by the new included subpopulations, it is removed.

- Stopping rule: For the current multi-objective problem, the algorithm finishes if during three consecutive iterations, the changes experimented in the *external_list* are negligible (in terms of the objective function values), i.e. the *Hausdorff distances* among three consecutive approximations of the Pareto-front are fewer than a tolerance *tol* (for this work, $tol = 10^{-7}$). Nevertheless, as a safeguard, another termination criterion based on the number of iterations executed by FEMOEA has been considered. Then, the algorithm stops if the previous condition holds *or* a maximum number of iterations has been fulfilled. This maximum value is represented by the input parameter L.

3 Computational Studies

All the computational studies in this paper have been run in the supercomputer Ben Arabi of the Supercomputing Center of Murcia, Spain, in particular, in Arabi, which is a Blade Cluster with 816 cores, organized in 32 nodes with 16GB of memory each, and 70 nodes with 8GB (102 nodes altogether). Each node has 8 cores, divided into 2 Intel Xeon Quad Core (E5450) to 3.0 GHz. For the interval branch-and-bound method (iB&B) introduced in [3], both the interval arithmetic in the PROFIL/BIAS library [7] and the automatic differentiation of the C++ Toolbox library [4], have been used. FEMOEA has been implemented in C++.

In order to have an overall view of the performance of the algorithms, different types of problems have been generated, varying the number n of demand points, the number m of existing facilities and the number k of those facilities belonging to the chain. For $n = 25, 50$ demand points the settings used were $(m = 2, k = 1)$, $(m = 5, k = 1, 2)$ and $(m = 10, k = 2, 4)$. For every setting, 10 instances were generated by randomly choosing the parameters of the problems uniformly within pre-defined intervals (see [11]). The searching space proposed in [11] has also been considered here for every problem.

As a general rule, the algorithm iB&B has been executed considering a tolerance of $eps = 0.03$ (the maximum width of a box in the solution list), which is not a negligible value. Even so, the algorithm ran out of memory when trying to solve several instances. In each of those cases, the value of eps was progressively increased until the algorithm was able to solve that particular problem. Regarding FEMOEA, and after extensive experiments, we found that a good parameter setting to deal with the current multi-objective optimization problem is: $L = 30$ and $R_L = 5e - 03$. The parameter R_1 coincides with the diameter of search space. Furthermore, FEMOEA has been executed twice varying the number of points in the approximation of the Pareto-front, $M = 200$ or 400.

To measure the performance of FEMOEA, two main aspects are under consideration, that of effectiveness and that of efficiency [5]. We start by saying that for stochastic algorithms, performance indicator values are also stochastic. For each random indicator, we approximate the expected value by taking the average over 5 runs.

As an effectiveness metric, we check whether the heuristic algorithm has successfully found an approximation of the Pareto-front. We say so when both the objective function values of the points in the *external_list* are included in the corresponding intervals provided by the iB&B method, and the points themselves are included in the corresponding solution boxes offered by iB&B. Additionally, for measuring the goodness of an approximation to the Pareto-front, the so-called hypervolume measure *Hyper* has also been computed [8, 12].

To measure the efficiency of the algorithms, one tries to compute the effort used to obtain the final result. Here, we measure the computing time to reach the result for iB&B (T), and the average computing time in the five runs $Av(T)$ for the algorithm FEMOEA.

Tables 1 and 2 summarizes the obtained results for a set of 10 instances with settings 25-5-2 and 50-10-4, respectively. At the end of each table, average values for the 10 problems have been computed. In those tables, the computing time and the hypervolume metric are shown for both iB&B and FEMOEA. The results for the rest of settings are not shown due to the lack of space, but similar conclusions can be inferred.

As can be observed in Tables 1 and 2, the iB&B algorithm is very erratic regarding computing time. Additionally, it is difficult for a given problem to determine the tolerance that, a priori, will be suitable to execute the algorithm. Of course, a large value of *eps* can be considered, but it will affect the quality of the obtained solution (the larger the value of *eps*, the greater the intervals containing the exact Pareto-front). For the cases where a different value of *eps* = 0.03 has been considered, this has been noted on the table, including the new value after the computing time spent by iB&B in brackets (see column *eps*). On the contrary, the evolutionary algorithm seems to be more regular, it always spends similar computing times for instances with the same settings. In average, FEMOEA has reduced the computing time by more than 99% (resp. 98%) for problems with settings 25-5-2 (resp. 50-10-4), and this considering a number $M = 400$ points in the approximation of the Pareto-front.

It is worth mentioning that FEMOEA approximates the Pareto-front with 100% success for all the problems (for both $M = 200$ and $M = 400$), i.e. its solutions

Table 1 Hypervolume and computing time for problems with setting 25-5-2

T	eps	$Av(T)_{200}$	$Av(T)_{400}$	$[lowH, uppH]$	$Av(Hyper)_{200}$	$Av(Hyper)_{400}$
209		262	537	[146.317,146.532]	146.320	146.332
489197	(.05)	347	1144	[1.326,1.328]	1.323	1.326
508028		273	596	[3.157,3.159]	3.146	3.158
97404		310	730	[112.931,113.380]	112.728	112.956
551537	(.05)	308	910	[1.751,1.764]	1.754	1.759
4536		274	562	[553.147,554.340]	553.151	553.163
569547		309	756	[2.282,2.283]	2.278	2.282
81687		266	596	[428.519,429.229]	427.964	428.800
389738	(.04)	338	1072	[1.342,1.344]	1.340	1.343
281464	(.04)	339	989	[1.762,1.763]	1.758	1.761
297340		302.6	789.2	[125.253,125.512]	125.176	125.288

Table 2 Hypervolume and computing time for problems with setting 50-10-4

T	eps	$Av(T)_{200}$	$Av(T)_{400}$	$[lowH, uppH]$	$Av(Hyper)_{200}$	$Av(Hyper)_{400}$
54099		1346	2764	[19.593,19.770]	19.601	19.660
81108		900	1877	[56.652,56.930]	56.797	56.867
47	(.07)	857	1715	[1.71e-05,0.005]	0.002	0.002
337078		785	1582	[133.242,133.723]	133.326	133.332
19419		817	1649	[184.410,184.873]	184.650	184.860
351722		931	1905	[163.254,164.069]	163.253	163.406
158215		796	1626	[123.899,124.458]	123.753	123.901
10695		873	1748	[103.307,104.152]	103.343	103.373
76578		820	1658	[1.267,1.290]	1.265	1.268
144828		925	1859	[2.354,2.376]	2.357	2.362
123380		905	1838	[78.797,79.164]	78.834	78.903

are always included in the intervals provided by the iB&B algorithm. Notice that FEMOEA increases its $Hyper$ value as the number of points in the approximation of the Pareto-front increases. In this sense, we should keep track of the number of generated points M in the set approximating the Pareto-front, since theoretically speaking, higher values of $Hyper$ represent better approximations. As can be observed in Tables 1 and 2, the hypervolume covered by the heuristic with 400 points is always included in the interval $[lowH, uppH]$, obtained from the lower-left and upper-right hand corners of the boxes provided by iB&B. On the other hand, with 200 points in the set approximating the Pareto-front, the hypervolume is only smaller than the lower limit for half of the problems (10 out of 20), but notice the number of points used to compute the hypervolume with iB&B (i.e., the number of boxes on the solution list) is usually much larger than M.

4 Conclusions and Lines for Future Research

In this work, a new general multi-objective optimization algorithm, FEMOEA, suitable for obtaining a good approximation of the complete Pareto-front of any continuous nonlinear multi-objective problem, has been proposed. Any decision-maker can adapt it to its particular problem with very little modifications.

To show its applicability, the method has been used to solve a bi-objective competitive location and design problem. Furthermore, it has been compared to an exact iB&B method. Results have shown that a FEMOEA use of the computing resources is more efficient. Moreover, the provided solutions by FEMOEA are competitive with respect to the ones given by iB&B: on the one hand, the solutions obtained by the heuristic algorithm are always included in the iB&B intervals and, on the other hand, with $M = 400$ they can cover practically all the area of the optimum Pareto-front (the hypervolumes obtained by FEMOEA are always included in the hypervolume intervals of iB&B).

In the future, we plan to compare the FEMOEA algorithm with other heuristic algorithms devised to cope with multi-objective problems on a set of benchmark problems. In particular, we will compare it with two standards widely referenced in literature, i.e. NSGA-II [1] and SPEA2 [12].

Acknowledgements. This work has been funded by grants from the Spanish Ministry of Science and Innovation (TIN2008-01117, ECO2011-24927), Junta de Andalucía (P08-TIC-3518 and P10-TIC-6002) and Fundación Séneca (The Agency of Science and Technology of the Region of Murcia, 00003/CS/10 and 15254/PI/10), in part financed by the European Regional Development Fund (ERDF). Juana López Redondo is a fellow of the Spanish 'Juan de la Cierva' contract program. The authors are indebted to the Supercomputing Center of Fundación Parque Científico of Murcia, Spain, for having allowed us to use its facilities and for its technical support.

References

1. Deb, K., Pratap, A., Agarwal, S., Meyarivan, T.: A fast and elitist multiobjective genetic algorithm: NSGA-II. IEEE Transactions on Evolutionary Computation 6(2), 182–197 (2002)
2. Fernández, J., Tóth, B.: Obtaining an outer approximation of the efficient set of nonlinear biobjective problems. Journal of Global Optimization 38(2), 315–331 (2007)
3. Fernández, J., Tóth, B.: Obtaining the efficient set of nonlinear biobjective optimization problems via interval branch-and-bound method. Computational Optimization and Applications 42(3), 393–419 (2009)
4. Hammer, R., Hocks, M., Kulisch, U., Ratz, D.: C++ Toolbox for Verified Computing I: Basic Numerical Problems: Theory, Algorithms, and Programs. Springer, Berlin (1995)
5. Hendrix, E.M.T., Toth, B.G.: Introduction to Nonlinear and Global Optimization. Springer, New York (2010)
6. Knowles, J.D., Corne, D.W.: The pareto archived evolution strategy: A new baseline algorithm for pareto multiobjective optimisation, vol. 1, pp. 98–105. IEEE Service Center (1999)
7. Knüppel, O.: PROFIL/BIAS - a fast interval library. Computing 1(53), 277–287 (1993)
8. Nebro, A.J., Luna, F., Alba, E., Dorronsoro, B., Durillo, J.J., Beham, A.: AbYSS: Adapting scatter search to multiobjective optimization. IEEE Transactions on Evolutionary Computation 12(4), 439–457 (2008)
9. Ortigosa, P.M., García, I., Jelasity, M.: Reliability and performance of UEGO, a clustering-based global optimizer. Journal of Global Optimization 19(3), 265–289 (2001)
10. Redondo, J.L., Fernández, J., Arrondo, A.G., García, I., Ortigosa, P.M.: Fixed or variable demand? does it matter when locating a facility? OMEGA-International journal of management science 40(1), 9–20 (2012)
11. Redondo, J.L., Fernández, J., García, I., Ortigosa, P.M.: A robust and efficient global optimization algorithm for planar competitive location problems. Annals of Operations Research 167(1), 87–106 (2009)
12. Zitzler, E., Laumanns, M., Thiele, L.: SPEA2: Improving the strength pareto evolutionary algorithm for multiobjective optimization. In: Giannakoglou, K.C., Tsahalis, D.T., Périaux, J., Papailiou, K.D., Fogarty, T. (eds.) Evolutionary Methods for Design Optimization and Control with Applications to Industrial Problems, Athens, Greece, pp. 95–100 (2001)

Improving Customer Churn Prediction by Data Augmentation Using Pictorial Stimulus-Choice Data

Michel Ballings[1], Dirk Van den Poel[1], and Emmanuel Verhagen[2]

[1] Faculty of Economics and Business Administration, Department of Marketing, Ghent University, Tweekerkenstraat 2, B-9000 Ghent, Belgium
http://www.crm.ugent.be, Michel.Ballings@UGent.be,
Dirk.VandenPoel@UGent.be
[2] Psilogy, Leopold De Waelstraat 17a, 2000 Antwerp, Belgium
Emmanuel.Verhagen@Psilogy.com

Abstract. The purpose of this paper is to determine the added value of pictorial stimulus-choice data in customer churn prediction. Using Random Forests and 5 times 2 fold cross-validation, this study analyzes how much pictorial stimulus – choice data and survey data increase the AUC of a churn model over and above administrative, operational and complaints data. The finding is that pictorial-stimulus choice data significantly increases AUC of models with administrative and operational data. The practical implication of this finding is that companies should start considering mining pictorial data from social media sites (e.g. Pinterest), in order to augment their internal customer database. This study is original in that it is the first that assesses the added value of pictorial stimulus-choice data in predictive models. This is important because more and more social media websites are focusing on pictures.

Keywords: Customer Relationship Management, Data Augmentation, Predictive Modeling, Customer Churn, Pictorial Stimulus-Choice Data, Pictures.

1 Introduction

In an increasingly competitive business environment, companies have come to realize that their most valuable asset is their customer base [1,2]. As a result, customer churn management has become the cornerstone of every customer intelligence department. From an analytical perspective, churn management consists of (1) predicting which customers are about to churn and (2) evaluating which marketing action is most effective in retaining those customers [3]. This study focuses on the former.

Because a customer's profitability increases over time, even small increases in retention can have substantial impact on a company's results [4]. [5] show that even an increase of 1%-point in retention can have a dramatic influence on contributions. Hence, companies are trying to improve customer churn prediction.

J. Casillas et al. (Eds.): Management Intelligent Systems, AISC 171, pp. 217–226.
springerlink.com   © Springer-Verlag Berlin Heidelberg 2012

In doing so, they adopt two main strategies: (1) improving data mining techniques and (2) enhancing the customer database [6]. The latter consists in adding other data types to the internal transactional database and will be the focus of this study. Some studies also assess the value of other data types in isolation. The internal transactional database is considered the baseline for database enhancement because it contains the top predictors in extant database marketing modeling: recency, frequency and monetary value (RFM) [7]. Since their identification [8], many studies have reported them as being the best predictors of customer behavior (e.g. [9]). In addition to the RFM variables, length of relationship (LOR) has also proven to be a top predictor [10]. Hence, studies that assess other data for analytical CRM purposes, in addition to RFM and LOR, are considered to have a database enhancement focus.

Table 1 provides a literature review of studies that enhance the customer database in order to improve predictive models.

Table 1 Literature review of data augmentation in analytical CRM

Study	Data-Variables
[11]	Geographical data (ZIP-codes)
[12]	Clickstream data
[13]	Consumer network data
[14, 15]	Call center e-mails
[6]	Commercially available survey data
[16]	Situational variables: weather, time, sales-person variables
[17]	Customer websites
[18]	Call center dialog transcripts
[19]	Company survey data: loyalty
[20]	Product features
[21]	RFID

Thanks to the increasing openness of social media platforms to data extraction and even the recent rise of a social network in which pictures are the centerpiece (Pinterest) new opportunities for data augmentation are created in the realm of pictorial content. Pictures contain a massive amount of information and user actions (e.g. 'liking' a picture) could potentially reveal part of that information. Hence, this study aims at contributing to the literature by assessing the added value of pictorial stimulus choice data to customer churn prediction over and above traditional customer data. Before detailing what pictorial stimulus choice data is, the next section will first classify traditional data types.

2   Customer Data Sources

According to the required level of investment companies have to make, and to which customer behavior the data represents, we differentiate four customer data sources that companies can tap into for customer intelligence purposes.

As [22] points out, smaller or startup companies have less financial abilities to invest in information and data gathering. Because they cannot (yet) stem the costs required to store, maintain and mine huge amounts of data they focus only on gathering data about necessities (what currently needs to be delivered and invoiced). We call this administrative data because it represents customer identification and contract specifics acquired through the administrative process. The second customer data source, operational data, is the entire history of all subscriptions or contracts and operations [23]. Companies that do have the means to make the necessary investments linked to the extra storage, maintenance, software and skill requirements are probably more mature and bigger than companies that only use administrative data. The third customer data source is complaints data [15]. This data type is related to the customer feedback process and requires a significant supplementary investment to mine given its often unstructured nature (e.g. email- mining). According to the exit-voice theory [24], complaining behavior is conceived of as one of two options when a customer is dissatisfied, next to leaving the provider, and can therefore be a valuable addition to predictive models. The fourth and final customer data source is surveys. While surveys are primarily aimed at uncovering insights (obtaining cross-sectional data for descriptive models), some companies take it one step further by using surveys as a customer data source for intelligence purposes (i.e. longitudinal data for predictive models) [6]. This data type can be conceived of as the final category in that customer intelligence departments are adding external data from other departments, to their internal data (i.e. administrations, operations and complaints data).

3   Methodology

3.1   Data

The customers of two Belgian newspaper brands were invited to participate in a study. Both brands can be considered similar, except their geographical targeting at the province/state level. One of the two brands consisted of different editions. 25,897 emails were sent, inviting customers to click on a link to go to an online questionnaire (an incentive in the form of a prize was offered). 6,661 (25.7%) customers opened the message and of those customers 4,360 (65.9%) clicked through to the questionnaire (many dropped out after the first page). 2,605 (59.7%) of them were subsequently used in the analysis. While we did this extra survey to obtain the pictorial stimulus- choice data we also included some traditional questions such as involvement [25], satisfaction [26], calculative commitment [27], affective commitment [28], normative commitment [29], renewal intentions [30] and value instrumentality [31].

Table 2 Sample characteristics (5 times 2 fold cross-validation)

Across folds	Average Number of accounts	Average Relative percentage	Standard deviation
Training/Test data	1302.5		
Non-Churners	1146.94	88.023%	0.306
Churners	156.073	11.978%	0.305

We merged the survey data with internal data from which we computed variables in the aforementioned categories. As such, we were able to assess the added value of pictorial stimulus-choice data. Table 2 displays the sample characteristics (see later for cross-validation)

3.2 Time Window

Depending on the length of the subscription and the promotional context, customers have to pay a certain price. They are sent a letter to remind them that they are approaching the end of the subscription and to ask them whether they want to renew their subscription, along with instructions on how to do that. Customers cannot cancel the subscription, as such churn prediction involves predicting whether the customer will or will not renew his or her subscription in the four-week period following the end of the subscription. During this period, customers keep on receiving newspapers, which they will have to pay if they renew.

Independent variables are computed from internal data from the period 03/01/1994-27/04/2010 preceding the period 01/05/2010-02/03/2011 in which the dependent churn variable is observed. Survey data is collected in the period 04/04/2010- 26/04/2010. Once the model is built in the first step, both the independent and dependent period can be shifted forward in time while respecting their relative positions in order to deploy the model.

3.3 Variables

3.3.1 Administrative Data

Administrative data represents all information regarding agreements made between the customer and the company at the time of the purchase decision. The data is acquired at the beginning of the relationship. This entails how much, where, when and to whom the newspaper needs to be delivered. It also comprises information about the price and possible promotions.

3.3.2 Operational Data

While administrative data only holds data about the current subscription level, operational data holds the entire customer history (at the subscriber's level). This means that variables are across subscriptions while they are per subscription in the administrative data and that data are acquired during the relationship, in contrast to at the beginning of the relationship.

This data also contains socio-demographic data, as opposed to customer identification data, and data about suspensions, forward interruptions, credit handling, and marketing actions, as well as response to such action: e.g. participation in games. Much of the data is not merely contractual as in the administrative data but represents a more complex commercial policy (such as credit and suspension processes) and the subsequent usage of the product (subscription) by the customer.

3.3.3 Complaints Data

This data type contains information about the topic of complaints and the solution and answer given. Feedback data and the Voice of Customers are acquired at customer initiated feedback moments and have been shown to be of value in predictive models [15].

3.3.4 Survey Data

Survey data can be conceived of as resulting from a company-initiated feedback process. Mind set variables (e.g. purchase intentions, commitment), customer life style information (e.g. interests and opinions) and product evaluation data (e.g. overall satisfaction, satisfaction drivers) are examples of this type of data and are impossible to collect from internal processes.

3.3.5 Pictorial Stimulus-Choice Data

The final type of data needs to be acquired through a choice process. This data was not available with the company so we collected it by sequentially presenting respondents with six picture sets (randomized between respondents) each containing nine pictures. The first five sets were accompanied with the following question. "Imagine a typical moment when you are reading [name of newspaper brand]. In general, how do you feel at that moment? Choose one of the nine pictures that best represents that feeling. Make your choice by clicking on the picture.". The sets respectively contained pictures of motivational scenes, a man's facial emotional expressions, a woman's facial emotional expressions, geometrical forms and colors. A sixth set contained pictures of couples in different relationship stages (such as being angry, an open relationship, engagement, marriage, having children, being grandparents) preceded with the question "Which of the following pictures best describes your relationship with [name of newspaper brand]?" For all pictures and sets, dummy-variables were then created indicating whether they were chosen by the respondent or not. This resulted in 48 dichotomous variables (6 sets x 9-1 pictures).

While we opted for an ad-hoc approach here (an online survey), companies could go online and mine social network data (e.g., which photo did a person like out of a particular photo set?).

3.4 Estimation Technique

For the estimation of the model we use random forests [32]. Random forests cope with the limited robustness and suboptimal performance [33] of decision trees by building an ensemble of trees (e.g. 1000 trees) and subsequently voting for the most popular class [32]. Each tree is grown on a bootstrap sample with a random subset of all available predictors.

The advantages of random forests are manifold. First, literature shows that random forests is one of the best performing classification techniques available [34] and is very robust and consistent [32]. Second, the method does not overfit [32], which is of particular importance for this study due to the relatively large number of predictors we want to test and the small sample size. Third, variable importance measures are available for all predictors [35]. Fourth, the algorithm has reasonable computing times [36]. Fifth, the procedure is easy to implement: only two parameters are to be set (number of trees and number of predictors) [23, 37]. We follow the recommendation of [32] by using a large number of trees (1000) and the square root of the total number of variables as the number of predictors.

3.5 Model Performance Evaluation Criteria

To evaluate the classification models' performance we'll use the area under the receiver operating characteristic curve (AUC or AUROC). AUC is argued to be an objective criterion for classifier performance by several authors [9, 38-40]. The receiver operating characteristic (ROC) curve is obtained from plotting sensitivity and 1-specificity considering all possible cut-off values [41]. AUC ranges from .5, if the predictions are not better than random, to 1, if the model predicts the behavior perfectly [6]. We use AUC instead of accuracy (Percentage of correctly classified, PCC) because AUC, in contrast to PCC, is the insensitive to the cut-off value of the 'a posteriori' probabilities. As such AUC is a more adequate performance measure (see [7]).

3.6 5 Times 2 Fold Cross-Validation

Research has shown that the usual method to compare classification methods (t-tests to confirm significant differences in the accuracies obtained from k-fold cross validation) results in increased type-I error [42]. [42] and [43] propose 5 times 2 fold cross validation to cope with the problem. 5 times 2- fold cross-validation randomly divides the sample (2,605 observations) in two parts (1,302 and 1,303) and repeats this process 5-times. Each part is used both as training and validation part. This process results in 10 AUC's per model [42]. In order to determine whether models are significantly different in terms of AUC, we follow the recommendation of [44] to use the Wilcoxon signed ranks test [45].

The Wilcoxon signed-ranks test [45] is a non-parametric test, that ranks, per data set, the differences in performance of two classifiers, while ignoring the signs, and compares the ranks for both positive and negative differences [44].

4   Results

4.1   Model Evaluations

Table 3 presents the added value per data type. The baseline data is the respective data source without pictorial stimulus-choice data.

Table 3 The added value per data type (5 times 2 fold cross validation)

Baseline Data	Baseline AUC	Baseline & pictorial stimulus- choice data AUC	Difference: % difference
Administrative data	0.8524	0.8599	0.0075:0.9%*
Administrative & Operational data	0.8727	0.8764	0.0037:0.4%*
Administrative & Operational & Complaints data	0.8775	0.8794	0.0019:0.2%
Administrative & Operational & Complaints & Survey data	0.8817	0.8826	0.0009:0.1%

*Significantly different p < 0.01

Wilcoxon signed-ranks tests of the null hypothesis that both models (baseline versus baseline plus pictorial stimulus-choice data) perform equally well are rejected for models based on administrative data (p < .01) and administrative & operational data (p < .01). Models based on administrative, operational & complaints data are not significantly different (p >.10). The same conclusion can be drawn for models based on administrative, operational, complaints & survey data (p > .10).

5   Conclusion

This study is the first to assess the added value of pictorial stimulus-choice data in predictive models.

We show that the augmentation of companies' databases with pictorial data can significantly improve customer churn prediction, even over and above traditional, internal predictors. We have shown that, as expected, companies with less data to deploy, administrative and administrative & operational data, in their customer intelligence projects would benefit most from this kind of data with AUC increases of respectively 0.9% and 0.4%.

Although traditional survey data also significantly improves the models' performances, it is clear that pictorial data, which can be freely mined from social networks (e.g. Pinterest), is a much more cost-effective and viable strategy to improve churn prediction for the entire customer database.

6 Limitations and Future Research

This study intends to gauge the added value of pictures in predictive models. Since a picture contains a huge amount of information, user actions (e.g. 'liking' a picture on a social media platform) could possibly say a great deal about future behavior. Whether this is the case depends on the informational content of the picture. A limitation in this study is then that we tested only six pictorial sets. A possible direction for future research is to test more pictorial sets.

A second limitation is that we used an ad-hoc approach (an online survey). Future research could use a structural approach by mining online social networks.

7 Practical Implications

More and more sources of data are accessible and managers' choice-options for data-augmentation are constantly growing. The practical implication of the findings in this study is that companies should start considering mining pictorial data from social media sites (e.g. Pinterest), in order to augment their internal customer database.

Because even small changes in retention rate (e.g. 1%-point) can have a large impact on contributions, including pictorial data in predictive models is a viable strategy.

References

1. Athanassopoulos, A.D.: Customer satisfaction cues to support market segmentation and explain switching behavior. J. Bus. Res. 47(3), 191–207 (2000)
2. Thomas, J.S.: A methodology for linking customer acquisition to customer retention. J. Marketing Res. 38(2), 262–268 (2001)
3. Hung, S.-Y., Yen, D.C., Wang, H.-Y.: Applying data mining to telecom churn management. Expert Syst. Appl. 31, 515–524 (2006)
4. Gupta, S., Lehmann, D.R., Stuart, J.A.: Valuing customers. J. Marketing 41, 7–19 (2004)
5. Van den Poel, D., Larivière, B.: Customer attrition analysis for financial services using proportional hazard models. Eur. J. Oper. Res. 157, 196–217 (2004)
6. Baecke, P., Van den Poel, D.: Data Augmentation by Predicting Spending Pleasure Using Commercially Available External Data. J. Intell. Inf. Syst. 36(3), 367–383 (2011)

7. Baesens, B., Viaene, S., Van den Poel, D., Vanthienen, J., Dedene, G.: Bayesian neural network learning for repeat purchase modelling in direct marketing. Eur. J. Oper. Res. 138(1), 191–211 (2002)

8. Cullinan, G.J.: Picking them by their Batting Averages' Recency – Frequency – Monetary Method of Controlling Circulation. Manual Release 2103, Direct Mail/Marketing Association, NY (1977)

9. Coussement, K., Van den Poel, D.: Churn prediction in subscription services: An application of support vector machines while comparing two parameter-selection techniques. Expert Syst. Appl. 34(1), 313–327 (2008)

10. Van den Poel, D.: Predicting mail-order repeat buying: Which variables matter? Tijdschr. Econ. Man. 48(3), 371–403 (2003)

11. Steenburgh, T.J., Ainsle, A., Engbretson, P.H.: Massively categorical variables, revealing the information in ZIP codes. Market Sci. 22, 40–57 (2003)

12. Hu, J., Zhong, N.: Web farming with clickstream. Int. J. Inf. Tech. Dec. Ma. 7, 291–308 (2008)

13. Hill, S., Provost, F., Volinsky, C.: Network-based marketing: Identifying likely adopters via consumer networks. Stat. Sci. 21, 256–276 (2006)

14. Coussement, K., Van den Poel, D.: Improving customer attrition prediction by integrating emotions from client/company interaction emails and evaluating multiple classifiers. Expert Syst. Appl. 36, 6127–6134 (2009)

15. Coussement, K., Van den Poel, D.: Integrating the voice of customers through call center emails into a decision support system for churn prediction. Inform. Manage. 45(3), 164–174 (2008)

16. Baecke, P., Van den Poel, D.: Improving purchasing behavior predictions by Data Augmentation with situational variables. Int. J. Inf. Tech. Dec. Ma. 36(3), 367–383 (2010)

17. Thorleuchter, D., Van den Poel, D., Prinzie, A.: Analyzing existing customers' websites to improve the customer acquisition process as well as the profitability prediction in B-to-B marketing. Expert Syst. Appl. 39(3), 2597–2605 (2012)

18. Gilman, A., Narayanan, B., Paul, S.: Mining call center dialog data. In: Zanasi, A., Ebecken, N.F.F., Brebbia, C.A. (eds.) Data Mining. V. WIT Press (2004)

19. Buckinx, W., Verstraeten, G., Van den Poel, D.: Predicting customer loyalty using the internal transactional database. Expert. Syst. Appl. 32(1), 125–134 (2007)

20. Lariviere, B., Van den Poel, D.: Investigating the role of product features in preventing customer churn, by using survival analysis and choice modeling: The case of financial services. Expert. Syst. Appl. 27(2), 277–285 (2004)

21. Wong, W.K., Leung, S.Y.S., Guo, Z.X., Zeng, X., Mok, P.Y.: Intelligent product cross-selling system with radio frequency identification technology for retailing. Int. J. Prod. Econ. 135(1), 308–319 (2012)

22. Alshawi, S., Missi, F., Irani, Z.: Organisational, technical and data quality factors in CRM adoption - SMEs perspective. Ind. Market Manag. 40(3), 376–383 (2011)

23. Larivière, B., Van den Poel, D.: Predicting customer retention and profitability by using random forests and regression forests techniques. Expert. Syst. Appl. 29(2), 472–484 (2005)

24. Hirschman, A.O.: Exit, Voice, and Loyalty–Responses to Decline in Firms.Organizations, and States. Harvard University Press, Cambridge (1970)

25. Zaichkowsky, J.: The Personal Involvement Inventory- Reduction, Revision, and Application to Advertising. J. Advertising 23(4), 59–70 (1994)

26. Fornell, C., Johnson, M.D., Anderson, E., et al.: The American customer satisfaction index: nature, purpose, and findings. J. Marketing 60(4), 7–18 (1996)

27. Gounaris, S.: Trust and commitment influences on customer retention: insights from business-to-business services. J. Bus. Res. 58(2), 126–140 (2005)

28. Gustafsson, A., Johnson, M.D., Roos, I.: The effects of customer satisfaction, relationship commitment dimensions, and triggers on customer retention. J. Marketing 69(4), 210–218 (2005)

29. Ros, M., Schwartz, S.H., Surkiss, S.: Basic individual values, work values, and the meaning of work. Appl. Psychol-Int. Rev. 48(1), 49–71 (1999)

30. Rossiter, J.: The C-OAR-SE procedure for scale development in marketing. Int. J. Res. Mark. 19(4), 305–335 (2002)

31. Lindeman, M., Verkasalo, M.: Measuring values with the short Schwartz's value survey. J. Pers. Assess. 85(2), 170–178 (2005)

32. Breiman, L.: Random forests. Mach. Learn. 45(1), 5–32 (2001)

33. Dudoit, S., Fridlyand, J., Speed, T.P.: Comparison of discrimination methods for the classification of tumors using gene expression data. J. Am. Stat. Assoc. 97(457), 77–87 (2002)

34. Luo, T., Kramer, K., Goldgof, D.B., Hall, L.O., Samson, S., Remsen, A., et al.: Recognizing plankton images from the shadow image particle profiling evaluation recorder. IEEE T. Syst. Man. Cy. B 34(4), 1753–1762 (2004)

35. Ishwaran, H., Blackstone, E.H., Pothier, C.E., Lauer, M.S.: Relative risk forests for exercise heart rate recovery as a predictor of mortality. J. Am. Stat. Assoc. 99(467), 591–600 (2004)

36. Buckinx, W., Van den Poel, D.: Customer base analysis: Partial defection of behaviourally-loyal clients in a non-contractual FMCG retail setting. Eur. J. Oper. Res. 164(1), 252–268 (2005)

37. Duda, R.O., Hart, P.E., Stork, D.G.: Pattern Classification. ch. 8. Wiley, NY (2001)

38. Provost, F., Fawcett, T., Kohavi, R.: The case against accuracy estimation for comparing induction algorithms. In: Shavlik, J. (ed.) Proc. of 15th International Conference on Machine Learning, ICML 1998. Morgan Kaufman, San Francisco (1998)

39. Langley, P.: Crafting papers on machine learning. In: Langley, P. (ed.) Proc. of 17th International Conference on Machine Learning, ICML 200. Stanford University, Stanford (2000)

40. De Bock, K.W., Coussement, K., Van den Poel, D.: Ensemble classification based on generalized additive models. Comput. Stat. Data An. 54(6), 1535–1546 (2010)

41. Hanley, J.A., McNeil, B.J.: The meaning and use of the area under a receiver operating characteristic (ROC) curve. Radiology 143(1), 29–36 (1982)

42. Dietterich, T.G.: Approximate statistical tests for comparing supervised classification learning algorithms. Neural Comput. 10, 1895–1924 (1998)

43. Alpaydin, E.: Combined 5 x 2cv F test for comparing supervised classification learning algorithms. Neural Comput. 11(8), 1885–1892 (1999)

44. Demšar, J.: Statistical Comparisons of Classifiers over Multiple Data Sets. Mach. Learn. Res. 7, 1–30 (2006)

45. Wilcoxon, F.: Individual comparisons by ranking methods. Biometrics 1, 80–83 (1945)

A Multiple-Agent Based System for Forecasting the Ice Cream Demand Using Climatic Information

Wen-Bin Yu[1], Hokey Min[2], and Bih-Ru Lea[1]

[1] Department of Business and Information Technology, Missouri University of
Science and Technology, Rolla, MO 65409
{yuwen,leabi}@mst.edu
[2] Department of Management, BAA 3008C, College of Business Administration,
Bowling Green State University, Bowling Green, Ohio 43403
hmin@bgsu.edu

Abstract. A multiple agent-based system is intended to capture complex behavioral patterns by utilizing a collection of autonomous computer systems (called
agents) that can interact with decision makers and then learn, perform, and delegate tasks on their behalf. With its ability to handle a large amount of information
from heterogeneous sources in dynamically changing environments, a multiple
agent-based system can significantly improve the company's business intelligence
and operational efficiency. Though rarely used in demand planning, this paper
proposes a multiple agent-based system for demand forecasting of ice cream
which poses unique challenges due to volatility and seasonality of ice cream consumption. To validate the usefulness of the proposed system for demand planning,
the forecasting outcomes of the proposed system was compared to those of traditional forecasting techniques. Our experiments showed that the proposed multiple
agent-based system outperformed its traditional forecasting counterparts in terms
of its accuracy and consistency.

1 Background

In the United States (U.S.) alone, approximately 1.52 billion gallons of ice cream,
including both hard and soft-serve, was produced in 2009. Indeed, ice cream is
one of the most favorite dairy products in the U.S. as evidenced by the fact that
more than 90% of U.S. households consumed ice cream every year [5]. To meet
this insatiable demand for ice cream, ice cream producers must make an accurate
prediction of ice cream consumption and then develop their production plans.
Due to the sensitivity of ice cream demand to weather (temperature), ice cream
production typically follows a seasonal production of ice cream starts in March
and April to fill retail and foodservice pipepattern [1,2]. For example, the prolines
in the late spring and early summer. Ice cream production usually peaks in June
and then remains strong through August to satisfy surging demand during the hot

J. Casillas et al. (Eds.): Management Intelligent Systems, AISC 171, pp. 227–238.
springerlink.com　　　　　　　　© Springer-Verlag Berlin Heidelberg 2012

summer season. Afterward, production rapidly declines through the end of the year. To make matters more complicated, ice cream consumption changes dramatically from one year to another for inexplicable reasons as shown in Figure 1. During a 20 year span between 1987 and 2006, Figure 1 shows a wild swing from a peak of 2002 to valleys of 1990 and 2004. As such, the use of traditional forecasting techniques (e.g., moving average, exponential smoothing, and regression) which rely heavily on historical data may not be suitable for predicting ice cream consumption.

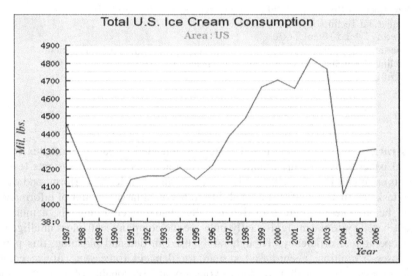

Fig. 1 Ice Cream Consumption Pattern in the U.S. (1987-2006)
Source: Gould, B. (2012), *Understanding Dairy Markets*, Unpublished Report,
http://future.aae.wisc.edu/data/annual_values/by_area/2348?tab=sales, Madison,
Wisconsin: Department of Agricultural and Applied Economics.

As a viable alternative to traditional forecasting techniques, we propose a multiple agent-based forecasting system which not only utilizes historical time series data, but also takes into account causal factors (e.g., weather, advertisement, promotional campaign, price discounts) affecting ice cream demands. Generally, an agent-based system is one of the distributed problem-solving techniques that divides a decision problem into sub-problems and solves those sub-problems using independent entities called agents [8]. Each agent can use different methodology, knowledge, and recourses to process given tasks. According to Reis [10], an agent refers to an autonomous entity that can take certain actions to accomplish a set of goals and can compete and cooperate with other agents while pursuing its individual goals. An agent is characterized by its ability to exploit significant amounts of domain knowledge, overcome erroneous input, use symbols and abstraction, learn from the decision environment, operate in real time, and communicate with others in natural language [9]. Exploiting such characteristics, the agent-based system was successfully utilized to develop aggregate demand planning and forecasting within the supply chain framework [7,11,12]. Unlike traditional forecasting techniques, an agent-based system is not predicated on the premise that future demand will follow the pattern of past demand. In particular, an agent-based system can be useful for predicting the future demand of new products and innovative services that were not extant in the past. Also, it is better suited to predict "erratic" (counter-intuitive) demand patterns because it can capture past (base-line agent), current (causal agent), and future (pattern agent) customer behaviors by combining human expertise and data mining techniques. Given dynamic and erratic patterns of climatic data, we utilize a multiple agent-based forecasting system (MABFS) to predict ice cream demand. By utilizing the concept of agent-based system, this research proposes a multiple agent-based framework that is comprised of several base-line forecasting agents and one causal agent that are designed to extract both historical and erratic patterns based on a dynamic weighting scheme.

2 An Architecture of the Multiple Agent-Based System

The architecture of the proposed MABFS aims at providing a framework for coordinating the interactions among multiple agents graphically displayed in Figure 2. In this architecture, agents are software entities that individually carry out their designated tasks. These agents include an interface agent, a data collection and monitoring agent, a coordination agent, and three forecasting agents. Especially, the forecasting agents fulfill forecasting duties by using various forecasting techniques. The forecasting agents consist of two base-line agents (EXP agent and BJ agent) that utilize exponential smoothing and Box-Jenkins methods, and one causal agent (TMP agent) that utilizes the climatic (temperature) information obtained from an outside source. While the base-line agents (i.e., EXP and BJ agents) determine the parameters for the proposed MABFS, a coordination agent integrates forecasting results obtained from the forecasting agents and then improves the overall forecasts using the agent-based system.

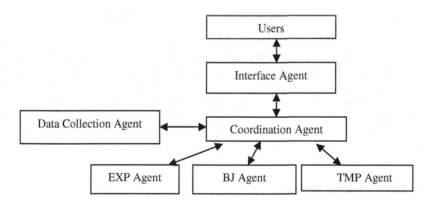

Fig. 2 The Basic Architecture of the MABFS for Ice Cream Demand Forecasting

Within the above architecture, the EXP agent found the best exponential smoothing parameter with a level smoothing weight of 0.59102 and a seasonal smoothing weight of 0.65209 using the Forecast Pro® program. Similarly, the BJ agent discovered the best ARIMA parameter with a multiplicative seasonal ARIMA (1, 1, 0) (1, 0, 0). In the meantime, the TMP agent using climatic data captures causal factors and sets the rules as follows.

(1) If the forecasted temperature is higher than 60 degree (°F), ice cream demand should be forecasted by using the following regression model:

Ice cream demand volume = -7.59 + 1.61 × temperature (°F).

(2) Otherwise, demand should be forecasted by using the following regression model:

Ice cream demand volume = 83.45 + 0.24 × temperature (°F).

After combining the forecasting results of three forecasting agents with proper weights given to each individual result with the help of a coordination agent, the proposed MABFS makes the overall forecast. The rationale for assigning a different weight to each forecasting agent being that a combination of different forecasting methods may produce more accurate results than does an individual forecast due to multiple sources of knowledge [3,4]. Also, the MABFS was designed to re-evaluate assigned weights of different agents periodically so that it can reflect dynamic changes in time series data over time. In particular, we developed a threshold weight-assigning procedure as part of the weighting scheme for the coordination agent. In this procedure, the mean squared error (MSE) is used to evaluate the performance of each forecasting agent and then will be used to determine the proper weights to be assigned to the agent. The weight to be assigned to each candidate agent is calculated as follows.

$$w_i = \frac{e*-mse_i}{\sum_{i \in C}(e*-mse_i)} \quad \text{for all } i \in C; \qquad w_i = 0 \text{ ,otherwise} \qquad (1)$$

where w_i is the weight for candidate agent i, mse_i is the MSE for candidate agent i, C is an index set for the candidate agent, and $e*$ is a threshold weight which is set to be larger than the MSE of the candidate agent. Using equation (1), the assigned weights are adjusted dynamically to improve forecasting accuracy. Since the MSE for each agent is likely to change when the new data arrive, the candidate set and the assigned weights need to be re-evaluated to reflect the best combination of the overall forecast. This dynamic re-evaluation of weights allows the MABFS to adapt to potential changes in time series data.

3 Model Experiments

To verify the usefulness and practicality of the proposed MABFS, we tested it with actual data. The actual data used in this study were obtained from the

International Dairy Foods Association, by the courtesy of Mr. Bill Bradbury. Ice cream sales data available from this source included weekly ice cream sales for the entire U.S. from July 1994 to Nov. 1996 as well as weekly frozen novelty sales (e.g., popsicles, etc.) for the same geographic areas. The ice cream sales data were primarily collected through the point-of-sale system (POS) at the checkout counters of supermarkets across the U.S. For illustrative purposes, however, the data that we used our experiments and analyses were limited to the regional data in the Louisville metropolitan area in Kentucky. We chose the Louisville market as the experimental target, because Louisville has a distinctive four seasons with varying temperatures throughout the year. Also, it is noted that we used the *Perceived Temperature* as climatic information instead of actual temperature. The rationale being that a variable called *Perceived Temperatures* provides a better insight into how the ice cream consumer actually perceives his/her sense of warm temperature and influences how much he/she craves for ice cream, as observed by Bradbury [1]. The *Perceived Temperatures* combines heat index readings (for temperatures of 80 degree Fahrenheit and above) and wind chill readings (for temperatures of 79 degree Fahrenheit and below) on a single scale.

The collected data were first examined to determine whether any time lag effect existed, because causal factors might have a delaying influence on the ice cream demand. The extent of a time lag effect can be gauged by examining the correlation coefficient between successive lagged temperature data and the sales data. The highest correlation coefficient between temperature data and ice cream sales data was obtained at lag = 0 weeks, which indicates no significant lag effect. Two regression models are then developed to accommodate the differences in the causal effects emanating from different temperatures. The weather may have a different impact on the sales of ice cream varying from warm to cold days. Figure 3 shows the graphical representation of ice cream sale volumes plotted against temperature.

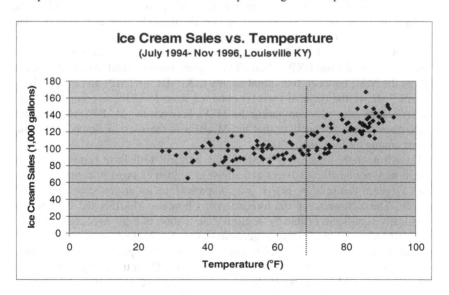

Fig. 3 Ice Cream Sales versus Temperature

Table 1 Result of the Regression Analysis for Temperature as a Causal Factor

	Temp < 60 °F	Temp ≥ 60 °F
Correlation Coefficient	0.198	0.782
Slope (in thousand gallons/°F)	0.244418	1.611384
Intercept	83.45115	-7.59337
R squared	0.039042	0.612367

Two distinct relationships between ice cream sales and the weather can be observed around temperature = 60 °F. Hence, two regression models are created: One for temperature equal to 60 °F and above to represent warmer weather conditions and another for temperature equal to 59 °F and below to represent colder weather conditions. The results are shown in Table 1. As expected, the correlation coefficient for ice cream sales and warm weather is significantly higher than the one with cold weather and the difference in correlation coefficient between warm and cold weather is statistically significant at $\alpha = 0.05$. This result indicates that the temperature is a more important causal factor during the summer than during the winter.

4 Results and Discussions

Results from three individual forecasting methods represented by agents BJ, EXP, and TMP and the MABFS are summarized in Table 2. The test results between each method (agent) and the MABFS are shown in Tables 3, 4 and 5. In Table 2, the result shows that MABFS has the smallest forecasting errors with a mean square error (MSE) of 54.53, and performed most consistently, measured by the standard deviation of squared errors, among all methods used in this study. The differences between MABFS and EXP, as well as MABFS and TMP, are statistically significant at $\alpha = 0.05$. Figures 4 through 7 summarize the ice cream demand forecasts obtained from EXP, BJ and TMP agents versus actual ice cream sales as well as the MSE between individual agents (EXP, BJ, or TMP) and the MABFS. Figures 7 shows the demand forecast obtained from MABFS as compared to actual ice cream sales. By comparing the MSEs between the MABFS and the other single agent, we can conclude that the MABFS is the most accurate forecasting system among all four methods.

Though a single forecasting method may perform well for a certain period of time, it may not be able to detect and adjust to sudden changes in consumer behavior. To see if this premise is true, we compared forecast errors between MABFS and BJ. The differences in the forecast errors between MABFS and BJ were not statistically significant (Table 5). A close examination of Figures 5a and 5b shows that ARIMA model, the BJ agent, generates exceedingly small forecast errors during the periods between July 1994 and May 1995, but produces very unstable forecasts after January 1996. Although the overall MSE of the BJ agent may seem to be as good as MABFS, ARIMA should not be chosen as the demand forecasting method because the demand pattern apparently deviates significantly from the

ARIMA model over time. Overall, the results of this study indicate that the MABFS enabled us to exploit the best features of existing forecasting techniques, resulting in the lowest overall forecast errors, and performed consistently well over time.

Table 2 A Comparison of Forecast Errors among Single Agents and MABFS

Methods	Ice Cream Sales, Louisville, Jul 94 – Nov 96			
	TMP	BJ	EXP	MABFS
MSE	122.89	65.83	72.68	54.53
Standard Deviation of Squared Errors	181.87	192.99	122.76	102.71
Max. Squared Errors	1352.66	1618.45	586.12	597.46
Min. Squared errors	0.01	0.00	0.00	0.01

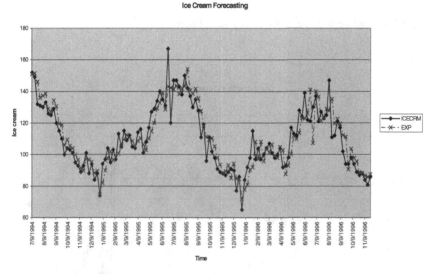

Fig. 4a Forecasting with the Exponential Smoothing Method (EXP agent)

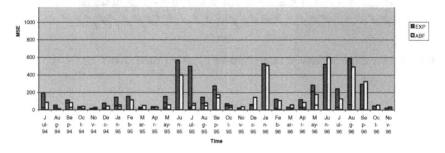

Fig. 4b Comparison of Forecast Errors between EXP and MABFS over Time

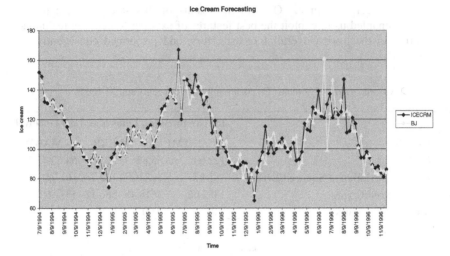

Fig. 5a Forecasting with the ARIMA method (BJ agent)

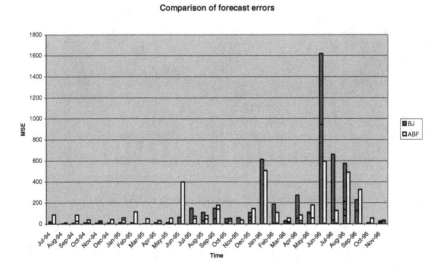

Fig. 5b Comparison of Forecast Errors between BJ and MABFS Over Time

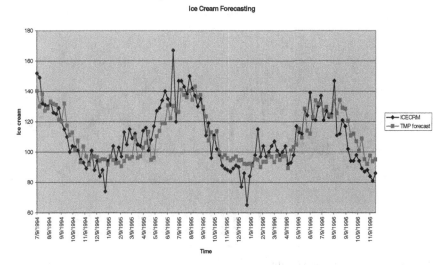

Fig. 6a Forecasting with the Causal Forecasting Method with Climatic Data (TMP agent)

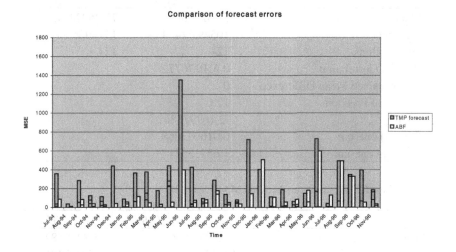

Fig. 6b Comparison of Forecast Errors between TMP and ABF over Time

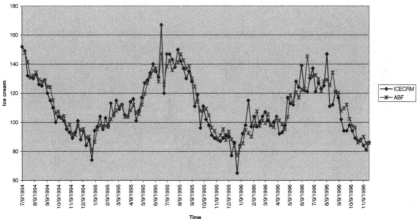

Fig. 7 Forecasting with the MABFS

Table 3 t-Test for forecast errors between MABFS and EXP

	MABFS	*EXP*
Mean	54.52508896	73.2597451
Variance	10549.51434	15149.2734
Observations	124	124
Pearson Correlation	0.650176772	
Hypothesized Mean Difference	0	
Degree of Freedom	123	
t Stat	-2.167975861	
P(T<=t) one-tail	0.016043068	
t Critical one-tail	1.657335815	

Table 4 t-Test for forecast errors between MABFS and BJ

	MABFS	*BJ*
Mean	54.52508896	65.8290587
Variance	10549.51434	37245.0406
Observations	124	124
Pearson Correlation	0.639540747	
Hypothesized Mean Difference	0	
Degree of Freedom	123	
t Stat	-0.840284338	
P(T<=t) one-tail	0.201189704	
t Critical one-tail	1.657335815	

Table 5 t-Test for Forecast Errors between MABFS and TMP

	MABFS	TMP
Mean	54.52508896	122.783489
Variance	10549.51434	33343.3654
Observations	124	124
Pearson Correlation	0.564263814	
Hypothesized Mean Difference	0	
Degree of Freedom	123	
t Stat	-5.041902091	
P(T<=t) one-tail	8.03559E-07	
t Critical one-tail	1.657335815	

5 Concluding Remarks and Future Research Directions

A demand forecast is the prediction of how much the company's products and services will be consumed by the customers. To make such prediction, the demand forecast typically relies on forecasting techniques that extrapolate historical demand (sales) data. However, these techniques may not be suitable for predicting ice cream demands, since ice cream demands often depend on the fluctuating weather condition and thus their historical data are no longer best representatives of the future ice cream sales.

Considering dependencies of ice cream demands on weather (temperature), this paper is one of the first to introduce the multiple agent-based forecasting system that combines the strengths of three popular forecasting techniques (exponential smoothing, Box Jenkins method, causal method) and then integrate the results of those techniques with the added capability to gather, filter, and synchronize climatic (weather) data.

Unlike traditional forecasting techniques, the proposed agent-based forecasting system has the ability to delegate forecasting tasks to each agent and then revise weights assigned to each agent dynamically based on its learning mechanism. As such, it is designed to improve overall forecasting accuracy and consistency. Our model experiments actually validated the fact that the proposed agent-based forecasting system outperformed its traditional counterparts in terms of its forecasting accuracy and consistency over time. In addition, the proposed agent-based forecasting system requires much less time-series data than the traditional forecasting techniques and thus can be more effective in forecasting the demand of new products or products with a short life cycle.

Another major contribution of the proposed multiple agent-based forecasting system is its ability to handle wild demand variations in erratic consumer behaviors that is missing from traditional forecasting methods. The proposed system allows us to use one of the forecasting agents that best matches the demand data at an early stage of the time series, while utilizing other dominant agents that produce the least forecast errors at the later stages of the time series. In other words, the proposed agent-based forecasting system can identify and utilize the best forecasting methods at different stages of the time series by dynamically re-evaluating and re-assigning weights to different forecasting agents. On the other hand, the traditional forecasting techniques would have required a remodeling process to capture sudden changes in consumer behaviors and thus had no choice but to discard the earlier portion of the time series data that reflect the old consumer behavior. Thus, the traditional forecasting technique was likely to generate unreliable forecasts due to the truncated time series data. Despite aforementioned advantages of the proposed agent-based forecasting system over the traditional forecasting techniques, it can be further extended to deal with the different degree of "fuzziness" involved in weather related information. Perhaps, future research efforts can incorporate fuzzy neural network concepts into ice cream demand forecasts. Another future line of research may look into the potential inclusion of other intelligent agents that exploit visual sensory pattern recognition, rule-based reasoning, and real-time prediction.

References

1. Bradbury, B.: The weather factor: Does a 100(degree) day beat a TPR? Frozen Food Age 48(8), 8–9 (2000)
2. Cawthorn, C.: Weather as a strategic element in demand chain planning. The Journal of Business Forecasting Methods and Systems 17(3), 18–21 (1998)
3. de Menezes, L.M., Bunn, D.W., Taylor, J.W.: Review of guidelines for the use of combined forecasts. European Journal of Operational Research 120(1), 190–204 (2000)
4. Granger, C.W., Ramanathan, R.: Improved methods of combining forecasts. Journal of Forecasting 3, 197–204 (1984)
5. International Dairy Food Association. Ice cream sales & trends (2012), http://www.idfa.org/newsviews/media-kits/ice-cream/ice-cream-sales-and-trends/ (retrieved on February 16, 2012)
6. Makridakis, S.: Why combining works? International Journal of Forecasting 5(4), 601–603 (1989)
7. Liang, W., Huang, C.: Agent-based demand forecast in multiechelon supply chain. Decision Support Systems 42, 390–407 (2006)
8. Min, H.: Artificial intelligence in supply chain management. International Journal of Logistics: Research and Applications 13(1), 13–39 (2010)
9. Newell, A.: Putting it all together. In: Klahr, D., Kotovsky, K. (eds.) Complex Information Processing: The Impact of Herbert A. Simon. Lawrence Erlbaum Associates, Hillsdale (1989)
10. Reis, B.Y.: A multi-agent system for on-line modeling, parsing and prediction of discrete time series data. In: Mohammadian, M. (ed.) Computational Intelligence for Modeling, Control & Automation, pp. 164–169. IOS Press (1999)
11. Yu, W., Graham, J.H., Min, H.: Dynamic pattern matching using temporal data mining for demand forecasting. In: Proceedings of the 2nd International Conference on Electronic Business, Taipei, Taiwan, December 10-13, pp. 400–402 (2002)
12. Yu, W., Graham, J.H.: A multiple agent architecture for demand forecasting in electronic commerce supply chain systems. In: Proceedings of the 17th ISCA (The International Society for Computers and their Applications, Inc.) International Conference on Computers and their Applications, San Francisco, CA, April 4-6, pp. 462–466 (2002)

Manual Intervention and Statefulness in Agent-Involved Workflow Management Systems

Pavlos Delias[1], Stelios Tsafarakis[2], and Anastasios Doulamis[2]

[1] Kavala Institute of Technology, Greece
[2] Department of Production Engineering & Management, Technical University of Crete, Greece

Abstract. Lack of adaptability within WorkFlow Management Systems (WFMS) has been early identified as one of their limitations. WFMS suffer from disadvantages such as not supporting the dynamic incorporation/modification of process models and poor adaptability of process models at runtime. The static workflow definition and its passive interpretation does not allow WFMS to demonstrate flexible behavior and to deal with real-life situations, such as fast changing customer requirements and enterprise goal shifts. In this work we propose the design and development of two features (manual intervention and statefulness), which are expected to tackle this limitation. Our work considers and agent-based environment for the WFMS implementation.

Keywords: Workflow Management, Intelligent Agents, Information Systems.

1 Introduction

The term workflow signifies the automation of a business process, which is defined within a process definition. WorkFlow Management Systems (WFMS) are supposed to guarantee that during run time every process is executed according to its definition, typically with little or no human intervention. Nevertheless, there are circumstances that a strict, automatic execution of the definition does not produce the desired outcome. There are some exceptional circumstances where the user needs to override the initial definition and manually change the execution path of the process. For instance, the user may detect invalid data in the process input data, or new information may become available. In such a case the process needs to rewind and resume execution from a previous step. Moreover, in a business environment, special events emerge (e.g., an ad-hoc agreement with a special customer) that may lead to different process rules (e.g., a document is not delivering or a deadline is getting loose). Ideally, the workflow administrator should have some tools to handle these exceptional circumstances, and manually specify the activity node that the system should execute next.

This lack of flexibility and the non-existence of *manual intervention* support has been early identified as a limitation of WFMS [1]. Systems that didn't provide

J. Casillas et al. (Eds.): Management Intelligent Systems, AISC 171, pp. 239–249.
springerlink.com   © Springer-Verlag Berlin Heidelberg 2012

this functionality were noticed to irritate end users, who felt that the systems were merely enforcing rigid rules [2]. Manual intervention can be expressed by many ways: performing the tasks manually, skipping some tasks, modifying the control flow, rewinding and repeating some tasks, providing manually values to evaluate conditions etc.

On the other hand, *statefulness* refers to the capability of maintaining the status of a process, recognizing at any moment what has been accomplished and what is yet to come, or at least what is coming next. In the workflow management context, wrapping stateful behavior is an innate requirement, which becomes crucial in case of long lasting workflows.

Two general modes to integrate this workflow functionality are popular [3]:

- The system determines the next task by querying the data contained in the process instance itself. The system is unaware of both the tasks that are already realized and the tasks that may follow. All state information is contained within the process instance. Thus, the instance's data need to indicate who is assigned to that unit of work, and all history information about what happened in the past. Examples of this style of implementation in an Agent-involved WFMS context can be found in [4-6]
- The system knows everything about the process instance, and the instance itself doesn't contain any history or "stateful" information. In [7, 8] this general implementation style is followed.

In this work we address both these sought-after features with a different approach, under an Agent-involved WFMS context. The term "Agent-involved Workflow Management Systems" (AWFMS) was introduced in [9] to capture all the cases where software agents and WFMS are intertwined. It is actually an umbrella term used to refer to cases like:

- *Agent-based* workflow [10] where "the software agents take full responsibility for process provisioning, enactment and compensation, with each agent managing and controlling a given activity or set of activities",
- *Agent-enhanced* workflow [11], "a technique whereby intelligent, distributed, autonomous software agents are used to improve the management of business processes under the control of a workflow management system",
- or even Agent-enabled systems [8], where broker agents enable workflow instances in distributed WF engines and they are used as front-end and they communicate through APIs with the WF engines.

In a previous work [9], we advocated what are the advantages of using an agent approach, and what the potential drawbacks could be. The interested reader is prompted therein for a thorough discussion.

The rest of the paper is organized as follows: The next section presents the intuition and the basic theoretical foundations of the proposed methods. The business process of "direct mail campaign" which will be used for illustration purposes is presented in Section 3. In the same section the application of the methods on the pivot process are explained, i.e. the methodology to reach statefulness in

AWFMS as well as its application on the mail campaign process are described. In section 4, a complementary database schema is presented as well, in order to further support our argument. Finally, a brief discussion concludes the paper.

2 The Proposed Methods

2.1 Manual Intervention

In this paper, manual intervention implies that a user can choose a specific point of a process and start execution from that point. Moreover, he/she can also choose to execute just a special part of the process and not the entire workflow. To succeed in allowing this, the notion of "*state*" is incorporated. The concept of "*state*" is analogous to a milestone within a workflow. Typically, a milestone indicates the end of a stage and it goes together with some specific deliverables. Thus, if there is a need to check whether the milestone is reached, it is sufficient to check if the deliverables are okay. This abstract idea is adopted in the proposed system. In particular, the process designer indicates a limited number of states that roughly split the workflow process into phases. A state is actually the interval between two milestones: one indicating the starting point and the other the finishing point. Often the finishing point is the process end. Following the procedure, the designer associates a set of "*requirements*" with every state. If the requirements are indeed accomplished, the user may begin workflow execution from that particular state. As it will be described in section 3, a "*requirement*" is a synonym for file. This technique allows end users to:

- Skip any number of activities, by providing manually the expected deliverables.
- Rewind workflow execution to a previous step and repeat process execution for a number of times.
- Intervene to the outcomes of the workflow without obstructing the process execution, by manually modifying the requirements' files.
- Execute just a part of the workflow, asynchronously if allowed by the business logic.

2.2 Statefulness through Document-Centric Stigmergy

The proposed approach can be characterized as a "*document-centric stigmergy*", a novel term, introduced here. Firstly, the use of "*stigmergy*" is explained:

Stigmergy is formed from the Greek words "*στίγμα*" (stigma – sign) and "*έργον*" (ergon – action), and it was coined in the 1950's by Grassé, a French entomologist who used the term to describe the indirect communication taking place among individuals in social insect societies [12]. Stigmergy captures the notion that agents' actions leave signs in the environment. Thus, if all agents are capable to understand and interpret these signs, they will determine their subsequent actions in such a way that the emergent behavior of the system is the desired one.

Stigmergy has been used as an optimization tool by a plethora of researchers [13], exploited mainly as a simple yet effective mechanism for agents' coordination. Nevertheless, the approach proposed here does not follow the strict formulation, as described in [13]. It rather uses the conceptual initiative of stigmergy to construct an organic design for workflow management. Actually, although the mechanism of stigmergy is mostly popular in insects societies, its original concept has indeed been analyzed as a coordination framework for collaborative activities in other environments as well [14], (e.g., humans [15], or software agents [16]).

In general, in order to apply a stigmergy mechanism, the following elements should be considered [15]:

- An *environment*, which is described by a *state*.
- The *dynamics* of the environment, which governs the evolution of its state over time.
- The agents' *sensors* that allow agents to interpret the state of the environment.
- The agents' *actuators* that allow agents to modify the environment.
- A *method* that configures agents' actions based on the sensed state of the environment.

3   Scenario-Based Evaluation

3.1   A Pivot Process

The proposed AWFMS approach is domain abstract, meaning that it could be applied to any domain. However, for illustration purposes and to make our contribution more evident, a pivot process from the marketing domain is presented. In point of fact, marketing is a very convenient domain for workflow management applications: Marketing processes are far more flexible and versatile than production processes since the process flows are not rigidly defined, heterogeneous resources are involved, and high customization per customer is required. However, the regular activities required to carry out a marketing process (e.g., writing a report, extracting data from databases, organizing campaigns, schedule meetings, etc.) have good potentials to be monitored by information systems. To such a context, automation prospects are significant and tightly related with the workflow perspective.

3.2   Direct Mail Campaign Automation

Direct mail marketing refers to sending an advertisement, offer, announcement, reminder or other item to a prospective customer. Kotler [17, p. 536] identifies direct-mail marketing as a major marketing communication mode, and as an important mean to inform, persuade and remind consumers about the brand. In fact, direct-mail campaigns serve multiple communication objectives, such as producing prospect leads, strengthening customer relationships, informing and educating customers, reminding customers of offers, and reinforcing recent customer purchase decisions.

Direct mail marketing (as opposed to mass marketing e.g., advertisement) is a targeted communication and is based on a one-to-one, brand-customer basis. It is becoming increasingly popular, as it can be personalized, which constitutes a very import feature in demassified markets. Direct mail campaigns include a broad mixture of tools and activities such as budgeting, forecasting, managing digital assets, and dealing with complex scheduling requirements. Because of the proliferation of products and brands, even larger number of market segments, fierceness of competition, and overall acceleration of change, direct mail campaigns have become complex and their planning and administrative decisions must be made under increasing time pressure. Indeed, timing and sequencing activities within a campaign is one of the critical decision variables [18].

To support the management of direct mail campaigns, and provide organizations with automation potentials, some vendors (SAP Table 1, Microsoft[1]) provide marketing campaign blueprints so that charting a campaign project and monitoring its workflow is facilitated. In this work, the basic outline of a direct mail campaign process is maintained, resulting in the detailed workflows described in the next sections.

Table 1 SAP Business Workflow in Campaign Automation. *Source: http://help.sap.com/ saphelp_crm70/helpdata/EN/45/cbced6f771fae10000000a1553f6/content.htm*

Workflow templates for Campaign Automation
WS14000061 Transfer Target Group to Channel
WS14000062 Create Target Group
WS14000062 Create Target Group and Channel Transfer
WS14000064 Send E-Mail to Employee Responsible
WS14000065 Authorization by Employee Responsible
WS14000066 Adding a Business Partner to a Target Group
WS14000067 Deleting a Business Partner from a Target Group
WS14000068 Start Target Group Optimization
WS14000069 Transfer Respondent to Channel
WS14000070 Start Subsequent Step Without Executing
WS15100040 Start Media Campaign

The rough main activities of a marketing communication process (and thus of a direct mail campaign) have been analytically described in popular handbooks of marketing [17]. These are roughly the following: 1) *Identify target audience*, 2) *Determine Objectives*, 3) *Design Communications*, 4) *Select Channels*, 5) *Establish Budget*, 6) *Decide on Media Mix*, 7) *Measure results*. However, it is clear that a campaign can focus on some special steps or it can omit some others, it can

---

[1] http://ce.microsoft.com/en-us/templates/
TC012330891033.aspx?CategoryID=CT102115851033

execute the steps sequentially or parallelize the process, according to the campaign's special requirements. Moreover, each step may contain different activities in a variety of flows. Because of the above particularities, campaigns may significantly differ from one another.

3.3 Applying Manual Intervention

Consider the pivot process "directMail", described in section 3.2. The states identified are:

- "NOT_STARTED". The process instance has been created but it hasn't started execution yet. It may be used to signify that a process id has been assigned to the instance but no other action has been performed (e.g., workflow assignment)
- "ESTABLISH_MARKETS". This is the initial state of the workflow. The workflow has been assigned and it is ready to start execution. The whole process will be executed.
- "SEGMENTATION". The process instance will start execution from the segmentation point, that is, it skips the "Establish_Markets" step.
- "QUANTIFY_TAM". Starts the process from the quantification of the total available market point. The steps of "Establish_Markets" and "Segmentation" are skipped.
- "BUDGET_RF". Begins executing the budgeting of response factor. All the previous steps are skipped.
- "PREPARE_PIECE". This state refers to the second phase of the process and if selected, it orders to skip the entire marketing research phase (which includes the states described previously).
- "LAUNCH_CAMPAIGN". This state orders that the two first phases (marketing research and prepare piece) should be both skipped.
- "SINGLE_SOLICIT_DESIGN". While all the previous states indicate that the process instances should start execution from a specific point and continue until the whole workflow is completed, this state (along with others that hold a prefix "SINGLE_") indicates that just a part of the work should be executed. This particular state refers to soliciting vendors to design the artwork for one marketing piece.
- "SINGLE_REVIEW_DRAFT". A state that applies the reviewing of the artwork of one marketing piece and then terminates.
- "SINGLE_CREATE_JOB_SCHEDULE". This state refers to the CreateJobSchedules class that the product manager implements to create work schedules for every group of assistants.
- "SINGLE_ASSISTANT_LAUNCHING". This state is about the execution of a task by one assistant. The reason to create such a state is that assistants may execute their assigned task at a different time, and asynchronously publish the results of their work.

When a state is selected as the starting point of a workflow execution, a requirements check is performed. If this check returns a positive answer, then the user is able to intervene to the process by altering the process starting point. The system assures that all states are related to the correct process instances through a process id, which is passed as a formal parameter to all the workflows and sub-workflows that correspond to a state.

3.4 Implementing a Stateful Approach

In the proposed document-centric approach, and with respect to the direct mail campaign process, the pertinent elements are defined as follows:

- **Environment:** The environment should be directly related with the process instance, and its state should exhibit the current execution state. By setting the environment to the process instance itself, a milestone in the process definition can be used to declare the environment's state. For this purpose, the notion of "*state*" which was described in section 2 can be exploited.
- **Dynamics:** States follow one another according to the process definition. Yet, a state cannot begin unless its requirements are fulfilled. These requirements are the core of the document-centric approach. More specifically, a document (or file in general) is an atomic piece of work of a process. Every document corresponds to the results of one (or more) atomic activity, but the inverse does not necessarily happen, since there may be some intermediate activities which do not need to be stored to a file. However, storing results in a document is the only way of saving process instances' data permanently. Documents are saved during runtime (process execution) and usually they follow a particular template. Thus, every document is a partial deliverable of a process instance and has a specific time point when it is delivered. Each state comprises a set of documents as its prerequisites. These documents are state requirements, and they are specified by the process designer during build-time.
- **Sensors:** Documents' paths are stored to a database. Agents (workflow performers) query the database to learn which requirements are fulfilled for a particular process instance.
- **Actuators:** When an agent performs a workflow, upon successful implementation of some work units, it updates the database.
- **Method:** Agents perform a workflow according to its definition. They sense the environment, interpret the signs and begin execution from a particular point (state). They know what they should execute next, since they can interpret the process definition and realize the point at which the process instance exists.

4   A Complementary Database Schema

An important capability of workflows is that they can be persisted (saved and reloaded at a later time). Workflow persistence is especially important when developing applications that coordinate human interactions, since those interactions can take a long period of time. But persistence is also applicable to other types of applications. Without persistence, the lifetime of workflows is limited. When the application is eventually shut down, any workflow instances simply cease to exist. Workflow persistence means to save the complete state of a workflow to a durable store such as a database or SQL file.

Nevertheless, the database schema is an important aspect of the application. In this section, a schema that is capable to support the *document-centric stigmergy* approach is proposed (Figure 1). Except for the "monitor_details" table which is used for monitoring reasons (out of the scope of this paper), the rest seven tables are exactly the tables that are needed to store workflows according to the document-centric approach. In particular, each workflow model has a specific *process type*, which corresponds to its definition. Process types are stored in the process_type table which needs to contain just the name of the process type (and maybe a short textual description). As discussed in section 2, for every process type, the process designer indicates a few "milestones" within its definition. Each milestone corresponds to a "*state*". Thus, the state table is incorporated. Every state is related with a specific process type and a workflow class that should be initiated upon the state's activation. Workflow classes are actually the process definitions and they are stored to the workflows table, along with a hint of what is the

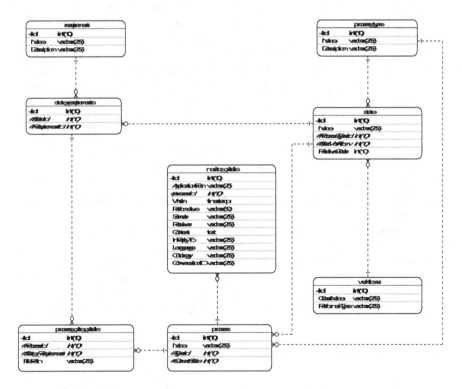

Fig. 1 The proposed database schema

appropriate performer type. An important notice is that the database does not need to store any additional information (e.g., regarding the flow of the activities, or the performers' types hierarchy) since this piece of information is hard-copied either into the body of the agents, or into the modular components of the application (e.g., workflow classes maybe deployed by their .jar files).

The process table refers to the process instance and it is used to track its execution details, which are actually stored in the process_data_details table. As mentioned in the previous section, the execution details (not referring to the monitored elements) are documents (files) that are delivered during the runtime. The process_data_details table is used to store the relative file paths. Every file is a "*requirement*", and as such it is defined within the requirement table. Finally, the state_requirements table is used to model an *m*-to-*n* relationship between the requirements and the states, that is every state may have zero or more requirements while a requirement may belong to one or more states.

The great advantage of this schema is that it is minimal respective to the application needs. It fully exploits agents' statefulness and the application's programming language to avoid storing large volume of data. Agents (as workflow performers) are fully conscious of what is the workflow they are executing, which activity follows next, what conditions will allow the transition to which activities, to whom they may delegate a piece of work, what is their type and role and where they should address in order to get informed about other agents or process related data.

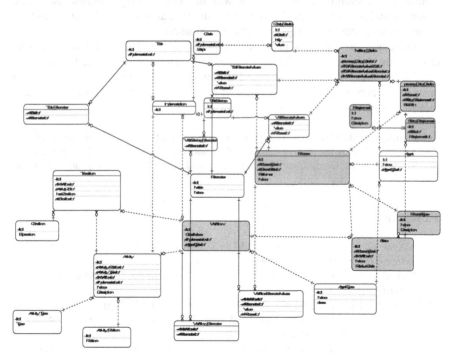

Fig. 2 A database schema which does not exploit application's features

For this advantage to become more evident, Figure 2 illustrates a database schema that would be needed if the agents' awareness was not exploited and process definitions were not hard-copied as JAVA classes, but they were stored to the database. The tables shaded in blue are the tables used also in the minimal schema. Although the schema of Figure 2 is not the only one that can respond to the issues mentioned in the previous paragraph, it becomes apparent that unless we exploit agenthood and a stigmergy approach, a significant overhead is added to the database, regarding process definition data, execution auditing activities, participants' hierarchy and workflow implementation details.

5 Conclusions

Manual intervention may provide the AWFMS with flexibility, but it incurs an added risk and cost. The risk associated with manual intervention is that when you override the process definition with a subjective – manual manner, there is no guarantee that the resulting process will be valid and sound. Moreover, when the requirements are fulfilled manually, there is also no guarantee that they have the appropriate content format or that they comply with the specified business rules. These factors make more error-prone the process instances that were manually mediated. The additional cost is related with the poor logging of manual activities. Since manual actions escape the system monitoring, auditing and backtracking become no longer possible for those particular instances.

Concluding, as business processes become more volatile, and as they start crossing the organization's boundaries, their interactions need a rather sophisticated supervisor. WFMS should find a way to manage the dynamic nature of business processes, and the two suggested features (manual intervention and statefulness) are parts of such a way.

Acknowledgement. This paper is supported by the project E-Park, "Exploitation of new Technological Trends for payment and handling public parking" approved under the Interreg III Programme, Greek-Cypriot cooperation and funded from European Union and Greek National funds.

References

1. Alonso, G., et al.: Functionality and Limitations of Current Workflow Management Systems. IEEE Expert 12(5) (1997)
2. Nutt, G.J.: The evolution towards flexible workflow systems. Distributed Systems Engineering 3, 276–294 (1996)
3. Eeles, P., Houston, K.A., Kozaczynski, W.: Building J2EE^{TM} Applications with the Rational Unified Process. Addison-Wesley Object Technology Series. Addison-Wesley Professional, Indianapolis (2003)
4. Barbara, D., Mehrotra, S., Rusinkiewicz, M.: INCAs: Managing Dynamic Workflows in Distributed Environments. Journal of Database Management 7(1), 5–15 (1996)

5. Budimac, Z., et al.: Lessons Learned From the Implementation of a Workflow Management System Using Mobile Agents. Novi. Sad. Journal of Mathematics 36(2), 65–79 (2006)
6. Cao, J., et al.: A dynamically reconfigurable system based on workflow and service agents. Engineering Applications of Artificial Intelligence 17(7), 771–782 (2004)
7. Tarumi, H., et al.: Work Web system–multi-workflow management with a multi-agent system. In: Supporting Group Work: The Integration Challenge. ACM Press, United States (1997)
8. Verginadis, Y., Mentzas, G.: Agents and workflow engines for inter-organizational workflows in e-government cases. Business Process Management Journal 14(2), 188–203 (2008)
9. Delias, P., Doulamis, A., Matsatsinis, N.: What Agents Can Do in Workflow Management Systems. Artificial Intelligence Review 35(2), 155–189 (2011)
10. Shepherdson, J., Thompson, S., Odgers, B.: Cross Organisational Workflow Coordinated by Software Agents. In: Cross-Organisational Workflow Management and Coordination. CEUR-WS.org, USA (1999)
11. Yan, Y., Maamar, Z., Weiming, S.: Integration of workflow and agent technology for business process management. In: The Sixth International Conference on Computer Supported Cooperative Work in Design. IEEE, London (2001)
12. Grassé, P.-P.: La reconstruction du nid et les coordinations interindividuelles chezBellicositermes natalensis etCubitermes sp. la théorie de la stigmergie: Essai d'interprétation du comportement des termites constructeurs. Insectes Sociaux 6(1), 41–80 (1959)
13. Dorigo, M., Bonabeaub, E., Theraulaz, G.: Ant algorithms and stigmergy. Future Generation Computer Systems 16, 851–871 (2000)
14. Schmidt, K., Wagner, I.: Ordering systems: Coordinative practices and artifacts in architectural design and planning. Computer Supported Cooperative Work 13(5-6), 349–408 (2004)
15. Van Dyke Parunak, H.: A Survey of Environments and Mechanisms for Human-Human Stigmergy. In: Weyns, D., Van Dyke Parunak, H., Michel, F. (eds.) E4MAS 2005. LNCS (LNAI), vol. 3830, pp. 163–186. Springer, Heidelberg (2006)
16. Ricci, A., Omicini, A., Viroli, M., Gardelli, L., Oliva, E.: Cognitive Stigmergy: Towards a Framework Based on Agents and Artifacts. In: Weyns, D., Van Dyke Parunak, H., Michel, F. (eds.) E4MAS 2006. LNCS (LNAI), vol. 4389, pp. 124–140. Springer, Heidelberg (2007)
17. Kotler, P., Keller, K.L.: Marketing Management, 12th edn. Pearson Prentice Hall, New Jersey (2006)
18. Roberts, M.L., Berger, P.D.: Direct Marketing Management, vol. 447. Prentice-Hall (1999)

A Statistical Approach to Star Rating Classification of Sentiment

Alexander Hogenboom, Ferry Boon, and Flavius Frasincar

Erasmus University Rotterdam, P.O. Box 1738, NL-3000 DR, Rotterdam, The Netherlands
{hogenboom,frasincar}@ese.eur.nl, ferry.boon@gmail.com

Abstract. Automated analysis of the ever-increasing amount of reviews available through the Web can enable businesses to identify why people like or dislike (aspects of) products or brands, yet to this end, a reliable indication of the intended sentiment of reviews is of crucial importance. This sentiment is typically quantified in universal star ratings, which are not always available. We propose and compare the performance of several statistical methods of automatically classifying star ratings of reviews represented by means of a binary vector representation, with features signaling the presence of sentiment-carrying words. A nearest neighbor classifier maximizes recall, whereas a naïve Bayes classifier excels in terms of precision, accuracy, and the root mean squared error of the assigned number of stars.

Keywords: Sentiment analysis, star ratings, nearest neighbor, naïve Bayes.

1 Introduction

The Web as it exists today encompasses a vast and ever-increasing amount of user-generated content. Popular Web sites like Twitter, Blogger, or Epinions enable anyone to write and publish short messages, blog posts, or reviews about anything at any time. Today's typical Web user exhibits a hunger for and reliance upon on-line advice and recommendations, yet in the wealth of user-generated content, explicit information on user opinions is often hard to find, confusing, or overwhelming [11]. Nevertheless, user-generated content does contain traces of people's sentiment. As recent estimates indicate that twenty percent of all tweets [6] and one third of all blog posts [8] discuss products or brands, automated information monitoring tools for consumer sentiment are crucial for today's businesses.

For such information systems, reviews form an important source of information for, e.g., marketing and reputation management. In reviews, users describe their experiences with a particular brand or product, while implicitly or explicitly expressing what they do or do not like about the subject of their respective reviews. The overall verdict of a review can typically be classified by means of universal star ratings, where the number of stars reflects the extent to which a reviewer intends to convey positive sentiment with respect to the review's subject. Such star classes,

J. Casillas et al. (Eds.): Management Intelligent Systems, AISC 171, pp. 251–260.

typically five, are defined on an ordinal scale, e.g., a piece of text that is assigned five stars is considered to be more positive than a four-star piece of text.

Star ratings can enable the extraction of valuable information from the multitude of available reviews, as they can facilitate analyses of, e.g., which aspects of an arbitrary product are mentioned in what context in reviews associated with particular ratings. Sentiment analysis techniques can be used to this end. Some of such techniques focus on identifying the subjectivity or objectivity of a text, whereas other techniques aim to determine the polarity of natural language text.

Typical sentiment analysis approaches involve scanning a text for cues signaling subjectivity or polarity, e.g., words, parts of words, or other (latent) features of natural language text, typically in statistics-based machine learning approaches. The use of sentiment lexicons – lists of words and their associated sentiment, possibly differentiated by Part-of-Speech (POS) and/or meaning [1] – has gained attention in recent research endeavors [2, 3]. Such lexicon-based methods have been shown to have a more robust performance across domains and texts than pure machine learning approaches [14]. Additionally, lexicon-based methods allow for intuitive ways of incorporating deep linguistic analysis into the sentiment analysis, for instance by accounting for structural or semantic aspects of text, but this comes at a cost of significant decreases in processing speed with respect to statistical approaches [2].

In order to be able to (semi-)automatically analyze user-generated content for clues as to, e.g., why people like or dislike (aspects of) products or brands, or how different aspects of products contribute to the overall user experience, a reliable indication of intended sentiment associated with this content is of crucial importance. Some Web sites offer users the possibility to assign scores to their reviews in order to express their intended sentiment, but such scores are not always available. For instance, opinionated blog posts or tweets are not typically assigned scores by their respective authors in order to signal their intended sentiment. Therefore, a major challenge is to automatically determine the star rating associated with reviews based on cues in the actual natural language content.

In this light, we propose and compare several statistical methods for classifying the star rating of reviews. In our current endeavors, we aim to contribute to combining the accuracy and processing speed benefits of statistics-based sentiment analysis approaches with the robustness of lexicon-based approaches.

The remainder of this paper is structured as follows. First, we discuss related work on sentiment analysis in Sect. 2. Then, we propose several statistics-based approaches to star rating classification of the sentiment associated with reviews in Sect. 3. An evaluation of our methods is presented in Sect. 4. Last, we conclude and propose directions for future work in Sect. 5.

2   Sentiment Analysis

The research area of sentiment analysis is related to natural language processing, computational linguistics, and text mining. The main goal of sentiment analysis is the extraction of subjective information from natural language text. Existing work focuses on several specific tasks. Some work aims to distinguish subjective text segments from objective ones or to identify the degree of subjectivity of text [17]. Other work is focused on determining the overall polarity of words, sentences, text segments, or documents [11]. This is typically treated as a binary classification problem, i.e., text is classified as either positive or negative, yet some research focuses on ternary classification by introducing a third class of neutral documents. Other work focuses on determining the degree of positivity or negativity of text.

In general, there are two main types of approaches to sentiment classification tasks. On the one hand, some approaches exploit (generic) sentiment lexicons when determining the subjectivity or polarity of natural language text. On the other hand, many state-of-the-art approaches rely on statistics-based machine learning techniques for sentiment analysis.

Lexicon-based approaches take into account the semantic orientation of individual words by matching words in a text with a list of words with their associated sentiment, possibly differentiated by POS and/or meaning. The overall semantic orientation of a text is then determined by aggregating (e.g., summing) the word scores, possibly while taking into account other aspects of content as well, e.g., negation [3, 5], intensification [13], or rhetorical roles of text segments [2, 4]. Lexicon-based approaches enable deep, yet computationally intensive linguistic analysis to be incorporated into the process of analyzing sentiment in natural language text [2] and have been shown to have a robust performance across domains and texts [14].

On the other hand, machine learning approaches have been shown to have great potential with respect to sentiment classification accuracy in specific domains for which they have been trained [14]. In such approaches, text is typically represented as a vector, which can be used to model the text as a bag-of-words, i.e., an unordered collection of words occurring in a document. Here, a binary representation of text, indicating the presence or absence of specific words [10] has been shown to be more effective than a frequency-based vector representation text [12]. Vectors may also contain features other than words, e.g., parts of words, word groups, or features representing other aspects of content such as semantic distinctions between words [16]. Features represented in vectors may be weighted as well [9].

Machine learning approaches have an attractive advantage over lexicon-based approaches in that they tend to perform better in terms of classification accuracy [14]. Additionally, lexicon-based methods tend to sacrifice computational efficiency when naturally incorporating deep linguistic analysis into the sentiment analysis process [2]. These properties render statistics-based machine learning techniques attractive approaches to sentiment analysis tasks. However, lexicon-based methods tend to be more robust across domains and texts [14]. Therefore, a statistics-based method in which sentiment lexicons are exploited as well appears to be a viable approach to our targeted multi-class sentiment analysis problem.

3   Star Rating Classification

In this paper, we aim to automatically determine the star rating of reviews by means of analyzing the sentiment conveyed by these pieces of natural language text. Rather than targeting a binary or ternary sentiment classification problem, we aim to distinguish five sentiment classes, i.e., one star, two stars, etcetera. These stars represent sentiment classifications ranging from very negative (one star), to neutral (three stars), and very positive (five stars).

As we hypothesize that the boundaries between classes may not be very clear-cut because of the different degrees of positivity and negativity represented by our star ratings, we assume that a statistics-based machine learning approach would be a better fit than unsupervised lexicon-based approaches for the problem we target in our current endeavors. Nevertheless, the robustness across domains and texts typically exhibited by lexicon-based approaches is an appealing feature. In this light, we propose to make a first step towards combining the classification accuracy and processing speed benefits of statistics-based sentiment analysis approaches with the robustness of lexicon-based approaches by means of linking vector representations of our texts to a sentiment lexicon.

In order to be able to apply statistical analyses on our data, we need a proper representation of our texts. We propose a novel *bag-of-sentiwords* representation, i.e., a vector with features representing the presence of sentiment-carrying words, retrieved from a sentiment lexicon. We include only sentiment-carrying words in our vector, as we assume these words to play a major, if not crucial role in conveying the overall sentiment of a text, as opinionated texts significantly differ from non-opinionated texts in terms of occurrences of subjective words [15]. We propose to use a binary representation, as we hypothesize that the sentiment conveyed by a text is not so much in the number of times a single word occurs in a text, but rather in the (number of) distinct words with a similar semantic orientation. Moreover, research has shown that such a binary representation is more effective for sentiment analysis purposes than a frequency-based vector representation of natural language text [12].

Statistical analyses and machine learning algorithms can be applied to the vector representations of text thus obtained in order to identify similarities between texts and to exploit these, such that the correct sentiment classification of a text can be identified. In this work, we consider two types of classifiers, both of which assume the availability of a set of training data, labeled with their corresponding sentiment classification, and a set of test data for which the sentiment needs to be classified based on the model built from the training data. The first type of classifier we consider is a nearest neighbor classifier. Additionally, we consider to use a naïve Bayes classifier for determining the star rating associated with a text.

In our nearest neighbor classifier, we compare an arbitrary unlabeled text with vector representations of each of our considered classes and subsequently assign to the text the label of the class with which the similarity is the highest. These vector representations of classes are typically representative documents or they represent the typical characteristics of documents in their respective classes. The similarity between two (vector representations of) documents can be measured in several ways.

First, we consider to compute a Jaccard similarity coefficient, by defining the similarity $s_{\text{jac}}(d_i, d_j)$ between documents d_i and d_j as the size – i.e., the number of ones in the vector representation – of the intersection of d_i and d_j in terms of the size of the union of these documents, i.e.,

$$s_{\text{jac}}(d_i, d_j) = \frac{|d_i \cap d_j|}{|d_i \cup d_j|}. \tag{1}$$

Alternatively, the similarity between two vector representations of natural language text could be computed by means of the cosine similarity $s_{\cos}(d_i, d_j)$ of document d_i to document d_j, i.e.,

$$s_{\cos}(d_i, d_j) = \frac{\sum_{f=1}^{n} d_{i_f} d_{j_f}}{\sqrt{\sum_{f=1}^{n} \left(d_{i_f}\right)^2} \sqrt{\sum_{f=1}^{n} \left(d_{j_f}\right)^2}}, \tag{2}$$

with d_{i_f} and d_{j_f} representing feature f out of n features for documents d_i and d_j, respectively.

Another design issue lies in the definition of a class, i.e., the determination of which vector representation(s) an unlabeled document should be compared with in order to determine its class. In our current endeavors, we consider three types of vector representations of a class.

The first vector representation of a class we consider is a centroid representation, where a class is represented by the document with the highest similarity to all other documents in its class. When using this representation, an unlabeled document is assigned the class of the centroid that is most similar to this document.

Second, we consider representing each class by means of all its associated documents. This implies that a new document can be classified by computing its similarity to each document in the training set and subsequently classifying it into the class associated with the highest similarity, averaged over its constituting documents.

Last, we consider to represent each class by merging all documents in each respective class into one vector representation per class. In this merger, a new vector is constructed for an arbitrary class by taking the union of all vectors this class is constituted by. When using this representation, an unlabeled text can be classified into the class of which the merged vector has the highest similarity to the vector representation of the unlabeled text.

As an alternative to our considered nearest neighbor methods, we consider a naïve Bayes classifier. In this classifier, a document d_i is assigned a class c_k, for which the probability $P(c_k|d_i)$ is maximized. This probability is defined as the product of the prior probability $P(c_k)$ of class c_k to occur – which can be estimated from the training data – and the probability $P(w_t|c_j)$ of each of its m distinct words w_t to occur in a document of class c_k, i.e.,

$$P(c_k|d_i) = P(c_k) \prod_{t=1}^{m} P(w_t|c_j). \tag{3}$$

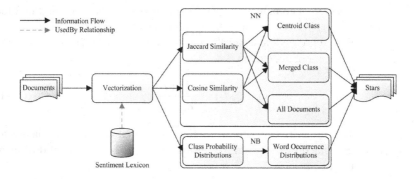

Fig. 1 Star rating classification of documents, represented by the occurrence of words retrieved from a sentiment lexicon, by means of nearest neighbor (NN) and naïve Bayes (NB) classifiers

Our star rating classification approaches are summarized in Fig. 1. The nearest neighbor classifiers use the Jaccard or cosine similarity measure, combined with class representations based on the centroids, all documents, or a merger of all documents of a class. Our naïve Bayes classifier models document similarity by means of probability distributions for star rating classes, given documents, where each class is modeled as the probability distributions for word occurrences in that class.

4 Evaluation

The statistical sentiment analysis methods proposed in Sect. 3 can be used for classifying the star rating of reviews based on cues in the actual natural language content of these reviews. These cues are constituted by the occurrence of specific sentiment-carrying words, derived from a sentiment lexicon, and are reflected in our novel binary vector representations of reviews. Our proposed methods can be applied to these vectorized reviews in order to identify similarities between reviews and to exploit these, such that their associated star ratings can be determined.

4.1 Experimental Setup

In order to evaluate and compare our proposed star rating classification methods, we assess their performance on a data set containing reviews posted on Amazon [7]. In this data set, the reviews have been annotated by their respective authors with a star rating between one and five stars. The reviews cover a multitude of products, including books, music, and movies, and hence span multiple domains. We randomly sample 10,000 reviews from this data set as our training set, and 10,000 reviews as our test set. The reviews in both sets are approximately normally distributed over five star classes, while being somewhat skewed towards the higher ratings.

The reviews in our data set need to be represented by means of vectors signaling the presence of sentiment-carrying words. In our current endeavors, we extract these sentiment-carrying words from the Multi-Perspective Question Answering (MPQA) corpus [18], which contains a large collection of subjective words collected from several news sources, covering a wide variety of subjects. We extract all subjective words and subsequently discard all duplicate entries while not accounting for POS or meaning. This process leaves us with 4,300 lexical representations of sentiment-carrying words, i.e., 4,300 features for our binary vector representation of reviews.

We evaluate and compare the performance of several star rating classification approaches on our vectorized data. In our experiments, we consider the nearest neighbor and naïve Bayes approaches proposed in Sect. 3. For the nearest neighbor classifier, we consider both the Jaccard and the cosine similarity measure. Additionally, we consider class representations based on the centroid reviews of each class, all reviews, and a merger, i.e., union, of all reviews constituting each respective class.

Each method is assessed by means of several performance measures. First, we assess the average precision, recall, and F_1 measure over all five classes. Precision is the proportion of the reviews classified as, e.g., one star, which in fact should have been classified as such. Recall is the proportion of the reviews with a particular classification which are also classified as such. The F_1 measure is the harmonic mean of precision and recall. We also assess the overall accuracy, i.e., the percentage of correct classifications. Finally, we assess the Root Mean Squared Error (RMSE) of the class numbers in order to evaluate how far off the classifications typically are.

4.2 Experimental Results

The experimental results presented in Table 1 suggest that, on our data set and with our vectorization method of the reviews in this data set, the Jaccard similarity measure typically yields better results for the nearest neighbor method than the cosine similarity measure does, especially in terms of precision. Furthermore, in terms of precision, recall, and F_1 measure, a class representation based on all reviews in a particular class appears to outperform the other considered class representations. However, a merger of all vector representations of reviews constituting a particular class appears to yield better results in terms of overall accuracy and RMSE of assigned class numbers than a class representation based on all reviews does.

However, the nearest neighbor classifiers are clearly outperformed by the naïve Bayes star rating classifier, especially in terms of overall accuracy and RMSE of assigned class numbers. The naïve Bayes approach is however outperformed in terms of recall by nearest neighbor classifiers with a class representation based on all reviews in a particular class, yet this is compensated for by the relatively high precision and, to a lesser extent, F_1 measure of the naïve Bayes approach as compared to the nearest neighbor star rating classifiers. All in all, the naïve Bayes approach appears to be superior to all considered nearest neighbor approaches.

Even though the performance of some of our considered methods seems rather promising, the algorithms leave room for improvement. An error analysis has

Table 1 Average precision, recall, F_1 measure, overall accuracy, and RMSE of assigned class numbers over all five star rating classes for the considered nearest neighbor (NN) and naïve Bayes (NB) star rating classifiers. The best performance is printed in bold for each performance measure.

Method	Precision	Recall	F_1	Accuracy	RMSE
NN (Jaccard, centroid)	0.241	0.235	0.219	0.300	1.879
NN (Jaccard, all)	0.294	**0.325**	0.261	0.323	1.673
NN (Jaccard, merged)	0.228	0.211	0.184	0.477	1.432
NN (cosine, centroid)	0.232	0.230	0.230	0.365	1.508
NN (cosine, all)	0.293	0.318	0.244	0.291	1.727
NN (cosine, merged)	0.227	0.229	0.223	0.392	1.567
NB	**0.328**	0.269	**0.269**	**0.508**	**1.296**

revealed that our considered approaches typically fail to correctly interpret more complex sentences, for instance those containing negation. Other errors are caused by occasionally sparse vectors due to a lack of identified sentiment-carrying words in some of the reviews in our data set. Another common source of errors appears to be off-topic noise in the reviews. People tend to discuss different aspects of their subjects and possibly even of other subjects before they arrive at their conclusions. The sentiment conveyed by the conclusions in such reviews appears to be a better proxy for the intended sentiment and thus for the overall verdict, quantified in a star rating. As such, a weighting scheme taking into account the position or role of words in a text may help improve the performance of our considered star rating classification methods.

5 Conclusions and Future Work

In this paper, we have proposed and assessed several statistical methods for classifying the star rating of reviews. The contribution of this work is two-fold. First, in an attempt to combine the classification accuracy and processing speed benefits of statistics-based sentiment analysis approaches with the robustness of lexicon-based approaches, we have proposed to represent the content of reviews by means of a binary vector representation, where the features represent the presence of sentiment-carrying words, retrieved from a general purpose sentiment lexicon. Second, we have compared the performance of several classifiers on these vector representations. A nearest neighbor classifier turns out to maximize recall, whereas a naïve Bayes classifier appears to excel in terms of precision, accuracy, and the RMSE of the assigned number of stars. These findings can help businesses in their marketing or reputation management efforts by providing a comparably reliable indication of intended sentiment in reviews. Such insights enable businesses to identify, e.g., why people like or dislike (aspects of) products or brands.

In future research, we plan to take more approaches into account in our comparisons of methods for star rating classification of sentiment. Furthermore, we plan to include additional features in our vector representations of content of reviews. Such features may be the frequencies, POS, and word senses of (sentiment-carrying) words. Additionally, we consider devising a weighting scheme for our vector representations in order to take into account the position or role of (sentiment-carrying) words in a text. Last, other machine learning algorithms, e.g., support vector machines, may be applied in order to possibly improve upon the performance of the star rating classification methods considered in our current endeavors.

Acknowledgements. The authors of this paper are partially supported by the Dutch national program COMMIT.

References

1. Baccianella, S., Esuli, A., Sebastiani, F.: SentiWordNet 3.0: An Enhanced Lexical Resource for Sentiment Analysis and Opinion Mining. In: 7th Conference on International Language Resources and Evaluation (LREC 2010), pp. 2200–2204. European Language Resources Association (2010)
2. Heerschop, B., Goossen, F., Hogenboom, A., Frasincar, F., Kaymak, U., de Jong, F.: Polarity Analysis of Texts using Discourse Structure. In: 20th ACM Conference on Information and Knowledge Management (CIKM 2011), pp. 1061–1070. Association for Computing Machinery (2011)
3. Heerschop, B., van Iterson, P., Hogenboom, A., Frasincar, F., Kaymak, U.: Analyzing Sentiment in a Large Set of Web Data while Accounting for Negation. In: 7th Atlantic Web Intelligence Conference (AWIC 2011), pp. 195–205. Springer (2011)
4. Hogenboom, A., Hogenboom, F., Kaymak, U., Wouters, P., de Jong, F.: Mining Economic Sentiment using Argumentation Structures. In: Trujillo, J., Dobbie, G., Kangassalo, H., Hartmann, S., Kirchberg, M., Rossi, M., Reinhartz-Berger, I., Zimányi, E., Frasincar, F. (eds.) ER 2010. LNCS, vol. 6413, pp. 200–209. Springer, Heidelberg (2010)
5. Hogenboom, A., van Iterson, P., Heerschop, B., Frasincar, F., Kaymak, U.: Determining Negation Scope and Strength in Sentiment Analysis. In: 2011 IEEE International Conference on Systems, Man, and Cybernetics (SMC 2011), pp. 2589–2594. IEEE (2011)
6. Jansen, B., Zhang, M., Sobel, K., Chowdury, A.: Twitter Power: Tweets as Electronic Word of Mouth. Journal of the American Society for Information Science and Technology 60(11), 2169–2188 (2009)
7. Jindal, N., Liu, B.: Opinion Spam and Analysis. In: 1st ACM International Conference on Web Search and Data Mining (WSDM 2008), pp. 219–230. Association for Computing Machinery (2008)
8. Melville, P., Sindhwani, V., Lawrence, R.: Social Media Analytics: Channeling the Power of the Blogosphere for Marketing Insight. In: 1st Workshop on Information in Networks, WIN 2009 (2009)
9. Paltoglou, G., Thelwall, M.: A study of Information Retrieval weighting schemes for sentiment analysis. In: 48th Annual Meeting of the Association for Computational Linguistics (ACL 2010), pp. 1386–1395. Association for Computational Linguistics (2010)
10. Pang, B., Lee, L.: A Sentimental Education: Sentiment Analysis using Subjectivity Summarization based on Minimum Cuts. In: 42nd Annual Meeting of the Association for Computational Linguistics (ACL 2004), pp. 271–280. Association for Computational Linguistics (2004)

11. Pang, B., Lee, L.: Opinion Mining and Sentiment Analysis. Foundations and Trends in Information Retrieval 2(1), 1–135 (2008)
12. Pang, B., Lee, L., Vaithyanathan, S.: Thumbs up? Sentiment Classification using Machine Learning Techniques. In: Empirical Methods in Natural Language Processing (EMNLP 2002), pp. 79–86. Association for Computational Linguistics (2002)
13. Taboada, M., Brooke, J., Tofiloski, M., Voll, K., Stede, M.: Lexicon-Based Methods for Sentiment Analysis. Computational Linguistics 37(2), 267–307 (2011)
14. Taboada, M., Voll, K., Brooke, J.: Extracting Sentiment as a Function of Discourse Structure and Topicality. Tech. Rep. 20. Simon Fraser University (2008),
   http://www.cs.sfu.ca/research/publications/
   techreports/#2008
15. van der Meer, J., Boon, F., Hogenboom, F., Frasincar, F., Kaymak, U.: A Framework for Automatic Annotation of Web Pages Using the Google Rich Snippets Vocabulary. In: Twenty-Sixth Symposium On Applied Computing (SAC 2011), Web Technologies Track, pp. 765–772. Association for Computing Machinery (2012)
16. Whitelaw, C., Garg, N., Argamon, S.: Using Appraisal Groups for Sentiment Analysis. In: 14th ACM International Conference on Information and Knowledge Management (CIKM 2005), pp. 625–631. Association for Computing Machinery (2005)
17. Wiebe, J., Wilson, T., Bruce, R., Bell, M., Martin, M.: Learning Subjective Language. Computational Linguistics 30(3), 277–308 (2004)
18. Wiebe, J., Wilson, T., Cardie, C.: Annotating Expressions of Opinions and Emotions in Language. Language Resources and Evaluation 39(2), 165–210 (2005)

Risk Assessment and Management

Non-parametric Statistical Analysis of Machine Learning Methods for Credit Scoring

V. García[1], A.I. Marqués[2], and J.S. Sánchez[1]

[1] Institute of New Imaging Technologies, Department of Computer Languages and Systems, Universitat Jaume I, Av. Sos Baynat s/n, 12071 Castelló de la Plana, Spain
{jimenezv,sanchez}@uji.es
[2] Department of Business Administration and Marketing, Universitat Jaume I, Av. Sos Baynat s/n, 12071 Castelló de la Plana, Spain
imarques@uji.es

Abstract. Various machine learning techniques have been explored for credit scoring and management, but no consistent conclusions have been drawn on which method shows the best behaviour. This paper presents an experimental analysis involving five real-world databases with several credit scoring models, including logistic regression, neural networks, support vector machines, decision trees, rule induction algorithms, Bayesian models, k nearest neighbours decision rule, and classifier ensembles. Particularly, we analyse the performance of this set of algorithms by means of a non-parametric statistical test and two post-hoc procedures for making pairwise comparisons.

1 Introduction

The recent international financial crisis has aroused increasing attention of financial institutions on credit and operational risk assessment, converting this into a key task because of the heavy losses associated with wrong decisions. One major risk for banks and financial institutions comes from the difficulty to distinguish the creditworthy applicants from those who will probably default on repayments. The decision to grant credit to an applicant was traditionally based upon subjective judgements made by human experts, using past experiences and some guiding principles. Common practice was to consider the classic five C's of credit: the character of the applicant, the capacity, the capital, the collateral and the economic conditions [2]. This method suffers, however, from high training costs, frequent incorrect decisions, and inconsistent decisions made by different analysts for the same application. These shortcomings have led to a rise in more formal and accurate methods to assess the risk of default. In this context, credit scoring and behavioural management have become primary tools for financial institutions to manage and evaluate credit risk, improve cash flow, reduce possible risks and make managerial decisions [23].

J. Casillas et al. (Eds.): Management Intelligent Systems, AISC 171, pp. 263–272.
springerlink.com © Springer-Verlag Berlin Heidelberg 2012

From the seminal reference to credit scoring in the introductory paper by Altman [4], many other developments have been subsequently proposed in the literature. The most classical approaches to credit scoring are based upon statistical and operations research models. However, the problem with applying statistical techniques to credit scoring is that some assumptions, such as the multivariate normality assumptions for independent variables, are frequently violated in practice, what makes them theoretically invalid for finite samples [16].

During the last decades, many research works have focused on the deployment of different machine learning techniques to design and implement credit scoring solutions. In contrast with statistical models, machine learning methods do not assume any specific prior knowledge, but automatically extract information from the training examples available. Although some researchers conclude that machine learning techniques are superior to statistical methods, their studies merely compare a reduced number of approaches to credit scoring [1, 5, 6, 7, 9, 11, 25]

The aim of this paper is to evaluate the performance of different statistical and machine learning techniques for credit scoring problems, in terms of accuracy, type-I error and type-II error by means of the use of Friedman's non-parametric test and post-hoc procedures. The models here evaluated comprise logistic regression, naïve Bayes classifier, support vector machine (with a linear kernel), multilayer perceptron and radial basis function neural networks, k-nearest neighbours classifier, random forest, and a rule induction algorithm. Apart from three common databases (Australian, German and Japanese credit data sets), this study uses two new credit data collected from Iranian and Polish financial institutions. Such an empirical analysis may contribute to a better understanding of the benefits of each credit scoring model, what entails important business implications for banks and financial institutions. In particular, this may help the development of the credit management process and provide credit analysts and decision makers with efficient and effective tools to assess credit risk more precisely.

2 The Credit Scoring Models

From a practical point of view, the credit scoring problem can be deemed as a binary classification problem where a new input sample (the credit applicant) must be categorized into one of the predefined classes based on a number of observed variables or attributes related to that sample. The input of the classifier consists of a variety of information that describes socio-demographic characteristics (gender, age, marital status, occupation, educational level) and economic conditions (loan amount, loan duration, monthly incomes, bank accounts) of the applicant, and then the classifier has to produce the output in terms of the applicant creditworthiness. In its most usual form, credit scoring aims at assigning credit applicants to either good (those who are liable to reimburse the financial obligation) or bad (those who should be denied credit because of the high probability of defaulting on repayments) classes.

The credit scoring problem can be formally described as follows. Given a data set of applicants $S = \{(x_1, y_1), (x_2, y_2), \ldots, (x_n, y_n)\}$, where each applicant x_i is characterized by m features or attributes, $x_{i1}, x_{i2}, \ldots x_{im}$, and y_i denotes the type of applicant (good/bad), then credit scoring viewed as a classification problem consists of constructing a model δ to predict the value y for a new applicant $\mathbf{x}$, that is, $\delta(\mathbf{x}) = y$.

Apart from the statistical logistic regression model, the credit scoring techniques included in this paper correspond to some of the most representative methods of different machine learning paradigms. The methods here evaluated are the statistical logistic regression model (LR) and seven different machine learning algorithms: the naïve Bayes classifier (NB), the multilayer perceptron (MLP) and the normalized Gaussian radial basis function (RBF) neural networks, the k-nearest neighbour (k-NN) decision rule, a support vector machine (SVM), the random forest (RF), and a single rule induction algorithm (RIPPER).

3 Experimental Protocol

Five real-world financial data sets have been taken to evaluate the performance of the strategies investigated in the present paper. The first three are from the UCI Machine Learning Database Repository [13] and they have been widely used in credit scoring research. The Iranian data set comes from a corporate client database of a small private bank in Iran [20] and the Polish data set contains bankruptcy information of 120 companies recorded over a two-year period [19]. Table 1 summarizes the main characteristics of these benchmarking databases: the number of attributes, the total number of samples per class and the a priori distribution of Bad/Good examples.

Table 1 Characteristics of the credit data sets used in the experiments

Data set	No. Attributes	No. Bad	No. Good	%Bad/%Good
Australian	14	383	387	55.5/44.5
German	24	300	700	30.0/70.0
Japanese	15	357	296	45.3/54.7
Iranian	27	50	950	5.0/95.0
Polish	30	112	128	46.6/53.5

The standard way to assess credit scoring systems is to use a holdout sample since large sets of past applicants are usually available. However, there are situations in which data are too limited to build an accurate scorecard and consequently, other strategies have to be used in order to obtain a good estimate of the accuracy rate. The most common way around this is cross-validation [3, Ch. 19] [23, Ch. 7].

A 10-fold cross-validation method has been adopted for the present experiments: each original data set has been randomly divided into ten stratified parts of (approximately) equal size. Stratification involves getting the correct proportion of examples in each class. For each fold, nine of the parts have been pooled as the training data, and the remaining block has been employed as an independent test set. Ten repetitions have been run for each trial. The results from classifying the test samples have been averaged across the 100 runs and then evaluated for significant differences between models using the Friedman's non-parametric and the Bonferroni-Dunn and Nemenyi post-hoc tests with $\alpha < 0.05$ [8].

All classifiers have been implemented using the Weka toolkit [14], which is a collection of statistical and machine learning algorithms for general data mining tasks. We have used the default parameter settings provided by this software as follows:

- LR: ridge parameter in the log-likelihood $= 1 \times 10^{-8}$.
- MLP: no. hidden layers $=$ (no. attributes $+$ no. classes)$/2$; learning rate $= 0.3$; momentum $= 0.2$; no. epochs $= 500$.
- RBF: no. clusters for K-means algorithm $= 2$; min. standard deviation for the clusters $= 0.1$; ridge parameter $= 1 \times 10^{-8}$.
- k-NN: no. neighbours $k = 1$; Euclidean distance.
- SVM: linear kernel; tolerance $= 0.001$; epsilon for round-off error $= 1 \times 10^{-12}$.
- RF: no. trees $= 10$; max. depth of trees $=$ unlimited; no. attributes randomly selected $= \log_2$ (no. attributes) $+1$.
- RIPPER: no. optimization runs $= 2$; min. total weight of the instances in a rule $= 2.0$.

3.1 Evaluation Criteria

For the experiments, the performance evaluation criteria correspond to standard measures used in credit scoring [24], including average accuracy, type-I error and type-II error. The definition of these metrics can be obtained from a 2×2 confusion or classification matrix as shown in Table 2, where each entry (i, j) contains the number of correct/incorrect predictions. For consistency with previous works in the topic of performance measures, the positive and negative classes correspond to good and bad applicants (or credit risk), respectively.

Table 2 Confusion matrix for the credit scoring problem

	Predicted positive (good)	Predicted negative (bad)
Positive class (good)	True Positive (TP)	False Negative (FN)
Negative class (bad)	False Positive (FP)	True Negative (TN)

Although many performance metrics (e.g., Gini coefficient, area under the ROC curve, geometric mean, true rate) can be defined from a confusion matrix [12, 22], most of credit scoring applications often utilize the accuracy rate (also called score of hits) as the criterion for performance evaluation. It represents the proportion of the correctly classified cases (good and bad) on a particular data set, and can be formally defined as follows:

$$\text{Accuracy} = \frac{TP + TN}{TP + FN + TN + FP} \tag{1}$$

However, in credit scoring, it is also very important to measure the error on each individual class:

$$\text{Type-I error} = \frac{FN}{TP + FN} \qquad \text{Type-II error} = \frac{FP}{TN + FP} \tag{2}$$

Type-I error defines the rate of good applicants being predicted as bad. When this happens, the misclassified good applicants are refused and therefore, the financial institution has opportunity cost caused by the loss of good customers. On the other hand, type-II error is the rate of bad applicants being categorized as good. When this happens, the misclassified bad applicants will become default. Therefore, if the credit granting policy of a financial institution is too generous, this will be exposed to high credit risk. Lee and Chen [18] stated that *"the misclassification cost associated with a type-II error is much higher than the misclassification cost associated with a type-I error"*.

3.2   Statistical Significance Tests

Probably, the most typical way to compare two classifiers over a set of problems is the Student's paired t-test. However, this appears to be conceptually inappropriate and statistically unsafe because parametric tests are based on a variety of assumptions (independence, normality and homoscedasticity) that are often violated due to the nature of the problems [8].

In general, the non-parametric tests (e.g., the Wilcoxon and Friedman tests) should be preferred over the parametric ones (the paired t-test and ANOVA), especially in multi-problem analysis, because they do not assume normal distributions or homogeneity of variance. In this work, we have adopted the Friedman test to detect significant differences among the classifiers.

The Friedman test is based on the average ranked performances of a collection of techniques on each data set separately. The Friedman statistic (χ_F^2) is distributed according to the Chi-square distribution with $K - 1$ degrees of freedom, when N (number of data sets) and K (number of algorithms) are big enough. The null-hypothesis being tested is that all strategies are equivalent and the observed differences are merely random. The main drawback of the Friedman and other related tests is that they only can detect significant differences over the whole set of comparisons, but they cannot compare a control technique with the $K - 1$ remaining algorithms.

When the null-hypothesis of the Friedman test is rejected, we can then use a post-hoc test in order to find the particular pairwise comparisons that produce statistically significant differences [8, 17, 21]. A post-hoc test compares a control algorithm opposite to the remainder techniques, making possible to define a collection of hypothesis around to the control method. The Nemenyi post-hoc test [17, Ch. 6], which is analogous to the Tukey test for ANOVA, states that the performances of two or more algorithms are significantly different if their average ranks are at least as great as their critical difference (CD) with a level of significance α:

$$CD = q_\alpha \sqrt{\frac{K(K+1)}{6N}} \tag{3}$$

where q_α corresponds to the critical value based on the Studentised range statistic but scaled by dividing it by $\sqrt{2}$ [8, 17].

A conservative post-hoc procedure is the Bonferroni-Dunn test, which allows comparisons controlling the family-wise error rate [8]. This test can be computed with Eq. 3, but using the critical values for $\alpha/(K-1)$. For the present set-up, the corresponding critical values, which have been obtained from the work by Demšar [8], are $q_{\alpha_{0.05}} = 3.031$ and $q_{\alpha_{0.05}} = 2.690$ for the Nemenyi and Bonferroni-Dunn post-hoc tests, respectively.

4   Results and Discussion

Table 3 reports the averaged results obtained for the eight learning algorithms in terms of accuracy, type-I error (TIE) and type-II error (TIIE). This table also shows the average ranks computed through the Friedman test. The best result for each data set has been underlined, whereas the lowest average rank value (the best performing technique) is highlighted in boldface.

A simple analysis on individual data sets could suggest that LR is the best algorithm in terms of accuracy since it achieves the highest rates in 3 out of 5 data sets. Analogously, the SVM and NB classifiers appear to be the best performing methods when using type-I and type-II errors, respectively. However, this kind of analysis only considers the performance of classifiers measured across an unique data set. From Table 3, the average ranking obtained by the Friedman test for the accuracy shows that RF is the best performing algorithm, whereas k-NN is the worst method. Although it has traditionally been claimed that SVM provides a good performance for credit scoring problems, it is significant worse than the RF and LR classifiers in our experiments.

When analysing the type-I error, we found that RF achieves the lowest average rank (2.70), followed by SVM (2.80). Once again, the k-NN classifier seems to be the worst algorithm. Unexpectedly, in the case of type-II error, the best technique corresponds to the NB classifier.

After applying the Friedman's non-parametric test, the Nemenyi and Bonferroni-Dunn post-hoc tests have been employed to report any significant differences with respect to the best performing classifier for each evaluation measure. The results of these tests are then depicted to illustrate the differences among the Friedman average ranks. Figure 1 plots the classifiers against average rankings, whereby all

Table 3 Average accuracy/error rates and the Friedman ranks for the eight algorithms considered

Accuracy								
Data set	NB	LR	MLP	RBF	SVM	k-NN	RIPPER	RF
Australian	77.04	86.39	82.91	82.97	85.51	79.80	84.78	86.25
German	75.48	76.91	71.09	72.78	76.57	67.23	73.00	74.53
Japanese	77.66	86.88	83.25	82.60	86.37	79.57	85.65	86.37
Iranian	23.42	94.06	94.06	95.00	95.00	92.90	94.33	95.11
Poland	68.83	73.38	75.21	67.88	71.71	76.67	73.08	75.75
Average Rank	6.80	2.50	5.30	5.50	3.20	6.00	4.40	**2.30**
Type-I Error								
Data set	NB	LR	MLP	RBF	SVM	k-NN	RIPPER	RF
Australian	0.42	0.13	0.19	0.27	0.07	0.22	0.14	0.13
German	0.14	0.11	0.19	0.15	0.11	0.24	0.15	0.10
Japanese	0.41	0.11	0.18	0.27	0.06	0.23	0.11	0.11
Iranian	0.80	0.01	0.02	0.00	0.00	0.04	0.01	0.01
Polish	0.14	0.25	0.25	0.29	0.31	0.25	0.25	0.18
Average Rank	5.80	3.20	5.50	5.60	2.80	6.30	4.10	**2.70**
Type-II Error								
Data set	NB	LR	MLP	RBF	SVM	k-NN	RIPPER	RF
Australian	0.08	0.14	0.16	0.20	0.19	0.16	0.14	0.14
German	0.49	0.50	0.51	0.52	0.54	0.55	0.54	0.63
Iranian	0.10	0.99	0.86	1.00	0.72	0.95	0.85	0.85
Polish	0.50	0.29	0.24	0.25	0.21	0.29	0.29	0.31
Average Rank	**2.40**	4.00	3.70	5.00	6.10	4.40	5.50	4.90

models are sorted according to their ranks. The horizontal line, which is at height equal to the sum of the lowest rank and the critical difference CD computed by each post-hoc test, represents the threshold for the best performing classifier at the significance level ($\alpha = 0.05$). This means that all algorithms above this cut line perform significantly worse than the best model.

In the case of the accuracy results, the Friedman ranking has suggested the random forest to be the best algorithm. However, according to Nemenyi's test, there do not exist significant differences between RF and the remaining techniques. When using the Bonferroni-Dunn test, only the NB classifier appears to be significantly worse than the random forest algorithm. Regarding to the type-I and type-II errors, both Nemenyi and Bonferroni-Dunn tests suggest that there are not significant differences among the eight classifiers.

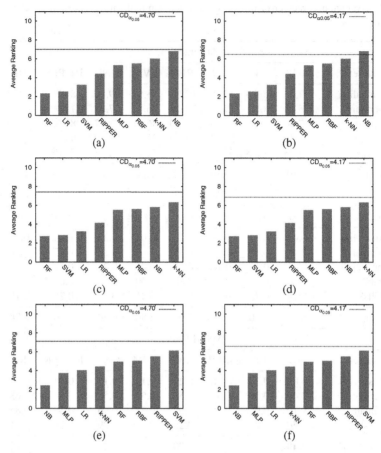

Fig. 1 Significance diagrams for accuracy (a, b), type-I error (c, d) and, type-II error (e, f). Left and right graphics correspond to Nemenyi and Bonferroni-Dunn tests, respectively

5 Conclusions

The categorization of good and bad credit applicants is of fundamental importance, and is indeed the ultimate objective of a credit scoring model. The need of an appropriate classification technique is thus evident. During the last decades, considerable research efforts have been addressed to identify more accurate methods for decision support systems in credit scoring, ranging from simple statistical models to more complex machine learning techniques. Several works have focused on comparing a reduced number of methods, but most of them employ an incomplete experimental protocol that is not easily reproducible.

This paper has investigated the performance of eight credit scoring models with three effectiveness evaluation metrics by adopting a 10×10-fold cross-validation strategy for the experiments. The methods here evaluated are the statistical logistic regression model and seven different machine learning algorithms: naïve Bayes classifier, MLP and RBF neural networks, k-NN decision rule, a SVM, random forest, and the RIPPER rule induction algorithm.

From a managerial perspective, the empirical analysis here carried out leads to the conclusion that there is no overall best credit scoring model for all data sets because it depends on a number of factors: the details of the problem, the structure of the data, the features of the application, the classification objectives, etc. Nevertheless, a thorough analysis of the results has revealed that random forest and logistic regression correspond to the methods with the highest performance and are also the most robust, whereas the naïve Bayes classifier and the k-NN decision rule have been identified as the models with the worst behaviour in terms of both accuracy rate and type-I error. Paradoxically, when using the type-II error, the naïve Bayes classifier and MLP neural network achieve the lowest Friedman ranks. Although differences are not statistically significant in most cases (according to the Nemenyi and Bonferroni-Dunn tests), the present experimental results can be deemed relevant in the context of credit scoring because an improvement in performance of even a fraction of a percent may translate into significant future savings for the financial institutions.

Future research is addressed to extend the present study to take care of the different misclassification costs associated to good and bad applicants. To this end, a promising direction is the deployment of cost-sensitive learning algorithms [10] in real-world credit scoring problems where different misclassification errors incur different penalties. In addition, other measures of performance might also be considered; for instance, measures such as the Gini coefficient, the area under the ROC curve, the Kolmogorov-Smirnov statistic and the information value that are commonly used to assess the classifier performance in the retail banking sector [15].

Acknowledgements. This work has partially been supported by the Spanish Ministry of Education and Science under grants CSD2007–00018 and TIN2009–14205.

References

1. Abdou, H.A.: An evaluation of alternative scoring models in private banking. The Journal of Risk Finance 10(1), 38–53 (2009)
2. Abrahams, C.R., Zhang, M.: Fair Lending Compliance: Intelligence and Implications for Credit Risk Management. Wiley, Hoboken (2008)
3. Alpaydin, E.: Introduction to Machine Learning. MIT Press, Cambridge (2010)
4. Altman, E.I.: Financial ratios, discriminant analysis and the prediction of corporate bankruptcy. Journal of Finance 23(4), 589–611 (1968)
5. Baesens, B., Gestel, T.V., Viaene, S., Stepanova, M., Suykens, J., Vanthienen, J.: Benchmarking state-of-the-art classification algorithms for credit scoring. Journal of the Operational Research Society 54(6), 627–635 (2003)
6. Bellotti, T., Crook, J.N.: Support vector machines for credit scoring and discovery of significant features. Expert Systems with Applications 36(2), 3302–3308 (2009)

7. Bensic, M., Sarlija, N., Zekic-Susac, M.: Modelling small-business credit scoring by using logistic regression, neural networks and decision trees. Intelligent Systems in Accounting, Finance and Management 13(3), 133–150 (2005)

8. Demšar, J.: Statistical comparisons of classifiers over multiple data sets. Journal of Machine Learning Research 7(1), 1–30 (2006)

9. Desai, V.S., Crook, J.N., Overstreet, G.A.: A comparison of neural networks and linear scoring models in the credit union environment. European Journal of Operational Research 95(1), 24–37 (1996)

10. Elkan, C.: The foundations of cost-sensitive learning. In: Proc. 17th Intl. Joint Conf. Artificial Intelligence, Seattle, WA, pp. 973–978 (2001)

11. Elsayad, A.M.: Implementing automated prediction systems for credit scoring. ICGST International Journal on Automatic Control and Systems Engineering 10(1), 11–19 (2010)

12. Ferri, C., Hernández-Orallo, J., Modroiu, R.: An experimental comparison of performance measures for classification. Pattern Recognition Letters 30(1), 27–38 (2009)

13. Frank, A., Asuncion, A.: UCI Machine Learning Database Repository (2010), http://archive.ics.uci.edu/ml

14. Hall, M., Frank, E., Holmes, G., Pfahringer, B., Reutemann, P., Witten, I.H.: The WEKA data mining software: an update. SIGKDD Explorations Newsletter 11(1), 10–18 (2009)

15. Hand, D.J.: Good practice in retail credit scorecard assessment. Journal of the Operational Research Society 56(9), 1109–1117 (2005)

16. Huang, Z., Chen, H., Hsu, C.J., Chen, W.H., Wu, S.: Credit rating analysis with support vector machines and neural networks: A market comparative study. Decision Support Systems 37(4), 543–558 (2004)

17. Japkowicz, N., Shah, M.: Evaluating Learning Algorithms: A Classification Perspective. Cambridge University Press, New York (2011)

18. Lee, T.S., Chen, I.F.: A two-stage hybrid credit scoring model using artificial neural networks and multivariate adaptive regression splines. Expert Systems with Applications 28(4), 743–752 (2005)

19. Pietruszkiewicz, W.: Dynamical systems and nonlinear Kalman filtering applied in classification. In: Proc. of 7th IEEE International Conference on Cybernetic Intelligent Systems, London, UK, pp. 263–268 (2008)

20. Sabzevari, H., Soleymani, M., Noorbakhsh, E.: A comparison between statistical and data mining methods for credit scoring in case of limited available data. In: Proc. of the 3rd CRC Credit Scoring Conference, Edinburgh, UK (2007)

21. Sheskin, D.J.: Handbook of Parametric and Nonparametric Statistical Procedures. CRC Press, Boca Raton (2011)

22. Sokolova, M., Lapalme, G.: A systematic analysis of performance measures for classification tasks. Information Processing & Management 45(4), 427–437 (2009)

23. Thomas, L.C., Edelman, D.B., Crook, J.N.: Credit Scoring and Its Applications. SIAM, Philadelphia (2002)

24. Yang, Z., Wang, Y., Bai, Y., Zhang, X.: Measuring scorecard performance. In: Proc. 4th Intl. Conf. Computational Science, Krakow, Poland, pp. 900–906 (2004)

25. Yobas, M.B., Crook, J.N., Ross, P.: Credit scoring using neural and evolutionary techniques. IMA Journal of Mathematics Applied in Business and Industry 11(4), 111–125 (2000)

Rule-Based Business Process Mining: Applications for Management

Filip Caron, Jan Vanthienen*, and Bart Baesens

Department of Decision Sciences and Information Management, KU Leuven,
Naamsestraat 69, 3000 Leuven, Belgium
Jan.Vanthienen@econ.kuleuven.be

Abstract. The abundance of available event data, originating from process-aware information systems, creates opportunities for enterprise risk management applications at the intersection of the business & management, artificial intelligence and knowledge representation research fields. This paper proposes a rule-based process mining approach for dealing with uncertainty and risk. The applicability of the approach is demonstrated using the updating and debugging process of a social security service provider.

1 Introduction

Contemporary organizations are increasingly exposed to uncertainties in the current dynamic business environments [1]. Each uncertainty will present either an opportunity or a risk. Consequently, adequate management of these risks becomes crucial in order to safeguard the organization's value creation abilities [6].

Recently, these organizations have been organizing their operations around business processes, frequently supported by process-oriented information systems. Process mining refers to the set of techniques that analyze event logs to acquire insights into the real processes [9]. Additionally, research has recognized the power of rules and ontology in governance, risk and compliance [11].

This paper contributes to the management science, research and practise by:

- proposing a knowledge representation based technique for enterprise risk management (including the identification and assessment of risks, the implementation of management control systems and to acquisition & communication of detailed information on the daily operations.
- providing an overview of the business rule types that are relevant in the context of both process mining and enterprise risk management.
- applying the proposed methodology on a case process of a social security service provider.

---

* Corresponding author.

J. Casillas et al. (Eds.): Management Intelligent Systems, AISC 171, pp. 273–282.

2   Rule-Based Business Process Mining

2.1   On Process Mining and the Position of Rule-Based Approaches

The process mining approaches can be roughly categorized into three classes. Firstly, the process discovery & visualization techniques enable the analyst to get an insight into the real process dynamics and discover potential issues [5, 9]. Secondly, conformance checking & delta analysis help to investigate inconsistencies between a prescriptive business process model and the corresponding real-life process [7]. These two classes promote the analysis of the process as a whole, without providing any traceability possibilities. Other issues might apply, such as too complex process visuals or incorrect prescriptive processes.

Rule-based business process mining techniques focus on mining business rules from event logs [4], as well as checking the event log against business rules [8]. The latter, which is highly related to the field of artificial intelligence and knowledge representation, will be of primary interest in this contribution. Consequently the analyst can provide precise and accurate information on certain behavior, while providing a full traceability back to the initial control objective(s) and risk(s) that needed to be mitigated.

The next subsection will provide a classification overview of all the business rule types that can be considered as relevant for business process mining.

2.2   Classifying Business Rules Type Relevant for Business Process Mining

In the context of business process mining we are interested in the operational business rules with a process orientation. These business rules directly represent the implementation of external directives and business policies [6] and can be classified based on **two dimensions**: the process mining perspective and the rule restriction focus.

The first dimension of the classification refers to the **process mining perspective (PMP)** that is used in the business rule. Four different perspectives on business process modeling were introduced in [2] and can also be used to classify the business rule types in this context.

- *Functional process perspective (PMP1)* that deals with the process elements (e.g. activities) that occur in a process instance, as well as the relevant process artifacts linked to these process elements (e.g. an invoice artifact for a pay activity).
- *Control-flow process perspective (PMP2)* that covers the process behavior in terms of when process elements occur in a process instance.
- *Organizational process perspective (PMP3)* that focuses on the organization behind the business process, which agent performs the different process elements

in a process instance taking into account factors such as timing, environmental conditions, etc.

- *Data process perspective (also known as informational perspective) (PMP4)* that represents the informational elements (e.g. event data, case date, etc.) that are used, produced or manipulated during the process, as well as relationships among them.

Secondly business rules can be classified along their main **rule restriction focus (RRF)**. Five new and distinctive business rule restriction focuses are identified:

- *Cardinality-based rules (RRF1)* are business rules that restrict the number of allowed instances of a specific process element type in a process instance.
- *Coexistence rules (RRF2)* can be defined as business rules that restrict the co-existence of process elements of different types over the execution of a process instance.
- *Dynamic data-driven rules (RRF3)* specify the influence of certain data elements (i.e. case or event data) and their value on the occurrence of process elements in a specific process instance.
- *Relative time rules (RRF4)* focus on specifying a time restriction on process elements relative to certain points in a process execution (e.g. start of a process, completion of a specific activity, etc.).
- *Static property rules (RRF5)* deal with specifying a specific property for a particular type of process element at a predefined process state.

While the first four rule restriction focuses deal with *dynamic* properties (i.e. history-based or future constraining), the last focus deals with *static* properties (i.e. properties in one specific process state).

2.3 Formulating Business Process Mining Rules

An **unambiguous interpretation** of the business rules can be obtained by formally specifying them. In this section we first develop a definition for processes, events and audit trails. Secondly, we provide proposals for formalizing both dynamic (RRF1 to 4) and static (RRF5) properties or business rules.

Definition 1: Business processes. A business process can be formally represented by a *process schema S*, which is defined by the tuple $(\mathscr{A}, \mathscr{O}, \mathscr{P}, \mathscr{BR})$ where

- The basic constructs include: $\mathscr{A} = \{a_1, a_2, a_3, ..., a_n\}$ that denotes the finite set of all activities, $\mathscr{O} = \{o_1, o_2, o_3, ..., o_n\}$ that represents the finite set of originators and $\mathscr{P} = \{p_1, p_2, p_3, ..., p_n\}$ that stands for the finite set of (all other) properties.
- $\mathscr{BR}$ the set of business rules specifying the relevant relations and constraints for a business process (e.g. precedence relations, required roles, etc.).

During the execution of a business process a multitude of business events can be observed. A business event is a relevant occurrence of something (e.g. start of a specific activity) that happens at a specific time and is of special interest to the business.

Definition 2: Business event. The events related to activities, which are considered as the states in a process instance or other relevant occurrences, are described as follows:

- An event is specified by the values of the related relevant data properties, as such an event can be denoted as $e : \mathscr{P} \rightharpoonup \mathscr{V}$ with $e \in \mathscr{E} = \{e_1, e_2, e_3, ..., e_n\}$ the set of all events and $\mathscr{V}$ the set of all possible values for the properties. Each of the rows in table 1 represents one event.

Contemporary information systems store a multitude of information about these events in a structured way. Therefore, the resulting event logs (denoted by α, β, etc.) precisely describes the execution of each process instances, within a certain timeframe.

Definition 3: Audit trail. An audit trail or trace $\sigma \in \mathscr{E}^*$ is an event sequence containing the events of a specific process instance, where $\mathscr{E}^*$ represents all traces composed of zero or more events of $\mathscr{E}$.

2.3.1 Formulating Business Process Mining Rules for Static Process Aspects

Static process mining rules focus on the properties in one specific state of a process instance, e.g. the performer of a specific activity (only for RRF5). Since the time aspect is not crucial here, the semantics of the first order logic expression language should suffice.

Definition 4: Business Process Mining Rule for Static Process Aspects. A rule for static process aspects is a first order logic formula that is interpreted for one specific state of a process instance. A first order logic formula p for an $e \in \mathscr{E}$ is a function $p : \mathscr{E} \rightarrow \{true, false\}$, with $e \vDash p$ denoting that event e satisfies formula p (i.e. $p(e) = true$) and $e \nvDash p$ denoting that event e does not satisfy formula p (i.e. $p(e) = false$).

2.3.2 Formulating Business Process Mining Rules for Dynamic Process Aspects

As process-aware information systems can be considered as reactive systems, rule-based business process mining could take advantage of the abilities of linear temporal logic (LTL) to interpret formulae over linear state sequences (in support of all PMP over RRF1 to 4, e.g. mutually included activities). However, a bounded version of the LTL (e.g. [3]) should be used to deal with the fact that business process instances are terminating in contrast to regular reactive systems.

Definition 4: Rule for Dynamic Process Aspects. A rule for dynamic process aspects is an LTL formula that is interpreted over the entire set of events (i.e. the different states of a single process instance. An LTL formula p over a subset of $\mathscr{E}$ is a function $p : \mathscr{E}^* \to \{true, false\}$, with $\sigma \vDash p$ denoting that trace σ satisfies formula p (i.e. $p(\sigma) = true$) and $\sigma \nvDash p$ denoting that trace σ does not satisfy formula p (i.e. $p(\sigma) = false$). For all LTL formulas p and q *true, false*, $\neg p$, $p \wedge q$, $p \vee q$, $\Box p$ (*p has to hold on the entire σ*), $\Diamond p$ (*p eventually has to hold in σ*), $\bigcirc p$ (*p must hold in the next state of σ*) and p U q (*p has to hold at least until q*) are LTL formulas as well. A full specification of LTL can be found in [3].

The main advantage LTL has to offer in the context of business process compliance checking, is the ability to express relative time properties between states. This becomes especially clear in controls that specify that something has to hold eventually or that something has to hold until.

3 Exploring the Applicability of Rule-Based Business Process Mining

The management of an organization will implement a risk management approach in order to align the strategy with the risk appetite, to enhance risk response & decisions, to reduce operational surprises & losses, seize opportunities, etc. Figure 1 represents the **dynamic process of interrelated components in risk management**, related to the COSO ERM framework [1]. The remainder of the section discusses the different components and elaborates on the identified links between the components and rule-based business process mining.

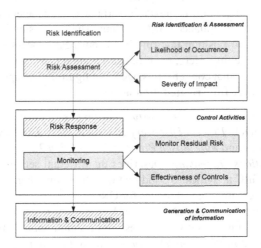

Fig. 1 Applicability of Rule-Based Business Process Mining for Enterprise Risk Management (filled up and shaded activity blocks represent respectively full and partial support)

Organizations generally accept that the environment is characterized by uncertainties that can result in both opportunities and risks. *Identifying these uncertainties and assessing the related risk* is a fundamental component of risk management. Rule-based business process mining techniques allow for a quantitative assessment of the likelihood of occurrence, based on company and process specific historic data.

After identifying and assessing the risks related to the process, management has to decide on the *risk response* strategy that will be used, i.e. either risk avoidance, reduction, sharing (e.g. insurance, outsourcing, etc.) or acceptance. Rule-based business process mining can be effectively used for the implementation of a wide variety of detective controls, corresponding to the acceptance response. Additionally, they can be employed for the *periodic or continuous monitoring* of the residual risk (i.e. risk remaining after the implementation of risk mitigating controls) and the effectiveness of the proactive and corrective controls.

Finally, rule-based business process mining outcomes enable the calculation of important statistics related to the risk and the implemented mitigation measures *(information and communication)*.

The next section positions this rule-based business process mining approach for enterprise risk management in the broader governance, risk and compliance setting of an organization.

4   Towards a Framework for Business Process Mining in Enterprise Risk Management

There is a growing need for a **clear guidance** as well as a common language for business process mining in the context of enterprise risk management and a broader governance, risk and compliance (GRC) setting. The conclusions made in the previous sections help to position this rule-based business process mining approach within this broader setting. Here, we propose a comprehensive process mining applicability framework that could fill this need, see figure 2. The applicability framework consists of **three interrelated dimensions**:

- *Process Mining Techniques Dimension*: Encompasses the techniques described in section 2, namely *Process Discovery & Visualization*, *Conformance Checking & Delta Analysis* and *Rule-Based Property Verification* 2.
- *Control Functions Dimension*: Describes the different sets of stakeholders involved in the GRC activities [6], and that therefore could benefit from the introduction of process mining techniques for GRC.
- *Control Function Activities Dimension*: Discusses the potential application areas of process mining for GRC activities: *Risk Identification & Assessment*, *Control Activities* and *Information, Documentation & Communication*. These where further elaborated in section 3.

The **rule-based business process mining approach** for enterprise risk management covers the application of rule-based mining techniques for all control function activities, with a focus on (but not limited to) the organization's management.

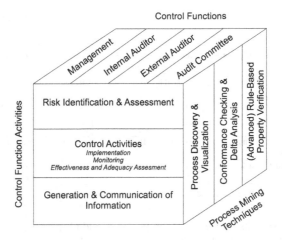

Fig. 2 Process Mining Applicability framework for Enterprise Risk Management and Audit Compliance Checking

5 Case Research: Risk Management in the Software Update and Debug Process of a Social Security Service Provider

Previous sections provided a situation of and legitimation for the use of rule-based business process mining, as well as a possible implementation. For the purpose of illustrating the possibilities of rule-based business process mining we here elaborate a case on risk management in a software development process. The event log describes the updates (including new features and resolving compliance issues) and bug repair instances of an online reporting application of a service provider in the social security sector.

5.1 Introduction to the Process Dynamics

The event log was retrieved from their project management tool, which has been used for tracking the evolution in software development process. In the context of this contribution we opted for the instances related to an important update or a debugging. In total the event log contains 463 events scattered over 158 cases. The activity subset $\mathscr{A}$ = {new feature business analysis, bug report analysis, information request, functional analysis, development, testing}, the originator set $\mathscr{O}$ encompasses 36 employees and the property set $\mathscr{P}$ = {case ID, timestamp, type of intervention, priority and summary}. An extract of the event log can be found in table 1.

Figure 3 represents the business process as it happened in reality (mined with the $\alpha++$ mining algorithm and plugin in ProM [10]). This figure already indicates certain deviations from the desired process model, which might suggest the existence

Table 1 Extract of the event log for the debugging and updating of the online reporting application

ID	Activity	Originator	Timestamp	Type of Intervention	Priority	Summary
28	Bug Report Analysis	John	22/11/2011 14:50	Bug	Critical	error code ...
20	Testing	Matt	24/11/2011 09:32:00	New Feature	Major	adapt to fiscal law art. ...
...	...	...	...	...	...	...
28	Functional Analysis	Anne	28/11/2011 14:45:00	Bug	Critical	error code ...
32	Development	Jack	28/11/2011 14:58:00	New Feature	Major	add ... query functionality to ...
...	...	...	...	...	...	...

of certain business risks. For example in points A and C (part of) the analysis was not performed. Additionally, for certain new features who were not identified as critical and did not result in a significant value addition, it was decided not to develop them after the functional analysis. These observations relate to the *risk identification* phase.

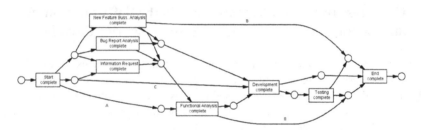

Fig. 3 Excerpt of retrieved process model specific for software updating and debugging by a social security service provider ($\alpha + +$ mining algorithm)

5.2 Excerpt of Rule-Based Business Process Mining Analysis

In this subsection we further elaborate on rule-based business process mining. Due to the limited amount of space, we restrict the example to a subset of control objectives related to the main objective of keeping customers satisfied. As the produced software is the main product of our service provider, the control objectives 'ensure that clients requirements are identified and communicated' as well as 'ensuring that these requirements are adequately addressed' become crucial. In order to achieve these objectives a multitude of internal controls can be devised, e.g. 'the existence of a new feature business analysis', 'a bug report analysis or an information request' and 'a separation of the development and testing responsibilities' (see table 2). However, one could also find for example service level agreements (i.e. business policies) describing the reasonable response times for executing critical, major or minor updating or debugging request.

Table 2 Risk management in software development case

	Rule 1: Activity existence	**Rule 2: Segregation of duties**
Objective	Ensure that the clients requirements (functional, legal, contractual, etc.) are identified and communicated.	Ensure that these requirements are adequately addressed in the developed software.
Risk	Non-compliance areas not identified, outdated requirements remain in effect, etc.	Financial losses and penalties, decreased customer satisfaction, increased likelihood of disputes, etc.
Internal control	Each process instance must contain a *new feature business analysis, bug report analysis or information request* activity.	*Person P* must not perform both the *development* and the *testing* activity for the same process instance
Business rule	$\Diamond(Activity(newfeaturebusinessanalysis)) \lor \Diamond(Activity(bugreportanalysis)) \lor \Diamond(Activity(informationrequest))$	$\Box((Activity(development).originator = P) \Rightarrow \Box(\neg(Activity(testing).originator = P)))$
Process mining perspective	Functional (PMP1)	Organizational (PMP3)
Rule restriction focus	Cardinality-based (RRF1)	Coexistence (RRF2)

Checking these business rules against the data contained in the event log results in the conclusions presented in table 3. Depending on the state of the enterprise risk management process, the risk analyst can make conclusions on the *likelihood of occurrence* (in risk assessment, monitoring residual risk or detective internal control) or on the *effectiveness of an implemented proactive or corrective control*. Finally, the results can be used for effective *communication*.

Table 3 Results of a rule-based business process analysis in the software development case

	Rule 1: Activity existence	**Rule 2: Segregation of duties**
Occurrence (total, percentage)	132 (158, 83.5%)	156 (158, 98.7%)
Occurrence Type = 'Bug' (total, percentage)	57 (65, 87.7%)	64 (65, 98.5%)
Occurrence Type = 'New Feature' x Priority = 'Critical' (total, percentage)	28 (35, 80.0%)	0 (35, 0.0%)

6 Conclusion

In this paper we proposed a rule-based process mining approach for enterprise risk management. This approach enables the management to effectively and efficiently identify & assess risks, implement management control systems and acquire and communicate detailed information on the business operations.

Firstly, the relevant rule types were identified and classified, while providing the necessary framework for formally expressing them. Followed by the identification of risk management components that could benefit from rule-based process mining support. Thirdly, we positioned our approach in a larger framework of GRC supporting techniques. Finally, the applicability of our approach was demonstrated by elaborating on the risk management components for the updating and debugging process of a social security service provider.

References

1. COSO. Enterprise risk management - integrated framework. Technical report, Committee of Sponsoring Organizations of the Treadway Commission (2004)
2. Curtis, B., Kellner, M.I., Over, J.: Process modeling. Communications of the ACM 35(9), 75–90 (1992)
3. Giannakopoulou, D., Havelund, K.: Automata-based verification of temporal properties on running programs. In: Proceedings of the 16th Annual Conference on Automated Software Engineering, pp. 412–416. IEEE Computer Society (2001)
4. Goedertier, S., Martens, D., Vanthienen, J., Baesens, B.: Robust process discovery with artificial negative events. The Journal of Machine Learning Research 10, 1305–1340 (2009)
5. Herbst, J.: A machine learning approach to workflow management. In: Lopez de Mantaras, R., Plaza, E. (eds.) ECML 2000. LNCS (LNAI), vol. 1810, pp. 183–194. Springer, Heidelberg (2000)
6. Pickett, K.H.S.: The Internal Auditing Handbook. Wiley (2010)
7. Rozinat, A., van der Aalst, W.M.P.: Conformance checking of processes based on monitoring real behavior. Information Systems 33(1), 64–95 (2008)
8. van der Aalst, W.M.P., de Beer, H.T., van Dongen, B.F.: Process mining and verification of properties: An approach based on temporal logic. In: Meersman, R. (ed.) OTM 2005. LNCS, vol. 3760, pp. 130–147. Springer, Heidelberg (2005)
9. Van der Aalst, W.M.P., Weijters, T., Maruster, L.: Workflow mining: Discovering process models from event logs. IEEE Transactions on Knowledge and Data Engineering 16(9), 1128–1142 (2004)
10. Wen, L., Wang, J., Sun, J.: Detecting implicit dependencies between tasks from event logs. In: Zhou, X., Li, J., Shen, H.T., Kitsuregawa, M., Zhang, Y. (eds.) APWeb 2006. LNCS, vol. 3841, pp. 591–603. Springer, Heidelberg (2006)
11. Yip, F., Wong, A.K.Y., Parameswaran, N., Ray, P.: Rules and ontology in compliance management. In: 11th IEEE International Enterprise Distributed Object Computing Conference, pp. 435–435. IEEE (2007)

A News-Based Approach for Computing Historical Value-at-Risk

Frederik Hogenboom, Michael de Winter, Flavius Frasincar,
and Alexander Hogenboom

Erasmus University Rotterdam, P.O. Box 1738, NL-3000 DR Rotterdam, The Netherlands
{fhogenboom,frasincar,hogenboom}@ese.eur.nl,
m.r.dewinter88@live.nl

Abstract. Within the field of finance, Value-at-Risk (VaR) is a widely adopted tool to assess portfolio risk. When calculating VaR based on historical stock return data, the data could be sensitive to outliers caused by seldom occurring news events in the sampled period. Using a data set of news events, of which the irregular events are identified using a Poisson distribution, we research whether the VaR accuracy can be improved by considering news events as additional input in the calculation. Our experiments show that when a rare event occurs, removing the event-generated noise from the stock prices for a small, optimized time window can improve VaR predictions.

1 Introduction

Despite its limitations in terms of interpretability and mathematical properties [2, 16], Value-at-Risk (VaR) is a widely adopted risk measure used by practitioners in the field of finance, quantifying the risk of loss on a portfolio of financial equities. It is defined as a threshold value and confidence level such that the probability that the loss on the portfolio over a given time horizon does not exceed a certain value at a given confidence level. It is generally assumed that there are no unexpected trend breaks. However, in reality we are faced with deviations from trends, mainly caused by emerging events. These events are usually reported in news and can greatly impact today's financial markets. For example, when Google announced a 29% increase in its 2011 Q3 net-income, within hours its shares went up by 7%.

According to the weak form of the efficient market hypothesis, news that contains information on an equity is not perfectly incorporated in the price when it is published. Studies have reported on the existence of such a delay [8], caused by initial over- or under-reactions to the news. Additionally, news events have an effect on the volatility of equities [15]. Hence, taking into account news events for VaR calculations (which are based on returns distributions) could be beneficial, as the volatility is the standard deviation of the distribution of returns.

As the usage of information extracted from text in a financial context has proven to be a vital strategy in many financial applications [6, 12], we hypothesize that

J. Casillas et al. (Eds.): Management Intelligent Systems, AISC 171, pp. 283–292.
springerlink.com   © Springer-Verlag Berlin Heidelberg 2012

we can improve VaR computations by introducing financial news events [3, 10] as an additional input. In our research, we employ the ViewerPro [18] software for the extraction of ticker data and news events. By using a Poisson distribution, we identify the irregular (and hence noisy) events. Subsequently, we cleanse the ticker data from event-generated noise, and aim to obtain a data set which is a more accurate representation of the expected returns distribution. In our experiments, we aim to optimize the time window for which the noise is removed by evaluating for different configurations the accuracies of the calculated VaR.

This paper is organized as follows. First, we describe related approaches to this research in Sect. 2. Then we introduce our framework in Sect. 3. Section 4 presents our implementation, our data set, and an evaluation of the framework on this data set. Last, in Sect. 5 we draw our conclusions and provide directions for future work.

2 Related Work

The existence of a relationship between the stock market and news events has been acknowledged by many previous studies [5, 7, 9]. Additionally, the number of news events and trading activity have proven to be correlated [15]. Even though the efficient market hypothesis supports that news information is fully and immediately processed into the value of shares, in practice this is not always the case [17]. Hence, for traders, timely and accurately reacting on news and estimating the VaR of portfolios correctly, is of utmost importance.

The three most widely used implementations for VaR calculations are the parametric method (assuming a specific distribution of equity returns), a Monte Carlo simulation-based method that predicts future returns by fitting a distribution based on historical data, and the historical method, which assumes that historical changes in the price accurately predict changes in the future. Common distributions for the parametric method are the normal and log-normal distributions, as they offer simplicity and robustness. However, in practice, equity returns are almost never normally distributed [1]. Assuming a specific distribution could therefore lead to a bias in the risk measure. Even though the Monte Carlo simulation overcomes this problem by randomly sampling the historical data multiple times to approximate its distribution, this method is rather slow as it is computationally intensive. As we aim for an application that is able to run real-time, Monte Carlo simulation-based methods are not suitable for our research. Similarly, the historical method also analyzes a set of historical returns instead of an assumed distribution. An advantage of the historical method over the Monte Carlo simulation-based methods is its simplicity, which fosters real-time computation. Therefore, in this paper we utilize the historical method for VaR prediction in which we implement event-based improvements.

Hull and White [11] improve the VaR calculation by updating the volatility in the historical method by means of GARCH/EWMA models in order to reflect the difference between the volatility at the time of the observation and the current volatility. While Hull and White analyze multiple equity portfolios, in our work we only observe single equity portfolios in order to prevent heteroscedasticity (i.e., interdependencies between variances, which is often the case with different financial equities in a portfolio). The authors propose a method to update the volatility in the appropriate time interval so that the volatility becomes a more dynamic factor in VaR calculation. Based on mean absolute percentage error (MAPE), their work is compared to another method, involving the assignment of weights to observations that are more recent [4]. The authors find that their method outperforms both the traditional historical method and second method for exchange rates, yet for stock indices, results are mixed.

Other work that aims to improve technical indicators with news was performed by Zhai et al. [19]. The authors make use of a simple text classification algorithm with a supervising learning method. Instead of only using company specific news, they are also integrating general market news in combination with technical indicators. It is concluded that technical indicators and news events alone are inaccurate as estimators, but that the combination of both could lead to better results. Based on a real-life market simulation, the authors show that by using their approach it is possible to make profit.

3   Framework

In order to be able to assess whether the incorporation of news into the calculation of the VaR of a specific equity improves the overall quality of the outcomes, we propose a framework that is based on two inputs, i.e., a list of stock prices and a list of financial events, which are extracted from several feeds such as Reuters using the ViewerPro application.

In a pre-processing phase, we cleanse the collected equity prices as follows. As stock markets are only open on specific dates and times, we filter the prices and keep those within market opening times. Also, in order to decrease computational complexity, the time intervals between individual prices are defined per hour instead of per second.

Subsequently, we read in news events, stemming from news items processed by ViewerPro using computational linguistics, semantic analysis, and formal logic. ViewerPro determines the positive and negative impacts of the information described in the news on the equities that are relevant to the user. Large amounts of news messages are filtered for equity-specific news, and the semantic component of ViewerPro analyzes each individual news message for economic impact. This yields a list of relevant annotated news events. Some general types of news events that are covered by the ViewerPro annotations are hiring and resignation of CEOs, acquisitions, profit announcements, etcetera.

An additional step is performed by identifying irregularly occurring event types from our event set, as these events are not likely to occur again and thus cause a significant noise in stock rates. As Poisson distributions are used in many fields to model the number of occurrences of events in a certain time interval (if the average rate of occurrences is known and we assume that events occur independently from each other), we apply a Poisson distribution F to a test set *test*, which is a function of the measured and expected number of occurrences in the test set, i.e., x and λ, respectively:

$$F(x;\lambda) = \frac{\lambda^x e^{-\lambda}}{x!}.$$ (1)

As depicted in Fig. 1, when using a threshold α of 0.05, for $x = 0$ (which means no event occurrences), $F(x;\lambda) < \alpha$ for $\lambda \geq 3$. For a training set *train* the expected number of occurrences λ' is obtained by scaling λ by the proportion of the set cardinalities, i.e., $\lambda' = \lambda \times \vartheta$, with $\vartheta = |train|/|test|$. Hence, we consider event types that occur $\geq 3 \times \vartheta$ as regular events, and events occurring $< 3 \times \vartheta$ as rare events.

The identified (rare) events are subsequently associated with times in which they occurred and also with the recorded stock rates. We adjust the collected prices for a time window to account for the generated noise by updating their values to the previously measured value, which is illustrated by Algorithm 1 that processes a list of chronologically ordered (hourly) recorded prices. For each stock price *price* in price list *prices*, we compare the stock price time with the time of each event *event* stored in event list *events* in order to check for event occurrences. If an event occurrence is identified, *impact* is set to the window size *window* (for which the optimization is given in Sect. 4), causing the value of the subsequent *price* items to be set to the current value. The value of *impact* is decreased with 1 every next *price* in price list *prices*, so that subsequent price values are updated up until the window size has been reached. In case of overlapping events, the *impact* counter is reset to the window size *window*. After processing all original prices stored in *prices*$_{hist}$, we obtain a new list of event-corrected prices, i.e., *prices*$_{event}$.

Both sets of original ("*hist*") and denoised ("*event*") prices are converted to sets with hourly returns. We compute the return set *returns* of a price set *prices* as the relative change between the price at time $t + 1$ and the previous price at time t, i.e.,

$$returns = \frac{prices_{t+1} - prices_t}{prices_t} \quad \forall t = 1, \ldots, N - 1.$$ (2)

where N represents the number of items in the list. A specific return *returns*$_t$ equals the profit that can be obtained if a share is bought at time t and sold at time $t + 1$.

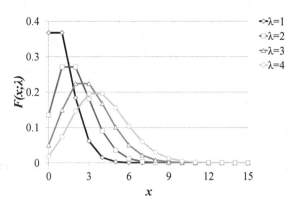

Fig. 1 Poisson distributions for various measured and expected occurrences, i.e., x and λ, respectively

---

Algorithm 1. News event processing (per equity)

Require: *prices* = array of stock prices and associated times
Require: *events* = array of events and associated times
Require: *window* = integer representing time window
 1: *previousprice.value* = *prices*.[1].*value*
 2: **for all** *price* in *prices* **do**
 3:     **for all** *event* in *events* **do**
 4:         **if** *impact* > 0 **then**
 5:             *impact* = *impact* − 1
 6:             *price.value* = *previousprice.value*
 7:         **end if**
 8:         **if** *price.time* = *event.time* **then**
 9:             *impact* = *window*
10:         **end if**
11:     **end for**
12:     *previousprice.value* = *price.value*
13: **end for**

We make use of the historical returns (both original and adapted) to estimate the future returns. The time horizon used for computing returns is 1 day. After sorting the return list *returns*, we calculate the Value-at-Risk, *VaR*, as

$$VaR = returns' \left[\lfloor \alpha \cdot \text{length}(returns) \rfloor \right] . \tag{3}$$

Here, *returns'* represents the ordered (sorted) list of returns and where the confidence level is denoted by α. Thus, in a data set with 20 historical returns – with the first element being located on position 1, and the last on position 20 – we select the first worst return (i.e., position 19) for a confidence level of 0.95. With (3), we calculate VaR_{event} and VaR_{hist} using our adjusted method and the traditional method (i.e., the historical method without the improvements proposed in [4, 11]), respectively.

4 Evaluation

In order to evaluate the performance of the proposed historical VaR calculation, our framework is implemented as a Java-based application that calculates the VaR of a single equity based on a data set containing news events and stock prices.

The data set used in our experiments stems from the ViewerPro software, and – after filtering – covers news events and stock data collected on an hourly basis for 363 heterogeneous equities on weekdays during the year 2010, and contains approximately 2,000 stock data points, 119 different event types, and 50 up to 75 associated events per equity. In order to evaluate the performance of the calculation, we predict the VaR_{event} and VaR_{hist} for 75% of our data set. The remaining 25% is used as a test set for comparing the predicted VaR with the actual VaR.

Even though many VaR analyses are currently performed using the Kupiec test [14], we employ a different set of measures. As explained by Kupiec in his original work, the test is statistically weak with sample sizes of one year. As our data set covers only the 2010, we need different measures that provide insight into the effectiveness of our proposed event-based approach.

In order to analyze for how many equities our adjusted event-based historical method provides better quality predictions in comparison to the traditional historical method, we measure each method's squared error. The squared error SE for equity e is defined as the squared difference between the equity's actual VaR ($VaR_{e,actual}$) measured in our test set and the predicted VaR ($VaR_{e,predicted}$) that has been predicted based on our training set, i.e.,

$$SE_e = \left(VaR_{e,actual} - VaR_{e,predicted}\right)^2 , \qquad (4)$$

where $VaR_{e,predicted}$ is one of VaR_{event} or VaR_{hist}.

Subsequently, the squared errors are combined into the mean squared error (MSE), yielding an MSE_{hist} and MSE_{event}. The MSE is calculated as the summation of the squared errors (SE) of all equities $e \in E$ divided by the number of equities, i.e.,

$$MSE = \frac{\sum\limits_{e \in E} SE_e}{|E|} , \qquad (5)$$

where $|E|$ denotes the total number of equities in set E, in our case 363.

Additionally, we evaluate the number of times both methods outperform one another, i.e., OPT (OutPerformed Total), by comparing the squared errors $SE_{e,hist}$ and $SE_{e,event}$ for each equity $e \in E$, yielding

$$OPT_{hist,event} = \sum_{e \in E} O(SE_{e,hist}, SE_{e,event}) , \qquad (6)$$

$$OPT_{event,hist} = \sum_{e \in E} O(SE_{e,event}, SE_{e,hist}) , \qquad (7)$$

$$O(X,Y) = \begin{cases} 1 & \text{if } X < Y \\ 0 & \text{else} \end{cases} . \qquad (8)$$

In our experiments, we compare the MSE and OPT for the traditional and event-based VaR calculation methods, both on the full event data set, as well as on a data set containing only the rare events, using an arbitrary time window of 8 hours (determined based on initial estimates). Subsequently, we determine the optimal time window size by observing plots of MSE and OPT values for the event-based VaR calculation method. Also, we take into account the number of overconfident predictions ($CONF$) of all equities $e \in E$, which is calculated as

$$CONF = \sum_{e \in E} C(VaR_{e,predicted}, VaR_{e,actual}) , \qquad (9)$$

$$C(X,Y) = \begin{cases} 1 & \text{if } X > Y \\ 0 & \text{else} \end{cases} , \qquad (10)$$

where $VaR_{e,predicted}$ represents the predicted VaR_{event} for equity e based on our adjusted data set (only containing the rare events).

Last, we perform a two-sample one-tailed t-test on the sets of individual squared errors SE_{hist} and SE_{event} (containing $SE_{e,hist}$ and $SE_{e,event}$ $\forall e \in E$, respectively) for our optimal configuration, in order to assess the significance of the measured difference between MSE_{hist} and MSE_{event}. For this, we use a significance level of 0.05 to reject the null hypothesis that there is no difference between the measured MSE values.

When comparing the results from both VaR calculation methods using our data set containing all (i.e., regular and rare) events and stock rates on an hourly basis, and when using a time window of 8 hours, we obtain the results depicted in Table 1, which shows the MSE and OPT values for both the traditional and the event-based historical VaR calculation methods (columns *hist* and *event*, respectively). We observe an improvement of 21.66% in terms of MSE when accounting for event-generated noise in our stock data. Additionally, the event-based VaR calculation method outperforms the traditional historical method (in terms of squared errors of predicted VaR values with respect to the actual VaR values) 232 times which is a share of 63.91% of all predictions, compared to 131 observations in which the traditional method outperforms our adapted method.

As presented in Table 1, repeating the same experiment on a filtered data set which only contains non-rare events (i.e., 345 in total) yields an additional improvement over the previous results. Now, in 71.88% of the cases (i.e., 248 out of 345), event-based historical VaR calculation outperforms the traditional method. Also, we see a performance gain in terms of MSE. The improvement in MSE values has increased from 21.66% to 26.29%. Both scores underline the added value of only considering the rare events.

Subsequently, we optimize the size of the time window by evaluating MSE and OPT values on the one hand, and the number of overconfident predictions ($CONF$) on the other hand. For this, we observe VaR prediction models with time windows ranging from 1 to 24 (i.e., 3 working days of 8 hours, which is the maximum effect of a news event [13]). As depicted by the graphs in Fig. 2, cleansing the data with a window of 10 hours yields the highest score for OPT (i.e., 249). However, the lowest MSE value is observed for a window of 14 hours. The number of overconfident predictions increases for each increase in window size, and hence we opt for a time window of 10 hours, as this maximizes the number of outperforming predictions while minimizing the number of overconfident predictions.

Table 1 Experimental results of the performance of traditional and event-based historical VaR calculation (columns *hist* and *event*, respectively), while employing a cleansing window of 8 hours

Measure	All events		Non-rare events	
	hist	*event*	*hist*	*event*
MSE	1.0590E−05	8.2965E−06	1.1220E−05	8.2700E−06
OPT	131	232	97	248

As shown in Table 2, utilizing a window of 10 instead of 8 hours on a data set with rare events yields an improvement both in terms of *MSE* and *OPT*. The *MSE* of our event-based historical VaR prediction models improves with 31.73% over the traditional historical VaR prediction method's *MSE*. This improvement is a lot higher than the measured improvement of 21.66% when using a cleansing window of 8 hours. Alternatively, we can also determine an optimal cleansing window for each event type separately by evaluating the percentile differences from the mean stock rate per equity in order to determine the impact of an event type. Large differences (e.g., > 50.00%) after an event occurrence indicate noise that should be cleansed, while a small difference (i.e., the smallest difference after an event) indicates that the market has returned to normal, hence not requiring any cleansing. This strategy for determining (individual) window sizes yields even higher improvements. For the measured *MSE* values we obtain a decrease of 35.47%, whereas 71.01% of the event-based VaR predictions outperform the traditional ones.

In order to assess the significance of the measured *MSE* improvement of 35.47%, we perform a paired two-sample one-tailed t-test based on SE_{hist} and SE_{event}, containing squared errors for all equities. We obtain a p-value of 0.0027, hereby

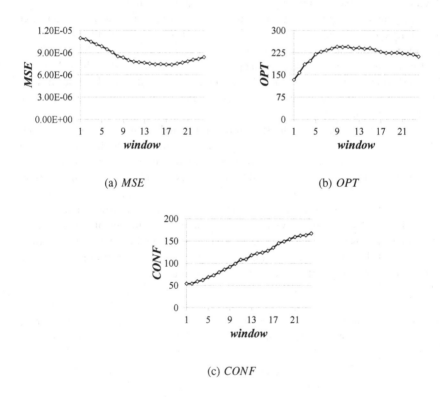

(a) *MSE*   (b) *OPT*

(c) *CONF*

Fig. 2 Performances of event-based VaR prediction models with various time windows

Table 2 Experimental results of the performance of traditional and event-based historical VaR calculation (columns *hist* and *event*, respectively) on non-rare events

	window = 10		window = event-based	
Measure	*hist*	*event*	*hist*	*event*
MSE	1.1220E−05	7.6600E−06	1.1220E−05	7.2400E−06
OPT	96	249	100	245

rejecting the null hypothesis that there is no difference between the measured *MSE* values when applying a significance level of 0.05. Hence, the proposed event-based historical VaR calculation method (using non-rare events and event-based window sizes) produces more reliable VaR predictions when compared to the traditional method.

5 Conclusions

When calculating Value-at-Risk (VaR) – a widely adopted tool in the field of finance to assess portfolio risk – based on historical stock returns data, the data could be sensitive to outliers caused by seldom occurring news events in the sampled period. In order to address this shortcoming, we have proposed a way to enhance the calculation and prediction of VaR based on historical data, by removing the event-induced noise. This would enable practitioners to make better predictions of risk in terms of distributions of expected future returns.

Using a substantial data set of stock rates and news events of 2010 stemming from the proprietary ViewerPro software, we have identified rare (and hence noisy) events using a Poisson distribution. Subsequently, the event-generated noise was removed from the stock rates. From our experiments, in which we evaluated various cleansing window sizes, we can conclude that the calculation and prediction of VaR can be improved with news (i.e., extracted events and stock rates) as an additional input. Our event-based method demonstrates a significant *MSE* improvement of 35.47% compared to the traditional historical method, and outperforms the latter in 71.01% of the cases.

For future work, we suggest to investigate accounting for the type of news events, which could affect the influence of an event on equity prices (e.g., mergers could generate a larger noise than quarterly profit announcements). Another direction is related to additionally accounting for general stock market events such as financial crises, instead of only just the company specific news. Last, we would also like to build a real-life market simulation for our improved historical VaR method.

Acknowledgement. The authors are partially sponsored by the NWO Physical Sciences Free Competition project 612.001.009: Financial Events Recognition in News for Algorithmic Trading (FERNAT) and the Dutch national program COMMIT. We would like to thank Mark Vreijling and Wijand Nuij from Semlab for fruitful discussions and technical support during this research.

References

1. Andersen, T.G., Bollerslev, T., Diebold, F.X., Ebens, H.: The Distribution of Stock Return Volatility. Journal of Financial Economics 61(1), 43–76 (2001)
2. Artzner, P., Delbaen, F., Eber, J.M., Heath, D.: Coherent Measures of Risk. Mathematical Finance 9(3), 203–228 (1999)
3. Borsje, J., Hogenboom, F., Frasincar, F.: Semi-Automatic Financial Events Discovery Based on Lexico-Semantic Patterns. International Journal of Web Engineering and Technology 6(2), 115–140 (2010)
4. Boudoukh, J., Richardson, M., Whitelaw, R.F.: The Best of Both Worlds: A Hybrid Approach to Calculating Value at Risk. Risk 11(5), 64–67 (1998)
5. Byström, H.: News Aggregators, Volatility and the Stock Market. Economics Bulletin 29(4), 2673–2682 (2009)
6. Chan, W.S.: Stock Price Reaction to News and No-News: Drift and Reversal After Headlines. Journal of Financial Economics 70(2), 223–260 (2003)
7. Engelberg, J.E., Parsons, C.A.: The Causal Impact of Media in Financial Markets. Journal of Finance 66(1), 67–97 (2009)
8. Fama, E.F.: The Behavior of Stock-Market Prices. Journal of Business 38(1), 34–105 (1965)
9. Goonatilake, R., Herath, S.: The Volatility of the Stock Market and News. International Research Journal of Finance and Economics 3(11), 53–65 (2007)
10. Hogenboom, A., Hogenboom, F., Frasincar, F., Kaymak, U., van der Meer, O., Schouten, K.: Detecting Economic Events Using a Semantics-Based Pipeline. In: Hameurlain, A., Liddle, S.W., Schewe, K.-D., Zhou, X. (eds.) DEXA 2011, Part I. LNCS, vol. 6860, pp. 440–447. Springer, Heidelberg (2011)
11. Hull, J., White, A.: Incorporating Volatility Updating into the Historical Simulation Method for Value-at-Risk. Journal of Risk 1(1), 5–19 (1998)
12. Ikenberry, D.L., Ramnath, S.: Underreaction to Self-Selected News Events: The Case of Stock Splits. Review of Financial Studies 15(2), 489–526 (2002)
13. Kalev, P.S., Liu, W.M., Pham, P.K., Jarnecic, E.: Public Information Arrival and Volatility of Intraday Stock Returns. Journal of Banking & Finance 28(6), 1441–1467 (2004)
14. Kupiec, P.H.: Techniques for Verifying the Accuracy of Risk Measurement Models. Journal of Derivatives 3(2), 73–84 (1995)
15. Mitchell, M.L., Mulherin, J.H.: The Impact of Public Information on the Stock Market. Journal of Finance 49(3), 923–950 (1994)
16. Rockafellar, R.T., Uryasev, S.: Conditional Value-at-Risk for General Loss Distributions. Journal of Banking & Finance 26(7), 1443–1471 (2002)
17. Rosenberg, B., Reid, K., Lanstein, R.: Persuasive Evidence of Market Inefficiency. Journal of Portfolio Management 11(3), 9–16 (1985)
18. Semlab: ViewerPro (2011), http://viewerpro.semlab.nl/
19. Zhai, Y.Z., Hsu, A., Halgamuge, S.K.: Combining News and Technical Indicators in Daily Stock Price Trends Prediction. In: Liu, D., Fei, S., Hou, Z., Zhang, H., Sun, C. (eds.) ISNN 2007. LNCS, vol. 4493, pp. 1087–1096. Springer, Heidelberg (2007)

Various Applications

Gaussian Mixture Models vs. Fuzzy Rule-Based Systems for Adaptive Meta-scheduling in Grid/Cloud Computing

R.P. Prado[1], J. Braun[2], J. Krettek[2], F. Hoffmann[2], S. García-Galán[1], J.E. Muñoz Expósito[1], and T. Bertram[2]

[1] Telecommunication Engineering Department in University of Jaén, Alfonso X el Sabio, 28 Linares, Jaén. Spain
{rperez,sgalan,jemunoz}@ujaen.es

[2] Control System Engineering, University Dortmund, D-44221 Dortmund, Germany
{jan.braun,johannes.krettek,frank.hoffmann,
torsten.bertram}@tu-dortmund.de

Abstract. Adaptive scheduling strategies are about considering the state of computational grids to obtain efficient and reliable schedules and to prevent the system performance deterioration. In this work, emerging adaptive strategies in grid computing, namely Fuzzy Rule-Based Systems (FRBS) -based strategies and a new adaptive scheduling approach, gaussian scheduling founded on Gaussian Mixture Models (GMMs) are compared. Both types of strategies focus on modeling the state of resources and select the most convenient site of the grid at every scheduling step given the current conditions. FRBSs provide a fuzzy characterization of the grid state and the inference of a suitability index based on their own knowledge given in the form of fuzzy IF-THEN rules. Besides, a GMM can be trained to model a complex probability density distribution indicating the suitability of every site in the grid to be the target of the schedule with the current conditions of its resources. This way the GMM scheduler assigns a probability to every state of the site where a higher probability is associated to a higher suitability of selection. Simulations based on real grid facilities are conducted to test the FRBS and GMM-based models and results are analyzed in terms of accuracy and convergence behaviour of their associated learning processes.

1 Introduction

Cloud computing is a new paradigm of distributed computing that provides companies with services such as the execution of applications (SoS), platforms for the development of applications (PaaS) and/or an infrastructure (IaaS) with high computing and storing capacity [7]. The final aim of these services consists of the improvement of productivity that companies can reach by externalizing their computational needs and hiring cloud services. On the other hand, a computational grid is a collection of autonomous, heterogeneous and geographically distributed

J. Casillas et al. (Eds.): Management Intelligent Systems, AISC 171, pp. 295–304.

computing resources that cooperate and share capacities to achieve a common goal [3]. Hence, an improvement in the features of a grid system represents an improvement in the features of a cloud system that uses it as underlying infrastructure. A major challenge of these computing networks is determined by their efficient cooperation and coordination or scheduling which is a NP-complete problem.

Adaptive scheduling strategies propose to consider both current and future resources conditions to avoid and prevent the grid system performance degradation [11]. In this regard, schedule-based strategies base their decisions on a "known" current state of the grid system to allow a more precise schedule of jobs and satisfy diverse QoS (Quality of Service) specifications [6]. The "known" current state makes reference to available resources domain capabilities, computational demands, number of jobs, etc. A key aspect is how to design a scheduling strategy able to simultaneously integrate and combine the diverse criteria describing the current conditions of the grid and to provide with a mechanism that founded on this state, can select the most suitable resource or site for the actual schedule.

In this work, emerging adaptive scheduling strategies based on FRBSs, fuzzy rule-based schedulers [8, 9] and a novel strategy derived from Gaussian Mixture Models (GMMs) [2, 12] are compared. Fuzzy Rule-Based Systems (FRBSs) [1] are expert systems derived from fuzzy logic and rule-based systems recently proposed as an alternative for the development of scheduling systems in grid computing on the basis of their adaptability to environments dynamism and capability to cope with uncertainty in systems information [8, 9]. Essentially, fuzzy rule-based schedulers suggest a fuzzy characterization of the states of sites and they associate a suitability index to every site according to their own knowledge of the system given in the form of IF-THEN rules. In this regard, given the dependence of these schedulers with their associated knowledge, learning strategies must be considered in their design such as Michigan and Pittsburgh approaches [1]. On the other hand, a scheduling strategy based on the consideration of probability density functions to model the state of the diverse resources domains making up the grid is presented. Specifically, the application of GMMs [12] is suggested for meta-scheduling in grid computing to model the state of resources and decide the most suitable resource selection at each stage. A GMM is a statistical model that uses a weighted sum of probability density functions of multiple Gaussian distributions to represent the distribution of a vector in the probability space [2, 12]. A GMM is trained to model a meta-scheduler that bases its decisions on a density distribution associated to the different resource domains (RDs) or sites and the state of these resources. Subject to a state of the grid, a probability of selection for the next schedule is obtained for every RD where the site showing a higher probability is selected as the most suitable target domain. Two approaches based on these models with evolutionary learning are proposed for meta-scheduling in grid computing and results are analyzed in terms of accuracy and convergence behaviour to fuzzy rule-based meta-schedulers with Michigan and Pittsburgh approaches as learning strategies.

The rest of the paper is summarized as follows. Sections 2 and 3 introduce an overview of fuzzy rule-based schedulers and present the novel adaptive approach, respectively. In Section 4, the simulation results are analyzed for both types of schedulers and finally, Section 5 outlines the main conclusions.

2 Fuzzy Rule-Based Schedulers

As introduced in the previous section, the state of resources must be considered in the scheduling strategies in order to provide QoS. In this sense, one of the biggest problems is given by the fact that the grids, unlike other classical distributed systems, are characterized by the dynamism and uncertainty in the state of theirs resources. Thus, a state-based scheduling that does not consider any uncertainty or imprecision in the handled information can make decisions founded on conditions that do not correspond with reality. The consideration of imprecision-tolerant techniques can be very beneficial in environments subject to uncertainty such as computational grids. FRBS-based schedulers have recently emerged as alternative scheduling systems to provide an efficient characterization of the state of sites in order to decide the suitability of the selection of the different sites in the schedule [8, 9]. Specifically, this decision is given by the application of expert knowledge in the form of IF-THEN rules that can be automatically optimized with diverse learning procedures in FRBSs based on evolutionary strategies such as Pittsburgh and Michigan approaches [1].

The structure of rules of a fuzzy rule-based scheduler generally follows the classical Mamdani model [1]:

$$R_i = IF\, x_1\, is\, A_{1g}\, and/or \ldots x_n\, is\, A_{nk}\, THEN\, y\, is\, B_z\, with\, w \tag{1}$$

where x_j denotes the component j of the antecedent, y indicates the consequent, and/or represents the possible connectives for the antecedents and w is the weight associated to the rule. Also, A_{jk} indicates the set k of the l possible fuzzy sets allowed for the component j of the antecedent and B_z represents the set z of the t possible fuzzy sets for the consequent. Fig. 1 shows the general structure of the fuzzy rule-based meta-scheduler within the grid environment. As illustrated, the general structure of the fuzzy meta-scheduler follows the classical schema of Mamdani fuzzy logic systems [1]. There exist three main components: the *fuzzification system*, the *inference system* and *defuzzification system*. The joint operation of these systems made up the reasoning strategy of the meta-scheduler. Specifically, as a result of the operation of the meta-scheduler, a RD selector factor y_o is obtained that shows the level of suitability for the selection of the RD under analysis to be selected in the current schedule. This operation is repeated for every available RD in the grid.

3 Gaussian Mixture Model Schedulers

Also, in this work, a new scheduling scheduling strategy is proposed to model the state of the diverse RDs integrated in the grid. Specifically, the selection of the most convenient site for a given schedule is to be done in regard of a suitability probability

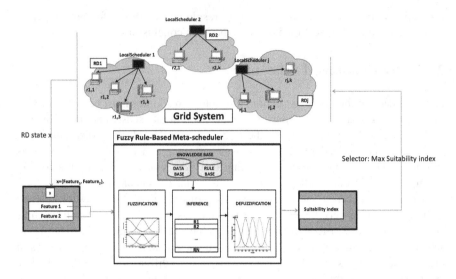

Fig. 1 General structure of a grid meta-scheduler based on FRBSs in a 2-dimensional space

which is calculated based on the state of the site and a weighted sum of probability density functions given by a Gaussian structure or GMM making up the scheduler decision core. A GMM is a statistical model that uses a weighted sum of probability density functions of multiple Gaussian distributions to depict the distribution of a vector in the probability space [2, 12]. A complete GMM is parametrized by mixture weights or priors α_i, mean vectors μ_i, and the covariance matrices Σ_i from all the M mixture components and can be denoted as:

$$\lambda = \{\alpha_i, \mu_i, \Sigma_i; i = 1, 2 \ldots, M\} \tag{2}$$

where $\mathbf{x}$ is a N dimensional feature vector to be modeled by GMM, $P_i(\mathbf{x})$, $i = 1 \ldots M$, are the probability density functions of $\mathbf{x}$ generated from the ith component of GMM which is denoted by λ_i and can be given as:

$$P_i(\mathbf{x}) = P_i(\mathbf{x} \mid \lambda_i) = \frac{1}{(2\alpha_i)^{N/2} \mid \Sigma_i \mid^{1/2}} exp\left\{-\frac{1}{2}(\mathbf{x} - \mu_i)'\Sigma_i^{-1}(\mathbf{x} - \mu_i)\right\} \tag{3}$$

A weighted sum of probability density functions of all the M mixture components is used to compute the probability that $\mathbf{x}$ belonged to model λ.

$$P(\mathbf{x} \mid \lambda) = \sum_{i=1}^{M} \alpha_i P_i(\mathbf{x}) \tag{4}$$

where α_i, $i = 1 \ldots M$ are the mixture weights and satisfy the constraint that $0 \leq \alpha \leq 1$, $\sum_{i=1}^{M} \alpha_i = 1$. The suggested scheduling strategy is based on the consideration of a GMM to model the scheduler decision module. Initially, the state of every RD participating in a given schedule is described through a finite set of features provided by the Grid Information System (GIS) as in the case of fuzzy rule-based meta-schedulers. To be precise, the consideration of a N dimensional feature vector **x**, representing the state of grid RDs, is to be modeled by GMM. I.e.,

$$\mathbf{x} = \{Feature_{1,j}, Feature_{1,j}, \ldots, Feature_{N,j}\}$$

for every RD j. Hence, at every scheduling step, the system retrieves the state of each RD j of the grid **x**, given by a set of features, and it computes the probability that **x** belongs to model λ, $P(\mathbf{x} \mid \lambda)$ for every site, considering a weighted sum of probability density functions as presented in Eq. 3. The RD obtaining the highest probability, is selected for the schedule and the process is repeated through the whole scheduling process. Fig. 2 represents an example of the scheduling GMM-based strategy with the RDs state described by two features. On the other hand, in order to obtain efficient GMM structures for the scheduling process, they are subject to an evolutionary learning. As regards the configuration of the scheduling system based on GMM, two different proposals are suggested.

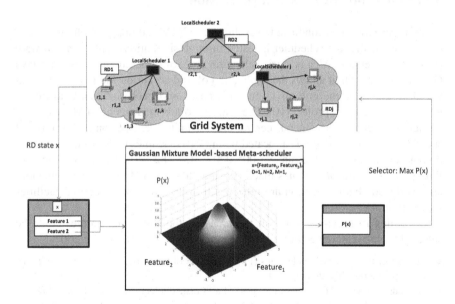

Fig. 2 Meta-Scheduling Gaussian structure in a 2-dimensional space for grid computing

3.1 Use of Simple Gaussian (GS) for the Whole Grid

The scheduling structure can be defined as a single N-dimensional Gaussian, with parameters λ, $\lambda = \{\mu, \Sigma\}$ and probability $P(\mathbf{x} \mid \lambda) = P(\mathbf{x})$. In this approach, the number of optimization parameters can be analyzed as follows. Considering Σ a diagonal matrix, the number of optimization parameters is $\#\lambda = \#\{\mu, \Sigma\} = \#\{N, N\} = 2 * N$

3.2 Use of Multiple Gaussian-Structure (GMM) for the Whole Grid

Another approach is to consider a scheduling structure defined as a D-GMM: D N-dimensional Gausssians make up the structure with λ_i, $\lambda_i = \{\alpha_i, \mu_i, \Sigma_i; i = 1, 2 \ldots, D\}$ and probability $P_i(\mathbf{x} \mid \lambda_i) = \sum_{i=1}^{D} \alpha_i P_i(\mathbf{x})$. In this second approach the number of optimization parameters can be determined as follows. Considering Σ_i as diagonal matrices and the same prior for all the component, the number of optimization parameters are $\#\lambda = \#\{\mu_i, \Sigma_i; i = 1, 2 \ldots, D\} = \#\{N * D, N * D\} = 2 * N * D$. As introduced above, the GMM parameters must be optimized in the learning of the GMM scheduler.

4 Simulations Results and Discussion

Several tests have been conducted to evaluate the GMM and fuzzy rule-based schedulers. Specifically, the scheduler is tested through simulations with Alea software [5] in its 2.1 version. Alea is a grid scheduling toolkit for simulation based on Java GridSim software that allows the application of grid scenarios and traces from real world. In our tests, the grid environment is based on Czech National Grid Infrastructure Metacentrum project [10]. The grid network is made up of 14 Metacentrum clusters with 806 heterogeneous central process units (i.e. Opteron and Xeon) and speed (i.e. 1,500-3,200 MHz) allocated within 210 machines under Linux with random access memory in the range 1,005,000 to 131,182,840 KB. Further, the configuration of the queues of sites machines, maintenance and reservation behaviour and workload characterization are retrieved from traces of Metacentrum facilities available at [4].

According to previous works in the area [8], the following features could be selected for the feature vector $\mathbf{x}$ describing the state of each RD in the grid:

- Number of free processing elements (FPE): Number of free processing elements within a participating resource domain, RD_j.
- Previous tardiness (PT): Sum of tardiness of all finished jobs in resource domain RD_j.
- Resource makespan (RM): Current makespan or finalization time of the last considered job in the RD_j.
- Resource tardiness (RT): Current tardiness of jobs assigned to the RD_j.
- Previous score in deadline evaluation (PS): Exceeding deadline time of already finished jobs in the RD_j.
- Resource score or number of delayed jobs (RS): Number of non delayed jobs so far in the RD_j.
- Resources in execution (RE): Number of resources executing jobs within the RD_j currently.

I.e., $\mathbf{x} = \{FPE_j, PT_j, RM_j, RT_j, PS_j, RS_j, RE_j\}$, for every RD j. Besides, the minimization of the latest job finalization time or *makespan* is pursued in this stage as scheduling goal as it is a general indicator of the grid productivity [11]. Two different approaches are considered for the GMM scheduling strategy. On the one hand, the proposal is evaluated using a scheduling structure is defined as a single 7-dimensional gausssian, i.e., $N = 7$, with parameters $\lambda = \{\mu, \Sigma\}$ with probability $P(\mathbf{x} \mid \lambda) = P(\mathbf{x})$. The number of optimization parameters in this approach, considering Σ a diagonal matrix, is: $\#\lambda = \#\{\mu, \Sigma\} = \#\{7,7\} = 14$. On the other hand, the scheduling structure is defined as a 3-GMM: three 7-dimensional Gaussians makes up the structure with $N = 7$ and $\lambda_i = \{\alpha_i, \mu_i, \Sigma_i; i = 1, 2 \ldots, 3\}$ and probability

$$P_i(\mathbf{x} \mid \lambda_i) = \sum_{i=1}^{3} \alpha_i P_i(\mathbf{x}) \qquad (5)$$

with $\alpha_i = 1/3; i = 1, 2 \ldots, 3.$, i.e., same probability for all the Gaussians. The number of optimization parameters, considering Σ_i as diagonal matrices, the number of optimization parameters is: $\#\lambda = \#\{\mu_i, \Sigma_i; i = 1, 2 \ldots, 3\} = \#\{21, 21\} = 42$. The Gaussians GS (1-GMM) and GMM models are evolved for 100 generations where 14 and 42 parameters, respectively, corresponding to mean and covariances matrices components are adjusted to find an optimum configuration, i.e., mean and covariance matrices are obtained though an evolutionary process to optimize the performance of the meta-schedulers in terms of makespan. Specifically, a genetic evolutionary process is considered. Furthermore, 30 runs are conducted for each GMM structure and results are compared to those of a fuzzy meta-schedulers based on Pittsburgh and Michigan approaches using the same description for the grid state, number of generations and Gaussian fuzzy sets for both the input and output.

Table 1 Simulation results for GS scheduler. Simulation results for 30 simulations in grid Metacentrum. Training Fitness makespan (s).

Parameter/Strategy	Michigan Fuzzy scheduler	Pittsburgh Fuzzy scheduler
Max	1,847,375.2	1,684,235.5
Min	1,654,505.4	1,625,058.2
Average	1,773,266.8	1,667,586.2
Standard Deviation	54,210.0	19,757.8
Confidence Interval (95%)	1,755,929.68, 1,790,604.09	1,660,515.96, 1,674,656.43

Table 1 presents the learning results for the meta-scheduler based on a 7-dimensional GS and 7-dimensional GMM. Specifically, it shows the mean result achieved by the strategies (Average) with the associated standard deviation (Standard Deviation) and 95% confidence interval (Confidence Interval-95%) and the best result (Min) and worst result (Max). As illustrated, the fuzzy meta-scheduler based on Pittsburgh evolution outperforms the GS and 3-GMM schedulers in terms of final *makespan* by 2.91% and 2.43% on average, respectively, what it is translated in a shorter time to perform the whole schedule. In addition, the best result

Table 2 Simulation results for GS scheduler. Simulation results for 30 simulations in grid Metacentrum. Training Fitness makespan (s).

Parameter/Strategy	GS-scheduler	GMM-scheduler
Max	1,781,326.0	1,781,326.0
Min	1,644,060.1	1,639,568.4
Average	1,717,557.7	1,709,183.5
Standard Deviation	51,057.0	50,924.8
Confidence Interval (95%)	1,698,492.80, 1,736,622.79	1,690,167.89, 1,728,199.20

(i.e., minimum *makespan*, Min) and worst result (i.e., maximum *makespan*, Max) achieved by the GS are 1.16% and 5.45% greater than the results obtained with the Pittsburgh fuzzy scheduler, respectively. Also, the best result (i.e., Min) and worst result (i.e., Max) achieved by the 3-GMM are 0.89% and 5.45% greater than the obtained with the Pittsburgh fuzzy scheduler. However, it can be observed that the GS and 3-GMM schedulers outperform the fuzzy meta-scheduler based on Michigan evolution in terms of final *makespan* by 3.14% and 3.61% on average, respectively. Furthermore, the best result (i.e., Min) and worst result (i.e., Max) achieved by the GS are 0.63% and 3.58% lower than the ones obtained with the Michigan fuzzy scheduler. On the other hand, the best result (i.e., Min) and worst result (i.e., Max) achieved by the 3-GMM are 0.90% and 3.58% more reduced than the results obtained with the Michigan fuzzy scheduler. Further, the standard deviation is also significantly higher with the following impact in the confidence interval. Finally, it is shown that the 3-GMM scheduler outperforms the GS model by 0.49% on average and that the best result (i.e., Min) is improved by 0.28%.

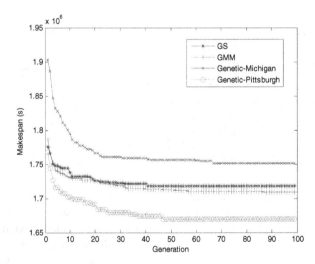

Fig. 3 Convergence of the GS scheduler in the training process

Also, the convergence of the GS, 3-GMM, Pittsburgh and Michigan schedulers in the learning process is presented in Fig. 3. As illustrated, the genetic Pittsburgh scheduler achieves the higher accuracy meanwhile the genetic Michigan scheduler presents the lower convergence speed and accuracy what shows the dependence of the fuzzy expert meta-schedulers with the learning strategy to achieve an efficient schedule. On the other hand, the GMM-based schedulers provide an intermediate solution. The GS and 3-GMM approaches follow a similar convergence behaviour. However, as discussed above, the 3-GMM achieves a higher accuracy. It is to be noted that the 3-GMM approach presents a higher number of optimization parameters than the GS model what allows a major flexibility for the systems but also a greater complexity in the simultaneous fixing. From these results it can be derived that the GMM schedulers could be competitive approaches in the design of state-based scheduling strategies for grid computing founded on the definition of probability density functions. However, fuzzy rule-based schedulers offer a greater scalability and accuracy and speed in their learning process in most configurations.

5   Conclusions and Future Work

The developments in the technical field related to Grid/Cloud Computing are increasingly becoming interesting and relevant for companies nowadays. Taking advantage of inactive time of computers to make a virtual supercomputer will have important effects on the organizations: reduction of infrastructure, less investments in hardware and software, the possibility to send surplus processing cycles, less risk with hardware problems, highest productivity of existing computers, etc. Thus, Grid/Cloud technologies have a growing impact over the governance structures and the self production support of companies. In this sense, improving the efficiency and scalability in the workload/resource scheduling is critical. In this paper, a comparison of FRBS schedulers for Grid/Cloud systems based on Michigan and Pittsburgh approaches and a new scheduling strategy based on GMM for grid computing has been presented. On the one hand, FRBSs models are flexible and they are more and more being adopted as scheduling systems for scheduling jobs in large-scale distributed networks based on the fuzzy characterization of the grid sites state to obtain an efficient schedule within an uncertain environment. On the other hand, the GMM scheduling strategy is founded on the association of density probability functions to every site in the grid which can provide a probability of suitability of selection at every scheduling stage on the basis of the characterization of sites state. To be precise, two different approaches based on GMM are proposed and the learning of the GMM scheduling structures are addressed through evolutionary processes. Simulation results based on existing grid infrastructure have shown that novel proposed approach, a Gaussian structure for all the sites, can be a compete alternative to some fuzzy state-based meta-scheduler systems such as FRBS Michigan schedulers although it is not able to outperform fuzzy Pittsburgh-schedulers in terms of accuracy and convergence behaviour. Also, fuzzy schedulers are more scalable. In future works, the definition of independent GMM structures for every involved site in the grid will be analyzed to increase the flexibility. In addition, a principal components analysis (PCA) will be introduced to achieve a reduction in system search space dimension and thus to accelerate the learning of the Gaussian schedulers.

Acknowledgements. This work has been financially supported by the Spanish Government (Research Project P07-TIC-02713).

References

1. Cordón, O., Herrera, F., Hoffmann, F., Magdalena, L.: Genetic fuzzy systems: Evolutionary tuning and learning of fuzzy knowledge bases. World Scientific Pub. Co. Inc. (2001)
2. Duda, R.O., Hart, P.E., Stork, D.G.: Pattern Classification, 2nd edn. Wiley Interscience (2000)
3. Foster, I., Kesselman, C.: The Grid 2: Blueprint for a New Computing Infrastructure. Morgan Kaufmann Publishers Inc., San Francisco (2003)
4. C.N.G. Infrastructure: Metacentrum data sets meta (2009), http://www.fi.muni.cz/~xklusac/index.php?page=
5. Klusáček, D., Matyska, L., Rudová, H.: Alea – Grid Scheduling Simulation Environment. In: Wyrzykowski, R., Dongarra, J., Karczewski, K., Wasniewski, J. (eds.) PPAM 2007. LNCS, vol. 4967, pp. 1029–1038. Springer, Heidelberg (2008)
6. Klusacek, D., Rudova, H.: Improving QoS in computational Grids through schedule-based approach. In: Scheduling and Planning Applications Workshop at the Eighteenth International Conference on Automated Planning and Scheduling (ICAPS 2008), Sydney, Australia (2008)
7. Mohammed, A.B., Altmann, J., Hwang, J.: Cloud computing value chains: Understanding businesses and value creation in the cloud. In: Neumann, D., Baker, M., Altmann, J., Rana, O. (eds.) Economic Models and Algorithms for Distributed Systems, Autonomic Systems, pp. 187–208. Birkhäuser Basel (2010)
8. Prado, R.P., García-Galán, S., Expósito, J.E.M., Yuste, A.J., Bruque, S.: Learning of Fuzzy Rule-Based Meta-schedulers for Grid Computing with Differential Evolution. In: Hüllermeier, E., Kruse, R., Hoffmann, F. (eds.) IPMU 2010. CCIS, vol. 80, pp. 751–760. Springer, Heidelberg (2010)
9. Prado, R., García-Galán, S., Yuste, A., Muñoz Expósito, J.: Genetic fuzzy rule-based scheduling system for grid computing in virtual organizations. Soft Computing - A Fusion of Foundations, Methodologies and Applications, 1–17 (2010)
10. Šustr, Z., Sitera, J., Mulač, M., Ruda, M., Antoš, D., Hejtmánek, L., Holub, P., Salvet, Z., Matyska, L.: MetaCentrum, the Czech Virtualized NGI (2009)
11. Xhafa, F., Abraham, A.: Computational models and heuristic methods for grid scheduling problems. Future Generation Computer Systems 26(4), 608–621 (2010)
12. Yu, G., Sun, J., Li, C.: Machine performance assessment using gaussian mixture model (gmm). In: 2nd International Symposium on Systems and Control in Aerospace and Astronautics, ISSCAA 2008, pp. 1–6 (2008), doi:10.1109/ISSCAA.2008.4776183

Reduced Large Datasets by Fuzzy C-Mean Clustering Using Minimal Enclosing Ball

Lachachi Nour-Eddine and Adla Abdelkader

Computer Science Department - Oran University - Algeria
{Lach_Nour,AekAdla}@Yahoo.fr

Abstract. Minimal Enclosing Ball (MEB) is a spherically shaped boundary around a normal dataset, it is used to separate this set from abnormal data. MEB has a limitation for dealing with a large dataset in which computational load drastically increases as training data size becomes large. To handle this problem in huge dataset used in different domains, we propose two approaches using Fuzzy C-mean clustering method. These approaches find the concentric balls with minimum volume of data description to reduce the chance of accepting abnormal data that contain most of the training samples. Our method uses a divide-and-conquer strategy; trains each decomposed sub-problems to get support vectors and retrains with the support vectors to find a global data description of a whole target class. Our study is experimented on speech information to eliminate all noise data and reducing time training. For this, the training data, learned by Support Vector Machines (SVMs), is partitioned among several data sources. Computation of such SVMs can be achieved by finding a core-set for the image of the data. Numerical experiments on some real-world datasets verify the usefulness of our approaches for data mining.

Keywords: Quadratic Programming (QP), Support Vector Machines (SVMs), Minimal Enclosing Ball (MEB), Core- set, kernel methods, Fuzzy C-Mean.

1 Introduction

Support vector machines (SVMs) [1], are a powerful technique for classification and regression. Training an SVM is usually posed as a quadratic programming (QP) problem to find a separation hyper-plane which implicates a matrix of density $n \times n$, where n is the number of points in the dataset. This needs more computational time and memory for large datasets, so the training complexity of SVM is highly dependent on the size of a dataset.

Here, we explore a technique that the training data, learned by Support Vector Machines (SVMs), is partitioned among several data sources. Computation of such SVMs can be achieved to find a core-set for the image of the data in a feature space. However, in the standard technique that uses two classes SVM, a certain kernel functions and kernel methods are used to handle large datasets.

J. Casillas et al. (Eds.): Management Intelligent Systems, AISC 171, pp. 305–314.

This paper develop an alternative method based on a equivalence between SVMs and Minimal Enclosing Ball (MEB) problems from which important improvements on training efficiency has been reported [2] [3] for large-scale datasets. We focus on multi-class problems where we explore two methods to extend binary SVMs to the multi-category setting which preserve the equivalence between the model and MEBs.

Although the SVM has proven to be a very suitable tool for classification problem, but later studies also found over fitting problem in SVM [4]. To overcome the limitation, also our study can be adapted to business risk management that is very complex and challenging task from the viewpoint of system engineering. It contains many processes, such as risk identification and prediction, modeling, control and management. In [5], a triple-phase SVM-based metamodel system for business risk identification has been developed. It implies that the proposed technique can provide a promising solution to business insolvency risk identification problem where *the partionner* in the generic SVM-based metamodel process in [5] can be replaced by our study.

Algorithms to compute SVMs based on the MEB equivalence are based on the greedy computation of a core-set, a typically small subset of the data which provides the same MEB as the full dataset. Then, we formulate new multiclass SVM problem using core-sets for reduce large datasets which can be considered optimally matched to the input demands of different background architectures of speaker systems. The core idea of these two approaches cited above is to adopt multiclass SVMs formulation and Minimal Enclosing Ball to reduce dataset without influence data noise.

2 Minimal Enclosing Balls (MEB) and Core-Set

Given a training dataset $S = \{(x_i, y_i)\}_{i=1}^{I}$ where $x_i \in \mathbb{R}^d$ and $y_i \in \{+1, -1\}$. To simplify the notation let us denote the pair (x_i, y_i) as $\tilde{z}_i$. Now the training dataset can be denoted as $S = \{\tilde{z}_i\}_{i=1}^{I}$. Let $\tilde{Z}$ be a space equipped with a dot product $\tilde{z}_i^T \tilde{z}_j$ corresponding to norm $\|\tilde{z}\|^2 = \tilde{z}^T \tilde{z}$. We define the ball $\mathcal{B}(c, R)$ of center $c \in \tilde{Z}$ and radius R in $\mathbb{R}$ as the subset of points $\tilde{z} \in \tilde{Z}$ for which $\|\tilde{z} - c\|^2 \leq R^2$. The minimal-enclosing ball [3] of a set of points $S = \{\tilde{z}_i : i \in I\}$ in $\tilde{Z}$ is in turn the ball $\mathcal{B}^*(S, c^*, R^*)$ of smallest radius that contains S, that is, the solution to the following optimization problem.

$$\min_{R,c} R^2$$
$$st: \|\tilde{z} - c\|^2 \leq R^2 \quad \forall \tilde{z} \in S \tag{1}$$

After introducing Lagrange multipliers we obtain from the optimality conditions the following dual problem

$$\min_{\alpha} \sum_{i,j \in I} \alpha_i \alpha_j \tilde{z}_i^T \tilde{z}_j - \sum_{i \in I} \alpha_i \tilde{z}_i^T \tilde{z}_i$$
$$st: \alpha_i \geq 0, \quad \sum_i \alpha_i = 1 \quad \forall i \in I \tag{2}$$

if we consider that $\sum_{i \in I} \alpha_i \tilde{z}_i^T \tilde{z}_i = \kappa$ a constant, we can drop it from the dual objective in (2), we obtain a simpler QP problem

$$\min_\alpha \sum_{i,j \in I} \alpha_i \alpha_j \tilde{z}_i^T \tilde{z}_j$$
$$st: \ \alpha_i \geq 0, \ \sum_i \alpha_i = 1 \quad \forall i \in I \tag{3}$$

This is a QP problem. In [3], it shows that the primal variables c and R can be recovered from the optimal α as $c = \sum_{i=1}^I \alpha_i \tilde{z}_i$, $R = \sum_{i,j \in I} \alpha_i \alpha_j \tilde{z}_i^T \tilde{z}_j$

2.1 Core-set for MEB

The algorithm of Bãdoiu and Clarkson [6] approximate the solution to this problem exploits the ideas of core-set and ϵ-approximation to the minimal enclosing ball of a set of points. If we consider S a set of I points in $\mathbb{R}^d$, R is the radius of $MEB(S)$, There exist a subset $C_S \subset S$ st:

- the size of C_S is less $\frac{2}{\epsilon}$
- the center $c(C_S)$ of $MEB(C_S)$ satisfies $d(z, c(C_S)) \leq (1 + \epsilon)R, \forall z \in S$

Such a subset C_S is a core-set of S for MEB. Then a core-set is a subset C_S of S such that:

- the size of C_S does not depend on I or d
- the solution for C_S is an approximation of the solution for S.

$\epsilon - coreset$: The solution for C_S is within ϵ of the solution for S.
Here we present the most usual version of the algorithm [6].

Algorithm 1. Bãdoiu-Clarson Algorithm

1: Initialize the core-set $C_{S,\epsilon.}$

2: Compute the minimal-enclosing-ball $\mathcal{B}(C_S, c, R)$ of the core-set $C_{S,\epsilon.}$

3: while A point $\tilde{z} \in S$ out of the ball $\mathcal{B}(C, c, (1 + \epsilon)R)$ exist **do**

4: Include $\tilde{z}$ in $C_{S,\epsilon.}$

5: Compute the minimal-enclosing-ball $\mathcal{B}(C_S, c, R)$ of the core-set $C_{S,\epsilon.}$

6: end while

3 Support Vector Machines (SVMs)

Support Vector Machines (SVMs) [1] address the problem of binary classification by building a hyperplane to represent the boundary between the two classes. This hyperplane $f(z) = (w^T z + b)$ is built in a feature space $Z = \phi(X)$ implicitly induced from X by means of a kernel function k which computes the dot products $z_i^T z_j = \phi(x_i)^T \phi(x_j)$ in Z directly on X. The so called $L2$-SVM chooses the separating hyperplane $f(z)$ by solving the following quadratic program:

$$\min_{w,b,\rho,\xi} \frac{1}{2} (\|w\|^2 + b^2 + C \sum_i \xi_i^2) - \rho$$
$$st : y_i f(z_i) \geq \rho - \xi_i \quad \forall i \in I \tag{4}$$

After introducing Lagrange multipliers, it can be shown that the latter problem is equivalent to solve.

$$\min_\alpha \sum_{i,j \in I} \alpha_i \alpha_j K_{ij}$$
$$st: 0 \leq \alpha_i, \ \sum_i \alpha_i = 1 \tag{5}$$

where $K_{ij} = y_i y_j k(x_i, x_j) + y_i y_j + \frac{\delta_{ij}}{c}$, δ_{ij} is the Kronecker delta function and $k(x_i, x_j)$ implements the dot-product $z_i^T z_j$.

The L2 implementation support a convenient reduction to a minimal enclosing ball (MEB) problem when the kernel used in the SVM is normalized by $k(x, x) = \kappa \ \forall x \in X$, where κ is a constant [3].

4 Multi-class Extensions

In a multi-class problem, examples $\{x_i\}$ belong to a set of L categories $c = \{c_k; k \in L \ / \ L \geq 2\}$. There are two types of extensions to build multiclass SVMs. One Against One strategy (OAO) that reduce the multi-class problem to a set of binary classification problems and One Against All strategy (OAA) that re-casts the binary objective functions to a multi-category problem.

The proposal multi-class extension of L2-SVMs is in [7] which show the reduction to a minimal enclosing ball problem characteristic of the binary L2-SVM that is the key requirement of our algorithms. The formulation associates each class $c_k, k \in L$ of the problem looks for a projector W operating on the feature space $Z = \phi(X)$ which should allow to recover a correct code for a given input z.

Let the training dataset be $S = \{(x_i, y_i)\}_{i=1}^l$ where $x_i \in R^d$ and $y_i \in R^L$ for some integers $d; L \geq 2$. i.e. we have I training points whose labels are vector valued. For a given training task having L classes, these label vectors are chosen out of the definite set of vectors $\{y_1, y_2, \ldots, y_l\}$. Now, for inputs $z = \phi(x)$ we can define the primal for the learning problem as

$$\min_\alpha \frac{1}{2} (\|W\|^2 + \|b\|^2 + C \sum_i \xi_i^2) - \rho$$
$$st: y_i^T (W^T z + b) \geq \rho - \xi_i^2 \geq 0 \quad \forall i \in I \tag{6}$$

Several selections are possible for the norm $\|W\|^2$. A common choice is the so called *Frobenius norm* $\|W\|^2 = trace(W^T W)$. Hence, the dual of the optimization problem obtained after introducing Lagrange multipliers is

$$\min_\alpha \sum_{i,j \in I} \alpha_i \alpha_j K_{ij}$$
$$st: 0 \leq \alpha_i, \ \sum_i \alpha_i = 1 \tag{7}$$

where $K_{ij} = y_i^T y_j k(x_i, x_j) + y_i^T y_j + \frac{\delta_{ij}}{c}$, δ_{ij} is the Kronecker delta function and $k(x_i, x_j)$ implements the feature dot-products $z_i^T z_j$.

5 MEB and Multi-class L2-SVMs Equivalence

Now, we suppose that the computing of the minimal enclosing ball is in feature space $\tilde{Z} = \phi(X)$ which has been induced from X by a mapping function $\phi: X \to \tilde{Z}$ where we can compute dot products in $\tilde{Z}$ directly from X by using a kernel function $\tilde{k}(x_i, x_j) = \phi(x_i)^T \phi(x_j) = \tilde{z}_i^T \tilde{z}_j$. Also, we suppose that the kernel is normalized, i.e., $\forall \ x \in X$, $\tilde{k}(x, x) = \kappa$ with $\kappa \in \mathbb{R}$ a constant.

As it has seen above, the optimization problem (1) is equivalent to solve the following quadratic program

$$\min_\alpha \Sigma_{i,j \in I} \alpha_i \alpha_j \tilde{K}_{ij}$$
$$st: \ \alpha_i \geq 0, \quad \Sigma_{i \in I} \alpha_i = 1 \quad \forall \ i \in I \tag{8}$$

where $\tilde{K}_{ij} = k(x_i, x_j)$. This problem coincides with the binary *L2-SVM* problem (5) and its multi-class implementation (7) is in the binary case if we set $\tilde{k}(x_i, x_j) = y_i y_j k(x_i, x_j) + y_i y_j + \frac{\delta_{ij}}{C}$, and in multi-category case if $\tilde{k}(x_i, x_j) = y_i^T y_j k(x_i, x_j) + y_i^T y_j + \frac{\delta_{ij}}{C}$. The key requirement of the latter equivalence is the normalization constraint on $\tilde{k}(x, x) = \kappa$.

6 Fuzzy C-Mean Clustering Algorithm

Fuzzy C-Mean (FCM) is an unsupervised clustering algorithm that has been applied to wide range of problems involving feature analysis, clustering and classifier design. This algorithm is examined to analyse different systems based on the distance between the various input data points. And it is based on clustering which allows one piece of data to belong to two or more clusters. It is based on minimization of the following objective function:

$$J_m = \Sigma_{i=1}^N \Sigma_{j=1}^C u_{ij}^m \|x_i - c_j\|^2 \tag{9}$$
$$1 \leq m \leq \infty$$

where m is any real number greater than 1, u_{ij} is the degree of membership of x_i in the cluster j, x_i is the ith of d-dimensional measured data, c_j is the d-dimension center of the cluster, and $\| * \|$ is any norm expressing the similarity between any measured data and the center. Fuzzy partitioning is carried out through an iterative optimization of the objective function shown above, with the update of membership u_{ij} and the cluster centers c_j by:

$$u_{ij} = \frac{1}{\Sigma_{k-1}^C \left(\frac{\|x_i - c_j\|}{\|x_i - c_k\|}\right)^{\frac{2}{m-1}}} \tag{10}$$

$$C_j = \frac{\Sigma_{i=1}^N u_{ij}^m \cdot x_i}{\Sigma_{i=1}^N u_{ij}^m} \tag{11}$$

This iteration will stop when $\max_{ij}\left\{\left|u_{ij}^{(k+1)} - u_{ij}^{(k)}\right|\right\} < \xi$, where ξ is a termination criterion between 0 and 1, whereas k is the iteration steps. This procedure converges to a local minimum or a saddle point of J_m. The algorithm is composed of the following steps:

Algorithm 2. Computation of the Fuzzy-CMean clustering

1: Initialize $U = [u_{ij}]$ matrix, $U^{(0)}$

2: At k-step: calculate the centers vectors $C^{(k)} = [c_j]$ with $U^{(k)}$.

3: Update $U^{(k)}$, $U^{(k+1)}$.

4: If $\left\|U^{(k+1)} - U^{(k)}\right\| < \xi$ then STOP; otherwise return to step 2.

In this algorithm, data are bound to each cluster by means of a Membership function, which represents the fuzzy behaviour of the algorithm [8]. To do that, the algorithm have to build an appropriate matrix named U whose factors are numbers between 0 and 1, and represent the degree of membership between data and centers of clusters.

7 Reduced Data Approaches

7.1 Formulation

The key idea of our method is to cast an SVM as an MEB problem in a feature space $\tilde{Z} = \phi(X)$ where the training examples are embedded via a mapping ϕ. Hence, we first formulate an algorithm to compute the MEB of the images $\tilde{S}$ of S in $\tilde{Z}$ when S is decomposed in a collection of sub-sets S_j. Then we will instantiate the solution for classifiers supporting the reduction to MEB problems.

The algorithm is based on the idea of computing core-sets $\mathcal{C}_j$ for each set $\tilde{S}_j = \phi(S_j)$ and taking its union $\mathcal{C} = \cup_j \mathcal{C}_j$ as an approximation to a core-set for $\tilde{S} = \cup_j S_j$. The generic procedure is depicted as algorithm (3). In a first step the algorithm extracts a core-set for each sub-set S_i. In the second step the MEB of the union of the core-sets is computed.

Algorithm 3. Computation of the MEB of $\tilde{S} = \phi(S)$

Require: A partition of the set S based Fuzzy-CMean clustering (algorithm 2) in a collection of subsets S_j

1: **for** Each subset S_j, $j = 1,...,p$ **do**

2:     Compute a ϵ-core-set C_j for one of the two instantiation

3: **end for**

4: Join the core-sets $C = C_1 \cup ... \cup C_p$

5: Compute the minimal enclosing ball of C. This is the Minimal enclosing ball of $\tilde{S}$ that define the reduced datasets.

For the computation of the core-sets we use algorithm 1.

7.2 Instantiation for the OAO Multi-class Approach

From the previous section we have that training a binary L2-SVM on a dataset S is equivalent to build a minimal enclosing-ball of S if $\phi(x)^T\phi(x)$ is implemented using the kernel $\tilde{k}(x_i, x_j) = y_i y_j k(x_i, x_j) + y_i y_j + \frac{\delta_{ij}}{c}$. The OAO procedure to obtain a multi-category SVM works by combining one binary SVM for each pair of classes. An instantiation of algorithm (3) would hence consist in computing core-sets for the subset of examples belonging to each pair of classes, and then joining them and finally recovering the binary model for this pair. However, since each class participates in L models, core-sets for each pair of classes can be highly redundant overloading the data unnecessarily.

Algorithm 4. Computation of the MEB using OAO L2-SVMs

1: for Each subset S_n , $n = 1, ..., p$ **do**

2:   **for** Each Class $k = 1, ..., L-1$ **do**

3:   **for** Each Class $m = k+1, ..., L$ **do**

4:   Let S_n^{mk} the subset of S_i corresponding to class k and m.

5:   Label S_n^{mk} using the standard binary codes +1 And −1 for class k and m respectively

6:   Compute a core-set C_n^{mk} of S_n^{mk} Using the kernel $\tilde{k}(x_i, x_j) = y_i y_j k(x_i, x_j) + y_i y_j + \frac{\delta_{ij}}{c}$

7:   **end for**

8:   **end for**

9:   Take the union of the core-set inferred for each pair of classes $C_n = C_n^{mk} \cup ... \cup C_n^{mk}$

10: end for

11: Join core-set $C_S = C_1 \cup ... \cup C_p$.

12: Compute the minimal enclosing ball of C_S using the same kernel $\tilde{k}$

7.3 Instantiation for the OAA Multi-class Approach

The OAA implementation is defined by a single optimization which coincides with a MEB problem just by using the kernel $\tilde{k}(x_i, x_j) = y_i^T y_j k(x_i, x_j) + y_i^T y_j + \frac{\delta_{ij}}{c}$. The use of algorithm (3) is hence straight forward and consists in computing any dot product $\tilde{\phi}(x_i)^T \tilde{\phi}(x_i) = \tilde{k}(x_i, x_i)$ using this kernel. The instantiation is depicted as algorithm (5).

Algorithm 5. Computation of the MEB using OAA Multiclass L2-SVM

1: for Each subset S_n , $n = 1, \dots, p$ **do**

2: Label each example $x_i \in S_n$ with the code y_{ik} assigned to the class of x_i and let y_i such label

3: Compute a core-set C_n of S_n using the kernel

$$\tilde{k}(x_i, x_j) = y_i^T y_j k(x_i, x_j) + y_i^T y_j + \frac{\delta_{ij}}{C}$$

4: end for

5: Join the core-sets $C_S = C_1 \cup \dots \cup C_p$.

6: Compute the minimal enclosing ball of C_S using the same kernel $\tilde{k}$

8 Experiments

This section presents the performance of a speaker verification system based on the *Gaussian Mixture* described in [9]. We compare the performance of speaker

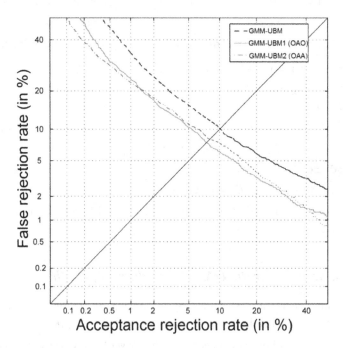

Fig. 1 DET curves for the speaker verification system using three UBMs

verification system with three UBMs, the first one was created directly from the corpus in [10] (formerly known as Speaker Verification), and the two last later is the reduced of the original UBM from the application of our two algorithms developed in section 7. The kernel used for the two algorithms is the Gaussian Radial Basis Function with à fixed value of σ with 0.50. We have trained a 512-mixture gender-independent from each UBM with diagonal covariance matrices. Speaker GMMs are trained by adapting only the mean vectors from the UBM using a relevance factor r of 16.

Figure 1 shows the detection error trade-off (DET) curves for the three systems. The system based reduced GMM-UBM2 from One Against All multiclass L2-SVM outperforms the GMM-UBM with an equal-error-rate (EER) of 8.17 %, compared to 10.13 % of the GMM-UBM2. The system based reduced GMM-UBM1 from One-Against-One L2-SVMs exhibits the best performance with an EER of 7.60 %.

9 Conclusion

In this paper we have treated the Core-set with SVMs for reducing huge dataset. Thus, we have explained the characteristics of SVMs that make them valuable from the huge dataset point of view: first, SVMs are discriminative models, thus more appropriate for classification problems; second they have the advantage of being capable to deal with samples of a very higher dimensionality; and third, they exhibit an excellent generalization ability that makes them especially suitable to deal with noisy data.

Then, we proposed two algorithms that compute an approximation to the minimal enclosing ball of a given finite set of vectors. Both algorithms are especially well-suited for large-scale instances of the minimal enclosing ball problem and can compute a small core-set whose size depends only on the approximation parameter.

We have explored two methods based on the computation of core-sets to train multi-category SVM models when the set of examples is fragmented. The main contribution has been to demonstrate through our experiments, that the methods proposed can reproduce the data with high accuracy where the noisy sample in huge dataset are eliminated, without complex and costly computation. SVMs based on core-sets have shown however important advantages in large-scale applications, which can hence be extended to business risk identification process as has been evocated in introduction for a generic SVM-based metamodel process.

References

1. Schölkopf, B., Smola, A.J.: Learning with Kernels: Support Vector Machines, Regularization, Optimization, and Beyond. MIT Press, Cambridge (2001)
2. Kocsor, A., Kwork, J., Tsang, I.: Simpler core vector machines with enclosing balls. In: ICML 2007, pp. 911–918. ACM (2007)

3. Cheung, P.M., Kwok, J., Tsang, I.: Core vector machines: Fast SVM training on very large datasets. Journal of Machine Learning Research (6), 363–392 (2005)

4. Tay, F.E.H., Cao, L.J.: Application of support vector machines in financial time series-forecasting. Omega, 309–317 (2001)

5. Lai, K.K., Yu, L., Huang, W., Wang, S.: A Novel Support Vector Machine Metamodel for Business Risk Identification. In: Yang, Q., Webb, G. (eds.) PRICAI 2006. LNCS (LNAI), vol. 4099, pp. 980–984. Springer, Heidelberg (2006)

6. Bădoiu, M., Clarkson, K.L.: Optimal core-sets for balls. Computing Geometry Theory Application 1(40), 14–22 (2008)

7. Asharaf, S., Murty, M., Shevade, S.K.: Multiclass core vector machine. In: ICML 2007, pp. 41–48. ACM (2007)

8. Al-Zoubi, M.B., Hudaib, A., Al-Shboul, B.: A fastfuzzyclusteringalgorithm. In: Proceedings of the 6th WSEAS Int. Conf. on Artificial Intelligence, Knowledge Engineering and Data Bases, Corfu Island, Greece, pp. 28–32 (2007)

9. Alkanhal, M., Alghamdi, M., Muzaffar, Z.: SpeakerVerification-based on SaudiAcceted Arabic Database. In: ISSPA 2007, 9th International Symposium on Signal Processing and its Applications, Sharjah, United Arab Emirate, pp. 1–4 (February 2007)

10. Speaker corpus in,
    http://www.ll.mit.edu/mission/communication/ist/corpora/
    SpeechCorpora.html

Impact of Initial Tuning for Algorithm That Solve Query Routing

Claudia Gómez Santillán, Laura Cruz Reyes, Gilberto Rivera Zarate,
Juan González Barbosa, and Marcela Quiroz Castellanos

Instituto Tecnológico de Ciudad Madero, México
{cggs71,lauracruzreyes,grivera984,
jjgonzalezbarbosa}@hotmail.com

Abstract. The algorithms are the most common form of problem solving in many science fields. Algorithms include parameters that need to be tuned with the objective of optimizing its processes. This work uses Hoeffding race techniques, with the objective to obtain the best initial combination of variables to use it as an input configuration. Hoeffding race quickly discard less promising candidates as soon as there are evidences enough to remove them from the competition. These evidences are based on the use of any statistical test that, at a given confidence level, would set a range of expected performance for configuration. All the experiment was applied in AdaNAS (Adaptive Neighboring-Ant Search), an algorithm that was developed to route queries through the Internet. Our results show that there is a significant gain in efficiency of the AdaNAS algorithm by using the simple, but powerful, technique of initial setting of parameters presented in this paper. In our experiments, the average efficiency was improved 50% by using a good initial configuration.

Keywords: Algorithms Optimization, Parameter Setting, Race Techniques, Ant Algorithms, Query Routing.

1 Introduction

All businesses need to share resources and collaboratively work, for these reasons their information exchange systems need adjust to the changing needs for resource management, this has impacted on turning them into complex systems in their structure, organization, distribution and access. Hence there is a necessity for creating algorithms that help to users find the information that they request within a reasonable processing time and with a higher quality of the obtained information.

New communication models have emerged in the Internet that manage information in a distributed manner and offer significant advantages for the business. Examples of such systems are peer-to-peer (P2P) networks that consist of a set of computers interconnected to offer its resources to other peers within the network.

J. Casillas et al. (Eds.): Management Intelligent Systems, AISC 171, pp. 315–323.
springerlink.com

The P2P systems together with the underlying communication network (Internet) form a complex system that requires autonomous operation through mechanisms of intelligent navigation [8]. To achieve this it is necessary to set appropriate values to its parameters.

Find the initial setting of parameters for an optimization algorithm is a nontrivial task, and it can consume, according to Adenso [1], to 90% of development time for solving a problem. Whenever we want to do a setting of parameters, we have two questions to answer: 1) How many runs of the algorithm will be needed? 2) How many instances should I run? It is commonly carried out 30 runs of the algorithm on the entire set of test instances. However, could be fewer runs or fewer instances?

In such scenarios is desirable to use a method to select a configuration of parameters without having to do all runs over all test instances but to ensure, with a certain level of confidence, that the chosen configuration is reliably the best.

In this paper we present a method based on statistical tests that complies with such features, called Hoeffding Race [2][3]. Hoeffding Race is a long studied technique, that even has been used in previous works [4][5] for parameter setting, however in recent years has lacked attention and its benefits have been denied in the area of automatic optimization. In this paper we present the use of Hoeffding Race applying it to an algorithm called AdaNAS.

AdaNAS [5][6] is based on the ACS (Ant Colony System) metaheuristic and NAS[7] algorithm, hybridized with local strategies such as: learning, characterization, and exploration. All the strategies were developed to resolve for the semantic query routing problem (SQRP) in peer to peer networks (P2P).

SQRP consists to discover routes as short as possible between a node that issues a query by the user, and a node (or several) that has the resources to satisfy that request. AdaNAS, as many other algorithms, has a lot of parameters that need to be properly tuned in order to yield a fully functioning algorithm. [1][4][8].

We describe the impact of initial tuning for AdaNAS algorithm by finding a good initial configuration through experimental evaluations statistically guided that reduce the number of experiments. In addition, the initial configuration helps to obtain a better efficiency from the beginning to the end of the execution of the algorithm and thus achieve the goal of SQRP, which is to maximize the amount of resources found in the P2P network and minimize the distance between the query node and the node with matching resources.

2 Parameter Setting

Determining the best combination of variables to use as input for a common practical problem is hard, since the input values have to be chosen in a way that the cost function is optimized. Parameter setting can be classified into: parameter tuning and parameter control. Michalewicz [9] and Angeline [10] define them as: Parameter Control is the setting done during the algorithm execution. This type of parameter setting supervises the local environmental changes and the current state of the algorithm to adapt locally the configuration to the local conditions; and Parameter tuning is the setting done before the algorithm runs, and provides a global initial configuration. It evaluates the general performance of the algorithm but, it does not assure that the values for the parameters will be the best in each instant of the run of the algorithm.

The parameter tuning can be applied through three techniques: a) by hand, doing a sequence of experiments with different values of the parameters, and choosing the configuration with the best performance, b) Meta-Evolution, using an auxiliary metaheuristic algorithm to improve the performance of the main meta-heuristic algorithm and c) Design of Experiment (DOE), this technique provides a great variety of statistics tests to make useful decisions [11][12]. DOE is a statistics tool set, useful on making plans, running and interpreting an experiment, while searching for valid and impartial deductions. An experiment can be defined, as a planned test which introduces checked changes in the process or system variables, with the aim of analyzing changes that could happen over the system outputs. Race is an algorithm family whose aim is to select from among a set of models, one that is considered best based on established criteria. When applied to the problem of parameter setting, racing algorithms can select from a set of configurations, the one whose associated cost is minimized by solving a set of instances [2][3][4].

2.1 Hoeffding Race

The Hoeffding race is a technique for finding out a good model for data by quickly discarding bad models and concentrating the computational effort at differentiating between the better ones [2][3][4].

The first stage of the method is to establish the necessary elements, such as: N points with which to test a given model, E_{true} is a real average error if we were to test a model on N points. But if you only want to test a model on n points ($n < N$), then you only have an estimate E_{est} of the average true error E_{true}. Hoeffding's bounds are useful when n points are tested with an identical independent distribution from the set of N original test point. In this case, you can say that the probability of E_{est} being more than ϵ away from E_{true} is: $\Pr (| E_{true} - E_{est} | > \epsilon) < 2e^{(-2n \epsilon^2)/B^2}$, where B bounds the greatest possible error that a model can make. Now, to calculate the parameter ϵ, which tell us how close the estimated mean is to the true mean after n points with confidence $1-\delta$, the Equation 1 is used.

$$\epsilon(n) = \sqrt{\frac{B^2 \log(2/\delta)}{2n}} \qquad (1)$$

The next stage of the process assumes the dataset has N data points and, for each iteration of the algorithm, a point from test set is randomly selected. Each data point is a box whose boundaries are $E_{est} + \epsilon$ (upper bound) and $E_{est} - \epsilon$ (lower bound), and the center is the average of the n points, E_{est}.

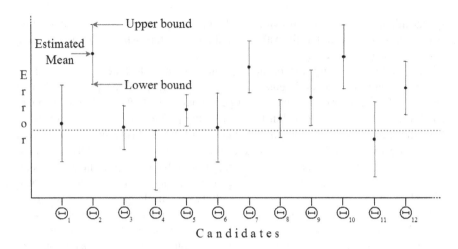

Fig. 1 Race Iteration example

For creating each box: 1) Compute E_{est}, depending on the n test points, 2) Compute ϵ, according to Equation 1, and 3) Calculate the width of the box.

Each box now has a bound within which the true average error lies. Then those boxes whose best possible error (lower bound) is still greater than the worst error of the best box (upper bound) can be eliminated. At each iteration, n is increased causing ϵ is decreased.

The algorithm continues picking test points until one condition occurs: a) All but one of the boxes has been eliminated or b) The algorithm can be stopped once ϵ has reached a certain threshold [2][3].

Figure 1 presents an example of race iteration, where θ_4 is the best configuration (with minimal error), because it has the minor estimated mean. In the worst case, θ_4 is still better than the best cases of θ_5, θ_7, θ_9, θ_{10} and θ_{12}. Because this, they will be eliminated from the competition, and the remainder will continue to compete.

3 Ant Algorithms for Semantic Query Routing

The basis of the ant algorithms can be found in a metaheuristic called ACO (Ant Colony Optimization). The ACO algorithm and its variants (e.g. ACS) were proposed to solve problems modeled as graphs. Each component of the network is represented by a *node* (or a vertex) and the interactions among them represent the *connections* (the edges). ACS needs to know information about all nodes in the network to select the destination node. Ant algorithms were inspired by the ant's behavior, while searching for food. Because when they perform the search, each ant drops a chemical called pheromone which provides an indirect communication among the ants [13].

However, in the Semantic Query Routing Problem (SQRP) the goal is to find one or more destination nodes for a query without having information from the complete network, requiring operating with local information. The SQRP consists in each peer deciding, based on a keyword in the query, to which neighboring peer to resend the text query. To avoid flooding, the goal is to maximize the number and quality of query results, while minimizing the use of the resources of the network. Existing approaches for query routing in P2P networks range from simple broadcasting techniques to sophisticated methods [8][14]. Due to the fact that P2P networks are based on non-central authorities and high-growing dimension, the challenge for query routing is the development of methods that adapt themselves to dynamic environments. Such intelligent adaptation must be based only on the local knowledge of each peer. Among the intelligent mechanisms successfully applied to several problems in distributed systems, lie the ant-colony methods.

3.1 Adaptive Neighboring Ant Search

AdaNAS is a metaheuristic algorithm, where a set of independent agents called ants cooperate indirectly and sporadically to achieve a common goal. The algorithm has two objectives: it seeks to maximize the number of resources found by the ants and to minimize the number of steps taken by the ants.

AdaNAS parameters

During the search process several parameters are used, which are a total of nine: 1) Local pheromone evaporation factor ρ, 2) Importance of local measures (Degree and distance) β_1, 3) Importance of pheromone β_2, 3) Relative importance between exploration and exploitation q, 4) Relative importance of the resources found during the time-to-live W_h, 5)Degree weight W_{deg}, 6) Distance weight W_{dist}, 7) Pheromone table initialization τ_0, 8) Initial value for distances D_0, and 9) Initial Time-to-live for the search agents TTL_{inic}.

Before using Hoeffding Race to find a valid assignment of parameters, we chose to perform a causal analysis [15][16] to identify which parameters significantly influenced the performance of the algorithm. PC algorithm was used [15][16] to perform causal analysis. This is used for identifying a causal order, in other words, the correct direction of the relationship between two variables and the intensity of the causal relationships found. To establish the right direction of causal relationships PC algorithm needs enough information, in this case, several runs of the algorithm using different values in its parameters. For this example, 800 runs were performed.

To acelerate the algorithm Hoeffding Racewe thought appropriate to identify significant parameters, because it cannot distinguish between two configurations when they only vary in the value of a non-significant parameter.

Non-significant parameters were β_2, W_h, W_{dist}, τ_0, D_0 [5].These parameters taking the values recommended by the literature[7][8][13]. The five remaining parameters were adjusted by Hoeffding race. Table 1 presents two possible parameter configurations for AdaNAS, the first (titled Non Tuned Values) presents the recommended values by the literature for these parameters; and the second one (titled Tuned Values) presents the obtained by Hoeffding Race. The methodology applied to select the best setting is shown in Figure 2.

Table 1 Configuration parameter of the AdaNas algorithm

Parameters	Non Tuned Values		Tuned Values	
	Value	Obtained By	Value	Obtained By
ψ_1	0.7	[4]	0.35	Hoeffding Race
ψ_1	2.00	[18]	2.00	Hoeffding Race
ψ_2	1.00	[4,18]	1.00	Recomended
q	0.9	[4,18]	0.65	Hoeffding Race
W_h	0.5	[4,18]	0.5	Recomended
W_{deg}	1.00	[18]	2.00	Hoeffding Race
W_{dist}	1.00	[18]	1.00	Recomended
ψ_0	0.009	[4,18]	0.009	Recomended
D_0	999	[18]	999	Recomended
TTL_{inic}	25	[4,18]	10	Hoeffding Race

4 Experiments and Results

In this section, we describe the experiments carried out on the AdaNAS algorithm. We use two versions of the same algorithm, is to say a) AdaNAS algorithm with initial configuration and b) AdaNAS algorithm without initial configuration. The objective of the first experiment is to examine the contribution of initial configuration in an instance, and in the second experiment AdaNAS algorithm is test over 90 different instances.

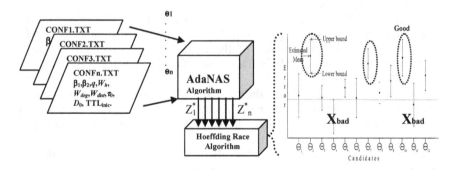

Fig. 2 Methodology for selecting the best configuration

4.1 Experiments Setup

In our implementation, an SQRP instance is determinate by three separate files: topology, repositories, and queries. The description of the instances used can be found at [5][6][7]. Each P2P simulation was run for 20,000 time units (queries).

The average performance was studied by computing performance measures each 100 units of time, called *Average efficiency*, defined as the *hits* (found matching resources) divided by the *hops* (distance between the query node and the node with resources).

The initial configuration of the algorithms is specified in a file containing a *global static configuration*. The configuration of the AdaNAS algorithm used in the experimentation is shown in Table 1.

4.2 The Experiment Results

The performance of the AdaNAS algorithm is analyzed experimentally in order to determine the contribution of the initial configuration that was obtained through Hoeffding Race Technique. For the evaluation of AdaNAS algorithm, the parameters of the algorithms were shown in Table 1.

The first experiment (see Figure 3) is an example of the behavior of the algorithm using the configuration found by Hoeffding race technique. In the second experiment (see Figure 4) are shown the algorithm performance on ninety different instances.

The Figure 3 shows the *average efficiency* reached during the execution of 14000 queries in an instance. For the first configuration –Non Tuned Values-, the algorithm started about at 2.1 average efficiency and at the end the average efficiency is 2.7 hits per hop; and for the second configuration –Tuned Values-, the algorithm start approximately at 2.9 average efficiency; at the end the average hit-rate increases to 3.5 hits per hop. We can see that the average performance always is better using Tuned AdaNAS.

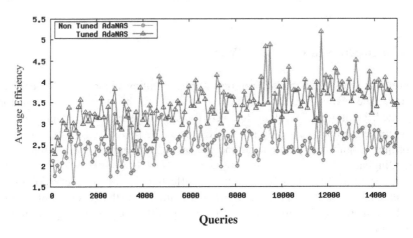

Fig. 3 Average Efficiency of the AdaNAS algorithm, during 14000 queries

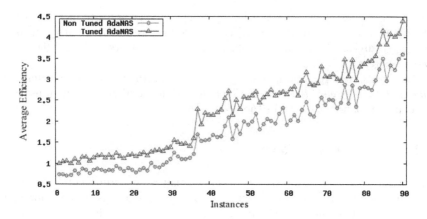

Fig. 4 Average Efficiency of the AdaNAS algorithm, during 90 different instances

Finally in the Figure 4 we show the results with 90 instances and, like in Figure 2, we can see that using the algorithm with the tuned parameters gives better results. It can be saw that, regardless of the hardness of the instances, Hoeffding race provided a better configuration.

AdaNAS with initial configuration –Tuned Values-, show the bigger contribution, giving an average efficiency of 1 hits per hop at the beginning and 4.5 hits per hop at the end, while the Non Tuned AdaNAS algorithm give an efficiency of 0.5 hits per hop at the beginning and 2.3 hits per hop at the end. Due to this result, it becomes relevant to study further the relations that exist between the problem characteristic and the algorithm parameter configuration in order to yield a bigger benefit.

5 Conclusions and Future Works

In previous works [5,6,7] we proposed a query routing algorithm that seeks to minimize the search time in the system and maximize the amount of information requested. To improve this algorithm were fundamental the parameter setting.

A simple methodology to do it was presented. The first step consists of finding algorithm's significant parameters, and the second one adjusts them through the Hoeffding race.

The experiment results demonstrated the effectiveness of this methodology as the version that uses the initial configuration outperforms the version with a non-tuned configuration. The improvement in average efficiency of the algorithm was up to 50%.

We are planning to analyze the parameter control of the AdaNAS algorithm, including more features of the problem and test instances, searching to further improve the performance of the algorithm.

References

1. Adenso-Díaz, B., Laguna, M.: Fine-Tuning of Algorithms Using Fractional Experimental Designs and Local Search. Operation Research, 99–114 (2004)
2. Maron, O., Moore, A.: Hoeffding Races: Accelerating Model Selection Search for Classification and Function Approximation. In: Advances in Neural Information Processing System, vol. 6, pp. 59–66 (1994)
3. Maron, O., Moore, A.: The Racing Algorithm: Model Selection for Lazy Learners. Artificial Intelligent Review 11, 193–225 (1997)
4. Birattari, M., Stutzle, T.: A Racing algorithm for Configuring Metaheuristics. Artificial Life, 11–18 (2002)
5. Gómez, C.: Afinación Estática Global de Redes Complejas y Control Dinámico Local de la Función Tiempo de Vida en el Proceso de Direccionamiento de Consultas Semánticas, Doctoral Thesis, Instituto Politécnico Nacional, México (2010)
6. Gómez, C., et al.: A Self Adaptive Ant Colony System for Semantic Query Routing Problem. Computación y Sistemas, Revista Iberoamericana de Computación 13(4) (2010), ISSN 1405-5546
7. Aguirre, M.: Algoritmo de Búsqueda Semántica en Redes P2P Complejas. Master Thesis, Instituto Tecnológico de Ciudad Madero (2008)
8. Michlmayr, E.: Ant Algorithms for Self-Organization in Social Networks. Doctoral Thesis, Women's Postgraduate College for Internet Technologies (WIT), Institute of Software Technology and Interactive Systems, Vienna University of Technology (2007)
9. Michalewicz, Z.: How to solve it: Modern Heuristics. Springer, NY (2000)
10. Angeline, P.: Adaptative and Self-Adaptative Evolutionary Computations. IEEE Computational Intelligence, 152–163 (1995)
11. Montgomery, D.C.: Design and Analysis of Experiments. John Wiley & Sons, New York (2001)
12. Barr, R., Golden, J., Kelly, M.: Designing and Reporting Computational Experiments with Heuristics Methods. Journal of Heuristics 1, 9–32 (1995)
13. Dorigo, M., Stützle, T.: Ant Colony Optimization. MIT Press (2004)
14. Sakaryan, G.: A Content-Oriented Approach to Topology Evolution and Search in Peer-to-Peer Systems, PhD Thesis, University of Rostock, Germany (2004)
15. Pérez, V.: Modelado Causal del Desempeño de Algoritmos Metaheurísticos en Problemas de Distribución de Objetos. Master Thesis, Instituto Tecnológico de Ciudad Madero (2007)
16. Quiroz, M.: Caracterización de Factores de Desempeño de Algoritmos de Solución de BPP. Master Thesis, Instituto Tecnológico de Ciudad Madero (2009)

Developing Anti-spam Filters Using Automatically Generated Rough Sets Rules

N. Pérez-Díaz, D. Ruano-Ordás, F. Fdez-Riverola, and J. R. Méndez

Dept. Informática, University of Vigo, Escuela Superior de Ingeniería Informática,
Edificio Politécnico, Campus Universitario As Lagoas s/n, 32004, Ourense, Spain
{npdiaz,drordas,riverola,moncho.mendez}@uvigo.es

Abstract. The huge amount of spam messages has limited the benefits introduced by e-mail communications. Therefore, spam filters are indispensable to fight against spam deliveries. However, the development of spam filters is very expensive whereas the usage of external filtering services can damage communications privacy. In such situation, we introduce an automatic procedure to integrate knowledge extracted by using rough-sets theory into spam filters to develop a low-cost filtering infrastructure.

1 Introduction and Motivation

Since the creation of the e-mail, their popularity and use have been widely explode due to four basic factors: (*i*) the low or nonexistent cost, (*ii*) their usability, (*iii*) the fast deliveries and finally, (*iv*) the versatility (allows attaching any file type). Nevertheless, this phenomenon also enables e-mail service to be used as a platform for the massively send of advertising messages (also called spam). This fact motivated the introduction of anti-spam frameworks able to discern between legitimate and spam messages.

This scenario has mainly affected the business sector, where the lack of a spam filtering system may cause the company to receive and send spam indiscriminately. This fact causes several drawbacks such as: (*i*) employees waste time removing spam from their inbox (reduced efficiency), (*ii*) the inefficient use of bandwidth, (*iii*) the failure of the corporate e-mail server, (*iv*) the loss of prestige and (*v*) judicial accusations for promoting spam.

Moreover, due to the continuous changes on techniques used to distribute spam, anti-spam filtering products become obsolete in a short time period. However, spam-filtering services are filters continuously updated to improve customer e-mails classification accuracy and handle the latest spam techniques. Behind these services, there is a team of experts examining your e-mail and updating filters behavior to detect the newest spam contents. Current filter frameworks as SpamAssassin [1] support filter customization by using a simple rule-based language.

Despite the suitability of this kind of services [2], providers cannot guarantee customers communications privacy because the domain primary MTA (Mail

J. Casillas et al. (Eds.): Management Intelligent Systems, AISC 171, pp. 325–334.
springerlink.com   © Springer-Verlag Berlin Heidelberg 2012

Transfer Agent) is replaced and all messages are delivered through it. Obviously, the replacement of this MTA is an ESMTP (Extended Simple Mail Transfer Protocol) [3] server belonging to filtering provider. Therefore most current filtering services are unsuitable when privacy is required.

The deployment of filtering services inside an enterprise is too expensive and involves hiring a team of experts to develop and maintain filtering rules everyday. In this work, we introduce a new proposal to automatically update filters through two global mailboxes to compile ham and spam messages respectively. Using messages compiled by these mailboxes we can use rough-sets theory [4, 5] to automatically generate rules and update spam filters.

The rest of the paper is structured as follows: Section 2 summarizes rough-sets theory and outlines its suitability to generate spam filtering rules while Section 3 shows current spam filtering middleware. Section 4 introduces our proposal to transform rules generated through using rough sets in filters following the syntax and restrictions of actual spam filtering middleware and Section 5 presents a case study of this technology. Finally, Section 6 shows the main conclusions.

2 Rough Sets for Spam Filtering

Rough-set theory [4, 5] provided a set of formal concepts and operations that can be successfully combined to induce categorization or classification rules from knowledge bases (datasets) in a target domain. Therefore, they have been successfully used to address manifold problems from different domains.

In order to allow the straightforward application of rough-sets theory to analyze a given e-mail corpus, the content from each message should be split into tokens. Accordingly, each e-mail is represented by a set of possible attributes $(a_1 ... a_{n-1})$, with their own values (usually in the form: includes the term 'Viagra', includes the term 'Cialis', etc.) and its corresponding *message class* (a_n). Therefore, this feature vector containing all the terms existing in the corpus plus the *class attribute* stands for the attribute set $A = \{a_1, a_2 ... a_{n-1}, a_n\}$. In addition, each attribute a_i is 1 when the message contains the term a_i and 0 otherwise. Similarly, the value for the *class attribute*, a_n, is 1 for spam messages and 0 for legitimate ones.

In this context, an information system can be defined as a pair $S = \langle U, A \rangle$, where U is a non-empty and finite set called the universe (representing all the messages comprising the whole corpus), and A is the non-empty and finite set of features previously defined. Using this characterization, an equivalence relation, also referred as an indiscernibility relation, is associated with every subset of attributes $P \subseteq A$. This relation is defined as shown in Expression 1.

$$IND(P) = \{(x, y) \in U \times U : \forall\, a \in P,\, a(x) = a(y)\} \tag{1}$$

Expression 1 establishes that two e-mails, x and y from U, are linked by considering the attributes included in P when they share the same values for those features. By using the indiscernibility relation $IND(P)$ from the set of attributes P, we can define the set of equivalence classes denoted by $U/IND(P)$.

Given any subset of features P, any concept $X \subseteq U$ can be defined approximately by the use of two sets, called lower and upper approximations. The lower approximation, denoted by $\underline{P}(X)$, is the set of elements in U which can be certainly classified as elements in the concept X using the set of attributes P, and the upper approximation, denoted by $\overline{P}(X)$, is the set of elements in U that can be possibly classified as elements in X. Expression 2 contains the definition of this concepts.

$$\underline{P}X = \cup \{Y \in U / IND(P) : Y \subseteq X\}$$
$$\overline{P}X = \cup \{Y \in U / IND(P) : Y \cap X \neq \varnothing\}$$

(2)

Through the utilization of upper and lower approximations, the boundary region of X (BN_P) is defined by computing the difference set from $\overline{P}(X) - \underline{P}(X)$. This set contains the objects (messages) that cannot be classified as members of X (i.e. spam e-mails) using the attributes of P.

In order to build a rule-based information system from any given set of features P, we should compute all reducts from P. A reduct is a subset of attributes $RED \subseteq P$ having the following properties: *(i)* the set of equivalence classes induced by RED is the same as computed using P, therefore $U/IND(P) = U/IND(RED)$ and *(ii)* the attribute set RED is minimal, therefore $U/IND(RED-\{a\}) \neq U/IND(RED)$ for any attribute a included in RED. This step is useful to remove superfluous features from input data reducing the computational overhead.

Moreover, the degree of dependency of a set of features P on a set of features R is denoted by $\gamma_R(P)$, $0 \leq \gamma_R(P) \leq 1$, being defined as shown in Expression 4.

$$\gamma_R(P) = \frac{card(POS_R(P))}{card(U)}$$

(4)

where

$$POS_R(P) = \bigcup_{x \in U / IND(P)} \underline{P}(X)$$

(5)

$POS_R(P)$ contains the objects of U which can be classified as belonging to one of the equivalence classes of $IND(P)$ using only features from the set R. If $\gamma_R(P) = 1$, then R functionally determines P.

By the execution of the above mentioned process and using different P sets that functionally determine the *class attribute*, we can apply rough sets conceptualization to generate the rules shown in Table 1.

As we can see from Table 1, attributes without an assigned value for a given rule (marked with a hyphen) indicate the irrelevance of their values regarding to the target consequent.

In order to treat messages included in boundary region, we found the following alternatives: *(i)* maintain them unclassified [6], *(ii)* classify them into a third category called 'suspicious' [7], *(iii)* use a further exploration to classify these messages [8] and *(iv)* classify as legitimate [9].

Table 1 Example of rules extracted from a knowledge base

	a_1 hello	a_2 work	a_3 student	a_4 show	a_5 more	a_6 Date	a_7 viagra	a_8 cialis	class
r_1	-	-	-	-	-	-	1	1	1
r_2	-	-	1	-	-	-	-	-	0
r_3	-	-	-	-	-	1	-	-	0
r_4	-	-	-	1	1	-	-	-	1

As shown, rough-sets theory has introduced the possibility of extract rules that summarize the knowledge included in datasets. In this work, we will use this information to automatically update spam-filtering services and implement the 3rd scheme to handle messages from borderline region. Next section introduces the technology operating behind spam-filtering services and outlines their most interesting features to support our automatically updating procedure.

3 Spam Filtering Middleware

Nowadays, SpamAssassin is one the most common and suitable framework to filter spam. This platform provides two great functionalities: (*i*) the possibility of modeling filters through rules and (*ii*) the facility to distribute them. Therefore, a wide variety of spam filtering products, such as Symantec Brightmail or McAffee SpamKiller, are based on the SpamAssassin features and filtering technology.

Under this platform, a filter is composed by a set of rules and a global threshold called *required_score*. Each rule contains an individual score and a trigger condition associated to an anti-spam technique (such as *Regular Expressions, Naïve Bayes, SPF, RBL...*). Each time a rule matches the e-mail, the filtering middleware sums the score of the rule to the message global score. Once all rules have been executed, the platform classifies the e-mail as spam only if the e-mail counter is greater or equal to the value of the *required_score*.

As shown on Figure 1, the first keyword of each rule definition has two types of modifiers *(body or header)*. This value indicates the location inside the target message where the rule is executed. Each time a rule matches an e-mail, the rule score is used to increment the message global score. Finally, all rules can optionally include a brief description in order to facilitate filter maintenance.

4 Integrating Rough Sets and Filtering Middleware

The challenge of this work is the improvement of SpamAssassin based filter using knowledge extracted from e-mails compiled by users through using rough sets. To this end, we developed an algorithm to transform RS-rules into framework rules.

```
00   #REGULAR EXPRESIONS
01   body BODY_DRUG_CIALIS /\bc.{0,2}i.{0,2}a.{0,2}l.{0,2}i.{0,2}s\b/i
02   describe BODY_DRUG_ CIALIS  Subject contains 'cialis'
03   score BODY_DRUG_CIALIS 2
04
05   body BODY_DRUG_VIAGRA /\v.{0,2}i.{0,2}a.{0,2}g.{0,2}r.{0,2}a.{0,2}/i
06   describe BODY_DRUG_VIAGRA  Subject contains 'viagra'
07   score BODY_DRUG_VIAGRA 2
08
09   header SUBJECT_DRUG_LEVITRA Subject =~ \l.{0,2}e.{0,2}v.{0,2}i.{0,2}t.{0,2}r.{0,2}a/i
10   describe SUBJECT_DRUG_LEVITRA  Subject contains 'levitra'
11   score SUBJECT_DRUG_LEVITRA 2
12
13   header SUBJECT_DRUGS  Subject =~ \d.{0,2}r.{0,2}u.{0,2}g.{0,2}s.{0,2} /i
14   describe SUBJECT_DRUGS  Subject contains 'drugs'
15   score SUBJECT_DRUGS 1.37
16
17   #META EXPRESIONS
18   meta BODY_SOME_DRUGS ( BODY_DRUG_CIALIS + BODY_DRUG_VIAGRA > 0 )
19   describe BODY_SOME_DRUGS Has some drugs on content message
20   score BODY_SOME_DRUGS 3
21
23   meta DRUGS_SUBJECT ( SUBJECT_DRUG_LEVITRA + SUBJECT_DRUGS > 0 )
24   describe DRUGS_SUBJECT Has some drugs on subject message
25   score DRUGS_SUBJECT 3
26
27   meta HAS_DRUGS ( BODY_SOME_DRUGS & SUBJECT_SOME_DRUGS )
28   describe HAS_DRUGS Message contains all references to drugs.
29   score HAS_DRUGS 40
```

Fig. 1 Regular expressions in current spam filtering middleware

Our transform scheme comprises 3 simple steps: (*i*) generating rules to detect the presence or absence of features (terms) in the target message, (*ii*) generating spam filter rules as a translation of RS-rules and, finally, (*iii*) generate scores for each rule. Figure 2 introduces a pseudo-code representation of the transformation process.

As we can see from Figure 2, in order to address the 3rd stage, a sufficient condition rule is introduced to implement each rule extracted through using rough-sets theory. In this paper, we introduce a simple method to implement this kind of rules in SpamAssassin. It is based on computing the *amount_subtract* and *amount_add* values as shown in Equation 6

$$amount\_add = \sum_{i=0..n}^{score(r_i \geq 0)} score(r_i)$$

$$amount\_subtract = \sum_{i=0..n}^{score(r_i < 0)} score(r_i)$$

(6)

```
00   T_RULESET=array of T_RULE;
01   T_RULE=RECORD
02        at_values: array [1..sizeof(rs_attributes)] of ENUM {1,0,-};
03        result: boolean;
04   END_RECORD
05
06   PROCEDURE RS2Filter (INPUT rs_rules: T_RULESET, INPUT rs_attributes: ARRAY of STRING)
07   BEGIN
08        FOREACH (ai ∈ rs_attributes) #first_stage
09             i=indexof(ai, rs_attributes);
10             print_filter_definition("header A"+i+"_SUBJECT Subject =~ /"+ai+"/i");
11             print_filter_definition("body A"+i+"_BODY /"+ai+"/i");
12             print_filter_definition("meta A"+i+"((A"+i+"_SUBJECT + A"+i+"_BODY)>0)");
13        END_FOREACH
14
15        FOREACH (ri ∈ rs_rules) #second stage
16             i=indexof(ri, rs_rules);
17             STRING rule="meta R"+i+" (";
18             FOREACH (aj: aj ∈ rs_attributes && ri.at_values[aj]=1)
19                  j=indexof(aj, rs_attributes);
20                  IF (rule[strlen(rule)-1]!='(' ) THEN
21                       rule+=" && ";
22                  END_IF
23                  rule+=aj;
24             END_FOREACH
25             rule+=")";
26             print_filter_definition(rule);
27        END_FOREACH
28
29        FOREACH (ri ∈ rs_rules) #third stage
30             IF (ri.result) THEN
31                  print_filter_definition("score R"+i+" "+convert_to_string(required_score-amount_subtract));
32             ELSE
33                  print_filter_definition("score R"+i+" "+convert_to_string(required_score-amount_add-1));
34             END_IF
35        END_FOREACH
36   END
```

Fig. 2 Pseudo-code representation of the process

where $R=(r_1, r_2, ..., r_n)$ are the rules previously included in a SpamAssassin filter. By computing this values, we can define a sufficient spam condition rule as a rule with a score of $scs(spam)$ and either, $scs(ham)$ is the score required to define a sufficient ham condition rule. The definition of these values is shown in Equation 7.

$$scs(spam) = required\_score - amount\_subtract$$
$$scs(ham) = required\_score - amount\_add - 1$$

(7)

In order to improve the readability of this paper and to complete the contents of this section, next section shows an example of the application of this technique.

5   Case Study

This section shows in detail how to transform rules included in Table 1 into a spam filter using the SpamAssassin framework and presents a performance comparison of our proposal and Naïve Bayes technique.

5.1   Converting RS-Rules to SpamAssassin

As mentioned, the transformation process involves the execution of 3 simple steps. During the first step, for each attribute a_i, we should generate two regular expressions (to detect the presence of the feature in the subject and body respectively) and one meta-rule to combine them (using OR). Following this scheme and considering the attribute a_1 from Table 1, we should generate the SpamAssassin rules showed in Figure 3.

```
header A1_SUBJECT Subject =~ /hello/i
body A1_BODY /hello/i
meta A1 ((A1_SUBJECT + A1_BODY) > 0)
```

Fig. 3 Example of rules generated by first step

During the second step, each rule r_i is transformed to a meta filter rule combining rules extracted in previous stage. Figure 4 shows rules created from the transformation of rules r_1 and r_2 from Table 1.

```
meta R1 (A7 && A8)
meta R2 A3
```

Fig. 4 Example of rules generated during second step

Finally, we should compute *amount_add* and *amount_subtract* to transform each rule identified in previous step in a sufficient condition rule. Figure 5 shows rule scores of r_1 and r_2 from Table 1 assuming *required_score* = 5, *amount_add* = 12 and *amount_subtract* = -4.

```
score R1 9
score R2 -8
```

Fig. 5 Scores assigned for each generated rule

Executing all steps from the whole amount of rules contained in Table 1, we generate all rules showed in Figure 6 following the conditions assumed to generate Figure 5.

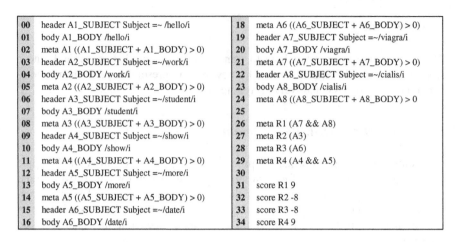

00	header A1_SUBJECT Subject =~ /hello/i	18	meta A6 ((A6_SUBJECT + A6_BODY) > 0)
01	body A1_BODY /hello/i	19	header A7_SUBJECT Subject =~/viagra/i
02	meta A1 ((A1_SUBJECT + A1_BODY) > 0)	20	body A7_BODY /viagra/i
03	header A2_SUBJECT Subject =~/work/i	21	meta A7 ((A7_SUBJECT + A7_BODY) > 0)
04	body A2_BODY /work/i	22	header A8_SUBJECT Subject =~/cialis/i
05	meta A2 ((A2_SUBJECT + A2_BODY) > 0)	23	body A8_BODY /cialis/i
06	header A3_SUBJECT Subject =~/student/i	24	meta A8 ((A8_SUBJECT + A8_BODY) > 0
07	body A3_BODY /student/i	25	
08	meta A3 ((A3_SUBJECT + A3_BODY) > 0)	26	meta R1 (A7 && A8)
09	header A4_SUBJECT Subject =~/show/i	27	meta R2 (A3)
10	body A4_BODY /show/i	28	meta R3 (A6)
11	meta A4 ((A4_SUBJECT + A4_BODY) > 0)	29	meta R4 (A4 && A5)
12	header A5_SUBJECT Subject =~/more/i	30	
13	body A5_BODY /more/i	31	score R1 9
14	meta A5 ((A5_SUBJECT + A5_BODY) > 0)	32	score R2 -8
15	header A6_SUBJECT Subject =~/date/i	33	score R3 -8
16	body A6_BODY /date/i	34	score R4 9

Fig. 6 Full SpamAssassin filter generated from rules included in Table 1

Finally, we should remark that this technique is able to successfully combine automatically generated rules with other techniques implemented in filter middleware.

5.2 Performance Comparison

With the goal of proving the effectiveness of rough sets in spam filtering domain a set of tests was executed for comparing their results. Naïve Bayes classifier is the most widely used algorithm in the spam filtering industry and it is the only content based intelligent technique included in SpamAssassin. Due to this, we compared the results obtained applying rough sets and naïve bayes approaches. A hundred features of SpamAssassin corpus were selected using Information Gain [10] method to perform the experiment. All tests were realized using 10-fold stratified cross-validation scheme [11].

Figure 7 shows the percentage of correct classifications, false positive and false negative errors. As we can see, rough sets achieved the best performance.

In order to analyze techniques from a cost-sensitive point of view, we used Total Cost Ratio (TCR) measure. TCR includes a λ parameter to weight false positive (FP) errors. When computing the score for a specific λ value, we assume that one FP error is λ times more costly than a FN one. Table 2 shows the performance comparison results.

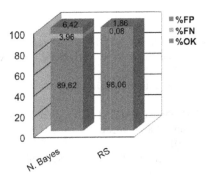

Fig. 7 Percentage comparison of models

Table 2 TCR scores for the analyzed proposals

	$\lambda=1$	$\lambda=9$	$\lambda=999$
RS	13,64640884	1,570247934	0,01420904
Naïve Bayes	2,465	0,416	0,004

As Table 2 shows, rough sets achieved better performance than Naïve Bayes.

6 Conclusions

In this paper, we introduce a novel technique to improve SpamAssassin filtering services through using rough-sets theory. This technology introduces the following benefits: (*i*) allow exploding changes in techniques used by spammers, (*ii*) automatic filter enhancement, (*iii*) reducing spam-filtering costs (increasing business profit) and (*iv*) improve business communications efficiency and effectiveness.

Moreover, we believe that this proposal introduces an affordable way to combine spam filters and current IA techniques to deploy spam-filtering services viable from a computational point of view and current ESMTP restrictions.

In order to improve this work, future developments should be focused in the implementation of definitive rules. This type of rules should allow the filter to abort the execution when the rule matches the e-mail. Under this scenario, the message will be automatically classified depending on the definitive value associated to that rule.

Consequently this concept aims to increase the classification speed without losing the effectiveness of the filter classification and therefore using less time and resources.

Acknowledgements. This work was partially funded by the projects Optimización de sistemas antispam (08TIC041E) and Deseño e validación de filtro antispam intelixente baseado en análise contextual ponderado do contido das mensaxes (09TIC028E) from Xunta de Galicia.

References

1. Garg, A., Battiti, R., Cascella, R.: May I borrow your filter? Exchanging Filters to Combat Spam in a Community. Adv. Inf. Netw. and Appl., 489–493 (2006), doi:10.1109/AINA.2006.1
2. AgentSpam: Spam & Virus filtering (2011), http://agentspam.com (accessed February 17, 2012)
3. Newman, C., et al.: ESMTP and LMTP Transmission Types Registration. RFC 3848, Network Working Group. (2004), http://tools.ietf.org/html/rfc3848 (accessed February 17, 2012)
4. Pawlak, Z.: Rough Sets: Theoretical Aspects of Reasoning about Data. Kluwer Academic Publishers, Boston (1991)
5. Pawlak, Z.: Rough sets. Int. J. of Parallel Program 11(5), 341–356 (1982), doi:10.1007/BF01001956
6. Glymin, M., Ziark, W.: Rough Set Approach to Spam Filter Learning. In: Int. Conf. of Rough Sets and Intel. Syst. Paradigms (2007)
7. Zhao, W., Zhu, Y.: Classifying Email Using Variable Precision Rough Set Approach. In: Wang, G.-Y., Peters, J.F., Skowron, A., Yao, Y. (eds.) RSKT 2006. LNCS (LNAI), vol. 4062, pp. 766–771. Springer, Heidelberg (2006)
8. Zhou, B., Yao, Y., Luo, J.: A Three-Way Decision Approach to Email Spam Filtering. In: Farzindar, A., Kešelj, V. (eds.) Canadian AI 2010. LNCS, vol. 6085, pp. 28–39. Springer, Heidelberg (2010)
9. Lai, G., Chen, C., Laih, C., Chen, T.: A Collaborative Anti-spam System. Expert. Syst. Appl. 3, 6645–6653 (2009), doi:10.1016/j.eswa.2008.08.075.
10. Méndez, J.R., Fdez-Riverola, F., Díaz, F., et al.: A Comparative Performance Study of Feature Selection Methods for the Anti-spam Filtering Domain. In: Industrial Conf. Data Min., pp. 106–120 (2006), doi:10.1007/11790853_9
11. Kohavi, R.: A Study of Cross-Validation and Bootstrap for Accuracy Estimation and Model Selection. In: 14th Int. Joint Conf. on Artif. Intel., pp. 1137–1143 (1995)

Author Index